Fifth Edition

Anatomy & Physiology

LABORATORY TEXTBOOK • ESSENTIALS VERSION

STANLEY E. GUNSTREAM
Pasadena City College

•

HAROLD J. BENSON
Pasadena City College

•

ARTHUR TALARO
Pasadena City College

•

KATHLEEN P. TALARO
Pasadena City College

Boston   Burr Ridge, IL   Dubuque, IA   New York   San Francisco   St. Louis
Bangkok   Bogotá   Caracas   Kuala Lumpur   Lisbon   London   Madrid   Mexico City
Milan   Montreal   New Delhi   Santiago   Seoul   Singapore   Sydney   Taipei   Toronto

 **Higher Education**

ANATOMY & PHYSIOLOGY LABORATORY TEXTBOOK: ESSENTIALS VERSION, FIFTH EDITION

Published by McGraw-Hill, a business unit of The McGraw-Hill Companies, Inc., 1221 Avenue of the Americas, New York, NY 10020. Copyright © 2010 by The McGraw-Hill Companies, Inc. All rights reserved. Previous editions © 2007, 2001, and 1996. No part of this publication may be reproduced or distributed in any form or by any means, or stored in a database or retrieval system, without the prior written consent of The McGraw-Hill Companies, Inc., including, but not limited to, in any network or other electronic storage or transmission, or broadcast for distance learning.

Some ancillaries, including electronic and print components, may not be available to customers outside the United States.

This book is printed on acid-free paper.

1 2 3 4 5 6 7 8 9 0 QPD/QPD 0 9

ISBN 978-0-07-352568-6
MHID 0-07-352568-5

Publisher: *Michelle Watnick*
Senior Sponsoring Editor: *James F. Connely*
Director of Development: *Kristine Tibbetts*
Senior Developmental Editor: *Lynne M. Meyers*
Marketing Manager: *Lynn M. Breithaupt*
Project Manager: *Joyce Watters*
Senior Production Supervisor: *Laura Fuller*
Associate Design Coordinator: *Brenda A. Rolwes*
Cover Designer: *Studio Montage, St. Louis, Missouri*
(USE) Cover Image: *© Brand X/Jupiter Images*
Senior Photo Research Coordinator: *John C. Leland*
Photo Research: *LouAnn K. Wilson*
Compositor: *S4Carlisle Publishing Services*
Typeface: *11.5/12 Times Roman*
Printer: *Quebecor World Dubuque, IA*

Photo Credits
3.1: Courtesy of the Olympus Corporation, Lake Success, NY; 5.2a-f, 18.1 and 18.2: © Ed Reschke; 18.3: © Victor P. Eroschenko; Remaining photos are Courtesy Harold Benson.

Some of the laboratory experiments included in this text may be hazardous if materials are handled improperly or if procedures are conducted incorrectly. Safety precautions are necessary when you are working with chemicals, glass test tubes, hot water baths, sharp instruments, and the like, or for any procedures that generally require caution. Your school may have set regulations regarding safety procedures that your instructor will explain to you. Should you have any problems with materials or procedures, please ask your instructor for help.

www.mhhe.com

CONTENTS

PREFACE

This *Essentials Version* of the *Anatomy and Physiology Laboratory Textbook* presents the fundamentals of human anatomy and physiology in a manner that is appropriate for students in allied health programs such as practical nursing, radiologic technology, medical assisting, and dental assisting. These students usually take a one-semester course in human anatomy and physiology and need a laboratory text that provides coverage of the fundamentals without the clutter of excessive details and unneeded terminology.

Overview of Changes to *Anatomy and Physiology Laboratory Textbook, Essentials Version, 5/e.*

- The art program has been substantially improved with new illustrations and, in addition, many existing figures have been updated. The text and laboratory reports have been rewritten to correlate with the figure changes. These improvements will enhance clarity and student understanding.
- Students are alerted to check with their instructor about using protective disposable gloves where appropriate.
- In Exercise 4, The Cell, coverage of the plasma membrane has been expanded to include the types and functions of proteins in the plasma membrane.
- In Exercise 19, The Nature of Muscle Contraction, the discussion of a motor unit has been expanded to distinguish between large and small motor units and their importance in muscle functions.
- In Exercise 23, Brain Anatomy: External, the discussion of cranial nerves has been expanded to more clearly define their functions.
- In Exercise 28, The Heart, coverage now includes coronary arteries and cardiac veins.
- Exercise 37, The Endocrine Glands has been reorganized with a reduction in the histology figures. Expanded coverage now includes thymosin and T-cell maturation, calcitonin and parathyroid hormone maintaining calcium balance, and adrenal sex steroids.

Exercise Format

Each of the thirty-eight exercises begins with a short list of objectives that outline the minimal learning responsibilities for the students. The exercises are basically self-directing, which minimizes the need for lengthy introductions by the instructor.

Instructors will find that the listing of required equipment and materials on the first page of each exercise facilitates laboratory preparation. The exercises use standard equipment and materials that are usually available in most biology departments.

Three exercises utilize electronic equipment. For investigating muscle contraction, instructors have a choice of using either the BIOPAC© Student Lab System in Exercise 20 or the Intelitool Physiogrip in Exercise 21. Exercise 32, Respiratory Physiology, offers options of using either nonelectronic equipment (Propper spirometers) or the Intelitool Spirocomp to determine lung volumes.

Each exercise topic is covered at a reduced level of difficulty appropriate for the student audience and presented in a direct, concise manner to facilitate student learning. Numerous illustrations and many photomicrographs are correlated with the text to improve understanding.

The necessary key terms are in bold print to aid students in building a vocabulary of anatomical and physiological terms.

Laboratory procedures are distinct from and follow the discussion of the exercise topic. They are presented in a concise, stepwise manner to guide the student. Activities consist of (1) labeling illustrations, (2) dissections, (3) study of specimens and models, (4) physiological experiments, and (5) microscopic studies. A major dissection specimen, such as the cat or fetal pig, is not included. Instead, a rat dissection is used in a single exercise to acquaint students with the basic organization of organ systems in mammals.

Pedagogical Design

The pedagogical design calls for students to develop an understanding of each exercise topic by (1) labeling the illustrations using information presented in the text and (2) completing corresponding portions of the laboratory report. Usually, students should complete these activities *before* coming to the laboratory or proceeding with the other laboratory studies.

The *laboratory reports* are designed to guide and reinforce student learning and provide a convenient site for recording data and making diagrams of observations. They are located at the back of the book to better control the pagination and to keep the book brief. The corresponding laboratory report should be removed from the back of the book whenever students *begin* working on an exercise. Removing the reports prevents page flipping and facilitates completing the laboratory report. Completed laboratory reports should be kept in the student's notebook. Instructors may choose to either post the answer keys from the *Instructor's Manual* so students can check their work or collect and grade the laboratory reports. The design of the laboratory reports makes them easy to grade.

Students in the target group typically have difficulty in learning the necessary anatomical and physiological terminology, especially the correct spellings. Experience has shown that students learn key terms and their correct spellings more easily if they write the terms. Thus, the laboratory reports are structured so that students are required to write the names of key terms rather than simply provide a matching letter or number.

Microscopic study is another difficult area for these students since they have trouble locating histological structures on slides when using diagrams as a reference. This problem is largely overcome by the inclusion of full-color photomicrographs of common tissues and of selected histological subjects. These are found in the Histology Atlas and in Exercise 37, The Endocrine Glands.

Instructor's Manual

The accompanying *Instructor's Manual* provides additional aid for the instructor: (1) a composite list of equipment and supplies, (2) operational suggestions, and (3) answer keys for the laboratory reports.

Adopters are encouraged to contact the authors regarding any problems and to submit constructive suggestions that will improve future editions. The Instructor's Manual can be found at www.mhhe.com/labcentral. Please contact your sales representative for additional information.

Acknowledgments

We also recognize the helpful suggestions provided by the following reviewers: Ann M. Findley, University of Louisiana at Monroe; J. Stephen Guffey, National Park Community College; Christine Holler-Dinsmore, Fort Peck Community College; Julie Huggins, Arkansas State University; Katharine K. Lawrence, Cochise Community College; Donald W. Linzey, Wytheville Community College; Karen Magatagan, Cochise College; Amy Griffin Ouchley, University of Louisiana at Monroe; Walt Sinnamon, Southern Wesleyan University.

TO THE STUDENT

This laboratory textbook has been designed to help you master the fundamentals of human anatomy and physiology, which provide the basis of the health-related professions. An understanding of anatomical structures, physiological processes, terminology, and techniques will give you the basic knowledge that is essential for success in your chosen field.

Each exercise begins with a list of learning objectives to guide your study. The sequence of activities in each exercise is established to facilitate the development of your understanding. Follow the sequence; don't skip around.

The activities consist of (1) labeling illustrations, (2) dissections, (3) study of specimens and models, (4) physiological experiments, and (5) microscopic study. Follow the directions for each activity with care to enhance your success in the laboratory. Work carefully and thoughtfully. Remember, the objective is to learn the material, not just to complete the exercise.

Each exercise has a *laboratory report,* located in the back of the book, that you are to complete as you work through the exercise. Remove it from the book when you *start* the exercise so that you can complete it without page flipping. Completed laboratory reports may be kept in your notebook.

Safety

A list of safety guidelines is included in the inside front cover.

Some of the exercises contain a Before You Proceed box following the material section. Be sure to check with your instructor about using protective disposable gloves where appropriate.

Labeling Illustrations

Most of the exercises are designed so that you will learn anatomy by labeling illustrations from the information provided in the text. This process helps you to understand the relationships between anatomical features, and it facilitates learning. Correctly labeled illustrations are then used as references when examining specimens and models. Use them for study and review purposes. Usually, the illustrations should be labeled *before* coming to the laboratory session.

Dissections

The dissections will include freshly killed rats and animal parts obtained from a slaughterhouse. The purpose of dissection is to expose anatomical parts for observation and study. Strive to cut as little as possible to achieve this goal. Use the scalpel sparingly. Most dissections can be accomplished with scissors, forceps, and probe.

Experiments

Before performing any experiment, read the directions completely through so that you have a good understanding of the experiment. Be sure that you have all of the equipment and materials required to complete the experiment before starting. Then, carefully follow the directions.

Microscopic Study

An examination of prepared slides enables you to visualize and understand the cytological or histological structure of specimens. Correlate your observations with the text, diagrams, and photomicrographs of the Histology Atlas. If drawings are required, make them with care and label the structures. In this way, you will understand the microscopic structures more quickly and better prepare yourself for lab practicums.

Laboratory Reports

Complete the laboratory reports independently to maximize your understanding. The laboratory reports are purposely designed so that you must write the key terms involved in the exercise, sometimes more than once. This format is used because writing the terms enables you to learn them more easily, especially their correct spellings.

General Operations

Success in the laboratory can be increased by following a few simple guidelines.

1. Be on time to class so that you will hear the comments and directions given by your instructor.

2. Follow your instructor's directions. Take notes on any changes in the equipment, materials, or procedures.
3. Keep your work area free of clutter. Extraneous items should be located elsewhere during the laboratory session.
4. Bring your textbook to the laboratory for reference.

5. Use equipment and materials with care. Report any problems or accidents to your instructor immediately.
6. Work independently, but be cooperative and helpful in team assignments.
7. Do not eat, drink, or smoke in the laboratory.

INTRODUCTION TO HUMAN ANATOMY

OBJECTIVES

After completing this exercise, you should be able to:

1. Correctly use directional and regional terms in describing the location of anatomical structures.
2. Identify the external body regions and body cavities on charts or a torso model.
3. Contrast the types of body sections used in anatomical studies.
4. Describe the organs and membranes of the body cavities.
5. Define all terms in bold print.

Materials

Human torso model with removable organs

Human anatomy is the study of the structures composing the human body and their interrelationships. The description of anatomical features requires the use of specific *anatomical terminology* that provides precise meaning. It is important that you learn the terminology presented in this exercise as quickly as possible because these terms will be used frequently throughout your study of the human body. In this exercise, you will be introduced to the anatomical terminology that is used to describe (1) relative positions of structures, (2) body sections or planes, (3) external features and body regions, and (4) body cavities and their membranes.

Directional Terms

The relative position of body structures is communicated through the use of **directional terms.** These terms typically occur as pairs, with the members of each pair having opposite meanings. For example, *anterior* (*ventral*) means toward the front of the body, and *posterior* (*dorsal*) means toward the back of the body. Most directional terms apply to organs of the body as well as to the body as a whole. Learn the meanings of the directional terms in Table 1.1.

When directional terms are used to describe the relative positions of body parts, it is assumed that the

TABLE 1.1
Directional Terms

Term	Meaning
Anterior (ventral)	Toward the front or abdominal surface of the body
Posterior (dorsal)	Toward the back of the body
Superior (cephalad)	Toward the head
Inferior (caudad)	Away from the head
Medial	Nearer the midline of the body
Lateral	Farther from the midline of the body
Proximal	Nearer the attachment of an extremity to the trunk
Distal	Farther from the attachment of an extremity to the trunk
External (superficial)	Toward or on the body surface
Internal (deep)	Away from the body surface
Parietal	Pertaining to the outer boundary of body cavities
Visceral	Pertaining to the internal organs

body is in a standard position called the *anatomical position.* In this position, the body is erect with arms at the sides and with the palms of the hands and the feet facing forward (anteriorly). The anatomical position is shown in Figure 1.1.

Body Sections

To observe the relative positions of internal structures, it is necessary to view them in sections that have been cut through the body. These sections, or planes, are applicable both to the whole body and to individual organs. See Figure 1.1.

Parasagittal sections are parallel to the longitudinal (long) axis of the body and divide the body into left and right portions. A **midsagittal section** is made along the **median line** and divides the body into *equal* left and right halves.

Frontal (coronal) sections divide the body into anterior and posterior portions and are perpendicular to sagittal sections but parallel to the longitudinal axis.

Figure 1.1 Body sections and surfaces. Note that the body is in the anatomical position.

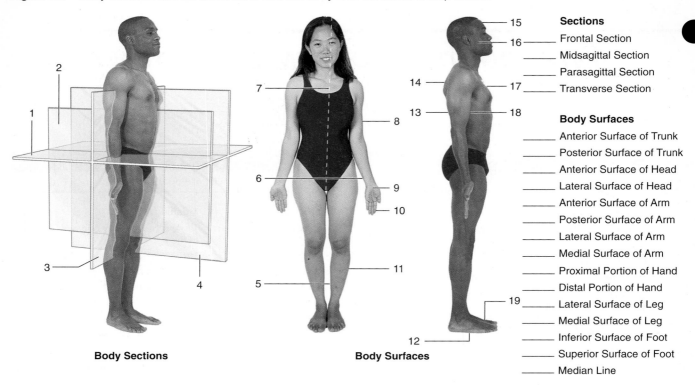

Sections
_____ Frontal Section
_____ Midsagittal Section
_____ Parasagittal Section
_____ Transverse Section

Body Surfaces
_____ Anterior Surface of Trunk
_____ Posterior Surface of Trunk
_____ Anterior Surface of Head
_____ Lateral Surface of Head
_____ Anterior Surface of Arm
_____ Posterior Surface of Arm
_____ Lateral Surface of Arm
_____ Medial Surface of Arm
_____ Proximal Portion of Hand
_____ Distal Portion of Hand
_____ Lateral Surface of Leg
_____ Medial Surface of Leg
_____ Inferior Surface of Foot
_____ Superior Surface of Foot
_____ Median Line

Body Sections   Body Surfaces

Transverse (horizontal) sections divide the body into superior and inferior portions and are perpendicular to the longitudinal axis of the body.

ASSIGNMENT

Check your understanding of directional terms, body sections, and body surfaces by labeling Figure 1.1. Place the correct number of each section and surface in the space in front of the labels listed. Record your responses for Figure 1.1 on Laboratory Report 1, which begins on page 235.

Regional Terminology

Specific anatomical terms are used to identify the various regions of the body. The **trunk** is the main portion of the body to which the **head, upper extremities** (arms), and **lower extremities** (legs) are attached. These major divisions are subdivided into smaller regions that are identified by specific anatomical terms. These anatomical regions are shown in Figure 1.2.

Trunk

The anterior portion of the trunk consists of an upper **thoracic region** (chest), which is subdivided into two **pectoral regions** separated by a central **sternal region.** The **abdominal region** is inferior to the thoracic region. Two **inguinal** (groin) **regions** lie at the junction of the trunk with the lower extremities. The **pubic region** lies between the inguinal regions.

The posterior surface of the trunk contains two **scapular regions,** which are located over the shoulder blades. An elongate **vertebral region** lies along the vertebral column. The small **sacral region** lies inferior to the vertebral region and between the two **gluteal regions** (buttocks). A **lumbar region** is located on each side of the vertebral region below the ribs.

The **axillary regions** (armpits) and the **coxal regions** (hips) lie on the lateral surfaces of the trunk.

Upper Extremity

An **acromial region** (shoulder) is formed by the attachment of each upper extremity to the trunk. The (upper) arm composes the **brachial region,** and the forearm composes the **antebrachial region.** The posterior surface of the elbow forms the **cubital region,** and the anterior surface of the elbow is the **antecubital region.** The **carpal region** (wrist), the **manual region** (hand), and the **digital region** (fingers) complete the regions of each upper extremity.

Lower Extremity

The proximal part of the lower extremity, the thigh, constitutes the **femoral region,** and the leg is the portion between the knee and ankle. The anterior portion of the leg is the **crural region;** the posterior portion is the **sural region.** The anterior surface of the knee is the

Figure 1.2 Regions of the body. Common names are noted in parentheses.

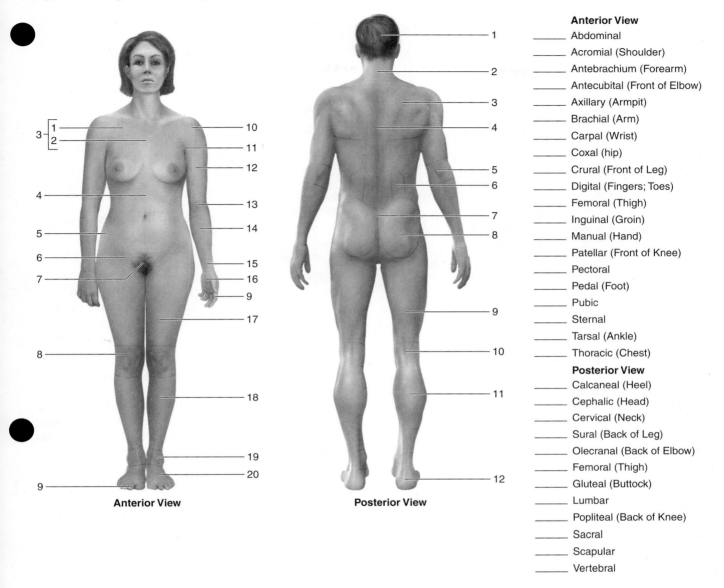

Anterior View   Posterior View

Anterior View
_____ Abdominal
_____ Acromial (Shoulder)
_____ Antebrachium (Forearm)
_____ Antecubital (Front of Elbow)
_____ Axillary (Armpit)
_____ Brachial (Arm)
_____ Carpal (Wrist)
_____ Coxal (hip)
_____ Crural (Front of Leg)
_____ Digital (Fingers; Toes)
_____ Femoral (Thigh)
_____ Inguinal (Groin)
_____ Manual (Hand)
_____ Patellar (Front of Knee)
_____ Pectoral
_____ Pedal (Foot)
_____ Pubic
_____ Sternal
_____ Tarsal (Ankle)
_____ Thoracic (Chest)
Posterior View
_____ Calcaneal (Heel)
_____ Cephalic (Head)
_____ Cervical (Neck)
_____ Sural (Back of Leg)
_____ Olecranal (Back of Elbow)
_____ Femoral (Thigh)
_____ Gluteal (Buttock)
_____ Lumbar
_____ Popliteal (Back of Knee)
_____ Sacral
_____ Scapular
_____ Vertebral

patellar region, and the posterior surface of the knee is the **popliteal region.** The **tarsal region** (ankle), **pedal region** (foot), **calcaneal region** (heel), and **digital region** (toes) complete the regions of each lower extremity.

Abdominal Divisions

Underlying internal organs may be located by dividing the abdominal surface into either nine regions or four quadrants. Anatomists prefer to divide the abdominal surface into nine regions formed by the intersecting of two vertical and two transverse planes as shown in Figure 1.3A.

The **umbilical region** is the central area that includes the umbilicus (navel). On each side of the umbilical region are the left and right **lumbar regions.** Just above the umbilical region is the **epigastric region,** which is between the left and right **hypochondriac regions.** Below the umbilical region is the **hypogastric,** or **pubic, region,** which is between the left and right **iliac regions.**

Health-care professionals often use a simpler system in which the abdomen is divided into quadrants by perpendicular vertical and transverse planes that intersect at the umbilicus. See Figure 1.3B. This division forms four quadrants: **upper right, upper left, lower right,** and **lower left.**

ASSIGNMENT

Label Figures 1.2 and 1.3 and record your responses on the laboratory report.

Figure 1.3 Abdominal regions and quadrants.

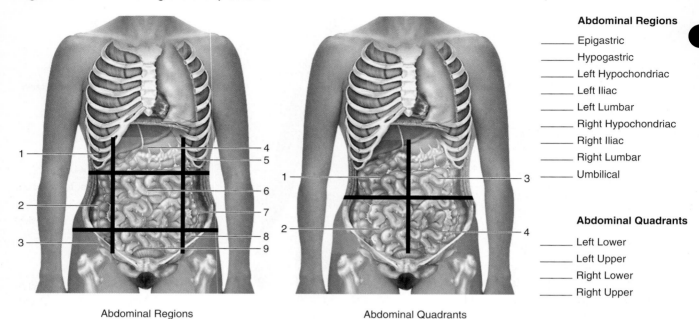

Abdominal Regions

Abdominal Quadrants

Body Cavities

Internal organs are located within body cavities. There are two major body cavities, the dorsal cavity and the ventral cavity, as shown in Figure 1.4. These cavities are named after the body surface nearest to them.

The **dorsal body cavity** consists of the **cranial cavity,** which contains the brain, and the **vertebral cavity,** which contains the spinal cord.

The **ventral body cavity** consists of the **thoracic cavity,** which houses the lungs, heart, and other thoracic organs, and the **abdominopelvic cavity,** which contains most of the internal organs. A thin, dome-shaped sheet of muscle, the **diaphragm,** separates the thoracic and abdominopelvic cavities.

Figure 1.4 Body cavities. (a) Midsagittal section; (b) frontal section.

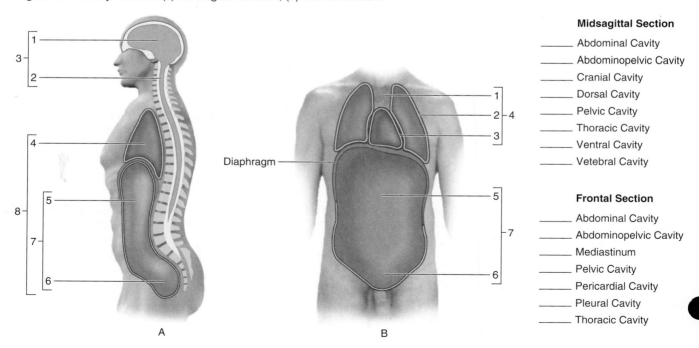

Diaphragm

A

B

Figure 1.5 Thoracic membranes. (a) Pleural membranes; (b) pericardial membranes.

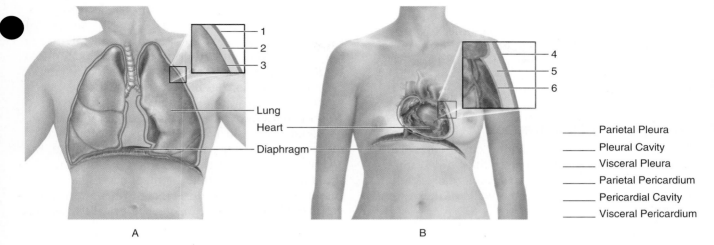

A   B

Lung
Heart
Diaphragm

_____ Parietal Pleura
_____ Pleural Cavity
_____ Visceral Pleura
_____ Parietal Pericardium
_____ Pericardial Cavity
_____ Visceral Pericardium

The thoracic cavity is divided into left and right portions by the **mediastinum,** a membranous partition that contains the heart, trachea, esophagus, and thymus gland. The **pericardial cavity,** containing the heart, is within the mediastinum. The right and left **pleural cavities** contain the lungs.

The abdominopelvic cavity consists of (1) the **abdominal cavity,** which contains the stomach, intestines, liver, gallbladder, pancreas, spleen, and kidneys,

and (2) the **pelvic cavity,** which contains the urinary bladder, sigmoid colon, and rectum. In females, it also contains the uterus, uterine tubes, and ovaries.

Membranes of Body Cavities

The body cavities are lined with membranes that provide protection and support for the internal organs. Refer to Figures 1.5 and 1.6 as you read this section.

Figure 1.6 Membranes of the abdominal cavity.

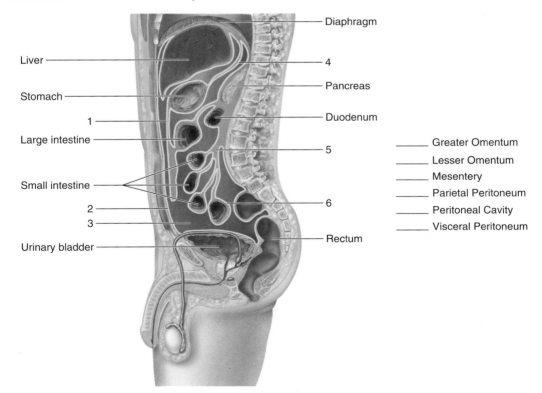

Liver
Stomach
1
Large intestine
Small intestine
2
3
Urinary bladder

Diaphragm
4
Pancreas
Duodenum
5
6
Rectum

_____ Greater Omentum
_____ Lesser Omentum
_____ Mesentery
_____ Parietal Peritoneum
_____ Peritoneal Cavity
_____ Visceral Peritoneum

Dorsal Cavity Membranes

The dorsal cavity is lined by three membranes that are collectively called the **meninges.** The outermost membrane is tough and fibrous; the other two membranes are more delicate. The thin innermost membrane closely envelops the brain and spinal cord. The meninges will be considered later when you study the nervous system.

Ventral Cavity Membranes

The membranes lining the ventral cavity are **serous membranes** that secrete a watery fluid called *serous fluid.* Their moist, slippery surfaces reduce friction between internal organs and the walls of the ventral cavity as the organs move.

Thoracic Cavity Membranes

The inner walls of the left and right portions of the thoracic cavity are lined by the **parietal pleurae;** the lungs are covered with the **visceral (pulmonary) pleurae.** The parietal and visceral pleurae are separated only by a thin layer of serous fluid. This moist, potential space between the pleural membranes is known as the **pleural cavity.** The parietal pleurae are continuous with the membranes forming the mediastinum.

The heart is enclosed within the **pericardium.** The **visceral pericardium,** or **epicardium,** is a thin serous membrane that is tightly attached to the exterior surface of the heart. Exterior to the epicardium is the loose-fitting, double-layered **parietal pericardium.** The inner layer of the parietal pericardium is a serous membrane, but the outer one is fibrous and very strong. The **pericardial cavity** is the potential space between the visceral and parietal pericardia.

Abdominal Cavity Membranes

The **peritoneum** lines the abdominal cavity. The **parietal peritoneum** lines the inner abdominal wall, and the **visceral peritoneum** covers the surface of the abdominal organs. The potential space between the visceral and parietal peritonea is the **peritoneal cavity.** See Figure 1.6. The kidneys are located in the **retroperitoneal region** (posterior to the peritoneum), so only part of their surfaces is covered with the parietal peritoneum.

Double-layered folds of the peritoneum, called **mesenteries,** extend from the cavity wall to the abdominal organs. The mesenteries support the abdominal organs, and they contain nerves and blood vessels serving the organs. A large mesenteric fold, the **greater omentum,** extends inferiorly from the stomach over the intestines and loops back up to join with the transverse colon (part of the large intestine). A smaller mesenteric fold, the **lesser omentum,** extends between the liver and the stomach.

The peritoneum does not line the pelvic cavity. It extends inferiorly only to cover the superior surface of the urinary bladder.

ASSIGNMENT

1. Label Figures 1.4, 1.5, and 1.6 and record the labels on the laboratory report.
2. Using a human torso model (a) locate the surface features and body cavities, (b) identify the major organs of each cavity, and (c) locate the organs within each of the nine abdominal regions and four abdominal quadrants.
3. Complete the laboratory report.

Exercise 2

BODY ORGANIZATION

OBJECTIVES

After completing this exercise, you should be able to:

1. List the organ systems and describe their general functions.
2. List the component organs for each organ system.
3. Locate and identify the organs in the ventral cavity of the rat.
4. Define all terms in bold print.

Materials

Freshly killed rat
Dissecting pan with wax bottom
Dissecting instruments and pins
Human torso model with removable organs
Protective disposable gloves

Before You Proceed

Consult with your instructor about using protective disposable gloves when performing portions of this exercise.

The human body is composed of a hierarchy of organizational levels. From simplest to most complex, they are chemical, cellular, tissue, organ, and organ system. Components of each organizational level work together to maintain **homeostasis,** the relative constancy of the body's normal internal environment. Homeostatic mechanisms enable the body to maintain its own stable internal environment.

Organ Systems

Most of the functions of the body are performed by organ systems. An **organ** is a structure that is (1) composed of two or more tissues, (2) has a definite shape, and (3) performs specific functions. The heart is an example of an organ. An **organ system** is a group of organs that function in a coordinated manner to perform specific functions. The cardiovascular system, which is composed of blood, heart, blood vessels, and spleen, is an example of an organ system.

In this exercise, you will study an overview of the components and general functions of the organ systems of the body in order to gain a general understanding of body organization and function. See Figures 2.1 and 2.2. Then you will dissect a freshly killed rat to observe the mammalian body organization, body cavities, and organs of the ventral cavity.

The Integumentary System

The skin, including hair, nails, associated glands, and sensory receptors, constitutes the integumentary system. The **skin** consists of two layers, an outer **epidermis** and an inner **dermis,** which protect the underlying tissues from mild abrasions, excessive water loss, microorganisms, and ultraviolet radiation. Perspiration secreted by sweat glands contains water and waste materials similar to dilute urine. The evaporation of perspiration cools the body surface.

The Skeletal System

The skeletal system forms the framework of the body and provides support and protection for softer organs and tissues. It consists of **bones, cartilages,** and **ligaments.** In conjunction with skeletal muscles, the skeleton forms lever systems that enable movement. In addition, **red bone marrow** produces blood cells.

The Muscular System

The contraction of muscles provides the force that enables movement. **Skeletal muscles** are attached to bones by **tendons** and constitute nearly half of the body weight. Their contractions move body parts during walking, eating, and other activities.

Two other types of muscle tissue occur in the body: smooth and cardiac. *Smooth muscle* is found in walls of hollow organs. *Cardiac muscle* occurs in the walls of the heart.

The Nervous System

The nervous system is a complex, highly organized system consisting of the **brain, spinal cord, cranial** and **spinal nerves,** and **sensory receptors.** These

Figure 2.1 Ten of the eleven human organ systems.

Integumentary

Components: skin, hair, nails, and associated glands
Functions: protects underlying tissues and helps regulate body temperature

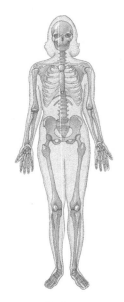

Skeletal

Components: bones, ligaments, and associated cartilages
Functions: supports the body, protects vital organs, stores minerals, and forms blood cells

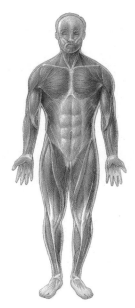

Muscular

Components: skeletal muscles, cardiac muscle of the heart, and smooth muscle in walls of internal organs
Functions: moves the body and body parts and produces heat

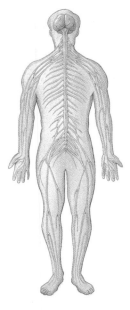

Nervous

Components: brain, spinal cord, nerves, and sensory receptors
Functions: rapidly coordinates body functions and enables learning and memory

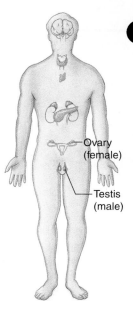

Ovary (female)

Testis (male)

Endocrine

Components: hormone-producing glands
Functions: secretes hormones that regulate body functions

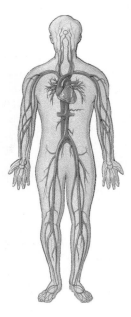

Cardiovascular

Components: blood, heart, arteries, veins, and capillaries
Functions: transports materials to and from body cells

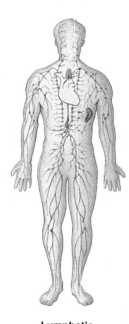

Lymphatic

Components: lymph, lymphatic vessels, and lymphatic tissue
Functions: collects, cleanses, and returns interstitial fluid to the blood; provides immunity

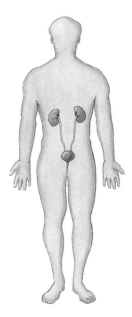

Urinary

Components: kidneys, ureters, urinary bladder, and urethra
Functions: regulates volume and composition of blood by forming and excreting urine

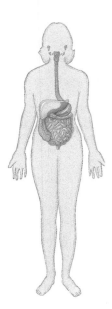

Digestive

Components: mouth, pharynx, esophagus, stomach, intestines, anus, liver, pancreas, and associated structures
Functions: digests food and absorbs nutrients

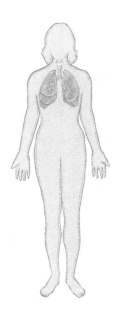

Respiratory

Components: nasal cavity, pharynx, larynx, trachea, bronchi, and lungs
Functions: exchanges O_2 and CO_2 between air and blood in the lungs

Figure 2.2 Male and female reproductive systems.

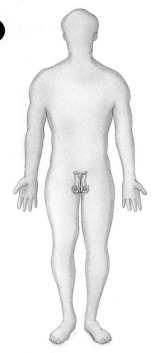

Male reproductive system
Components: testes, vasa deferentia, prostate gland, bulbourethral gland, and penis
Functions: produces sperm and transmits them into the female vagina during sexual intercourse

Female reproductive system
Components: ovaries, uterine tubes, uterus, vagina, and vulva
Functions: produces ova, receives sperm, provides intrauterine development of offspring, and enables birth of baby

components work together to enable rapid perception and interpretation of the environment and the almost instantaneous coordination of body functions. The human brain is responsible for the intelligence, will, self-awareness, and emotions characteristic of humans.

The Endocrine System

The endocrine system consists of small masses of glandular tissue that secrete hormones. Hormones, which are chemical messengers, are absorbed by the blood and transported throughout the body, where they bring about the chemical control of body functions. The larger masses of endocrine tissues occur in the **endocrine glands: pituitary, thyroid, parathyroid, thymus, adrenal, pancreas, pineal, ovaries,** and **testes.** Smaller masses of hormone-producing tissues occur in other organs, such as the kidneys, heart, and placenta, and in the digestive tract.

The Cardiovascular System

The **heart, arteries, veins, capillaries, spleen,** and **blood** constitute the cardiovascular system. These components work together to transport materials such as oxygen, carbon dioxide, nutrients, wastes, and hor-

mones throughout the body. Contractions of the heart circulate the blood through the blood vessels. Blood, consisting of cells and plasma, is the transporting agent and also provides the primary defense against disease organisms. The spleen serves as a blood reservoir and removes worn-out red blood cells from circulation.

The Lymphatic System

The lymphatic system consists of **lymphoid tissue** and a network of **lymphatic vessels** that collect fluid from interstitial spaces (spaces between cells) and return it to large veins under the collarbones. Tissue or interstitial fluid is called **lymph** after it enters a lymphatic vessel. En route, lymph passes through **lymph nodes,** nodules of lymphoid tissue, that remove cellular debris and microorganisms. Lymphoid tissue is found in many organs, such as the spleen, thymus, tonsils, adenoids, liver, intestines, and bone marrow. The lymphatic tissue produces immune cells and plays a central role in the defense against pathogens.

The Respiratory System

The exchange of oxygen and carbon dioxide between the atmosphere and the blood is enabled by the respiratory system. It consists of air passageways and gas-exchange organs. The passageways are the **nasal cavity, nasopharynx, larynx, trachea,** and **bronchi.** The gas exchange organs are the **lungs.**

The Digestive System

The digestive system converts large nonabsorbable food molecules into smaller nutrient molecules that can be absorbed into the blood. Digestion involves mechanically breaking food into smaller particles and mixing them with digestive fluids and chemically breaking large molecules into smaller molecules. The digestive tract consists of the **mouth, pharynx, esophagus, stomach, small intestine,** and **large intestine.** Major digestive glands are the **salivary glands, liver,** and **pancreas.**

The Urinary System

The nitrogenous wastes of metabolism and excess water and minerals are removed from the blood and body by the urinary system. The **kidneys** remove wastes and excess materials from the blood to form urine, which is carried by two **ureters** to the **urinary bladder** for temporary storage. Subsequently, urine is voided via the **urethra.**

The Reproductive System

Continuity of the species is the function of the male and female reproductive systems. The male reproductive system consists of a pair of sperm-producing **testes** located in the **scrotum;** two tubes, the **vasa deferentia,** which carry sperm to the **urethra;**

accessory glands that secrete fluids for sperm transport; and the **penis,** the male copulatory organ. The female reproductive system consists of a pair of **ovaries** that produce ova, two **uterine tubes** that carry the ova to the **uterus, accessory glands** that secrete lubricating fluids, and the **vagina,** the female copulatory organ and birth canal.

ASSIGNMENT

1. Complete the laboratory report for this exercise.
2. Locate the major organs of the organ systems on the human torso model.

Rat Dissection

You will work in pairs to perform this part of the exercise. Your principal objective in the dissection is to *expose the organs for study, not to simply cut up the animal.* Most cutting will be performed with scissors. The scalpel blade will be used only occasionally, but the flat blunt end of the handle will be used frequently for separating tissues.

Skinning the Ventral Surface

1. Pin the four feet to the bottom of the dissecting pan as illustrated in Figure 2.3. Before making any incision, examine the oral cavity. Note the large **incisors** in the front of the mouth, which are used for biting off food particles. Force the mouth open sufficiently to examine the flattened **molars** at the back of the mouth. These teeth are used for grinding food into small particles. Note that the **tongue** is attached at its posterior

Figure 2.3 Incision is started on the median line with a pair of scissors.

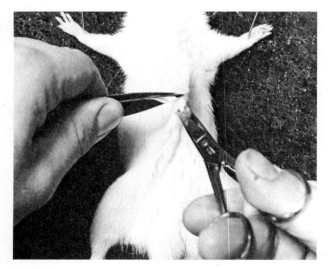

Figure 2.4 First cut is extended from the incision up to the lower jaw.

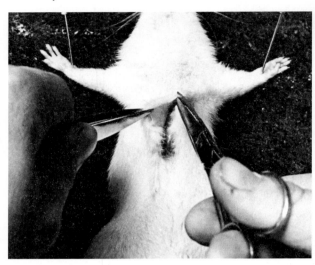

Figure 2.5 Completed incision from the lower jaw to the anus.

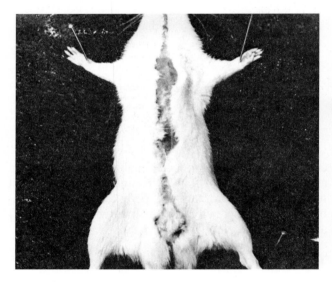

end. Lightly scrape the surface of the tongue with a scalpel to determine its texture. The roof of the mouth consists of an anterior **hard palate** and a posterior **soft palate.** The throat is called the **pharynx,** which is a component of both the digestive and respiratory systems.

2. Lift the skin along the midventral line with your forceps and make a small incision with scissors as shown in Figures 2.3 and 2.4. Cut the skin upward to the lower jaw, turn the pan around, and complete this incision to the anus, cutting around both sides of the genital openings. The completed incision should appear as in Figure 2.5.

3. With the handle of the scalpel, separate the skin from the musculature as shown in Figure 2.6. The fibrous connective tissue

Figure 2.6 Skin is separated from musculature with scalpel handle.

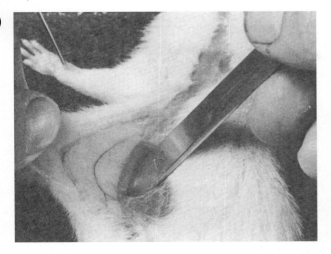

Figure 2.7 Incision of musculature is begun on the median line.

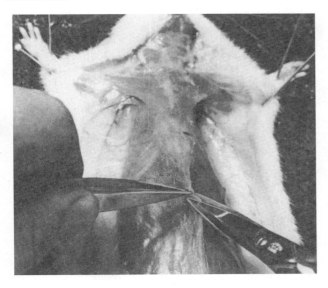

Figure 2.8 Lateral cuts at base of rib cage are made in both directions.

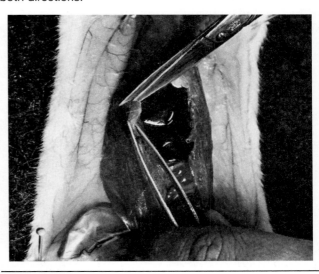

Figure 2.9 Flaps of abdominal wall are pinned back to expose viscera.

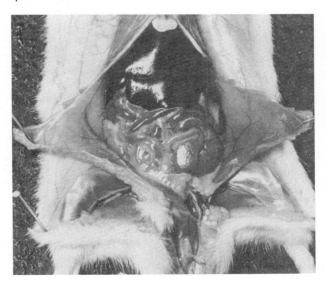

that lies between the skin and musculature is the **superficial fascia.**

4. Skin the legs down to the "knees" and "elbows" and pin the stretched-out skin to the wax. Examine the surfaces of the **muscles** and note that **tendons,** which consist of tough fibrous connective tissue, attach the muscles to the skeleton. Covering the surface of each muscle is another thin, gray feltlike layer, the **deep fascia.** Fibers of the deep fascia are continuous with fibers of the superficial fascia, so considerable force with the scalpel handle is necessary to separate the two membranes.

5. At this stage your specimen should appear as in Figure 2.7. If your specimen is a female, the mammary glands will probably remain attached to the skin.

Opening the Abdominal Wall

1. As shown in Figure 2.7, make an incision through the abdominal wall with a pair of scissors. To make the cut, you must hold the muscle tissue with a pair of forceps. **Caution:** Avoid damaging the underlying viscera as you cut.

2. Cut upward along the midline to the rib cage and downward along the midline to the genitalia.

3. To completely expose the abdominal organs, make two lateral cuts near the base of the rib cage—one to the left and the other to the right. See Figure 2.8. The cuts should extend all the way to the pinned-back skin.

4. Fold out the flaps of the body wall and pin them to the wax as shown in Figure 2.9. The abdominal organs are now well exposed.

Figure 2.10   Viscera of a female rat.

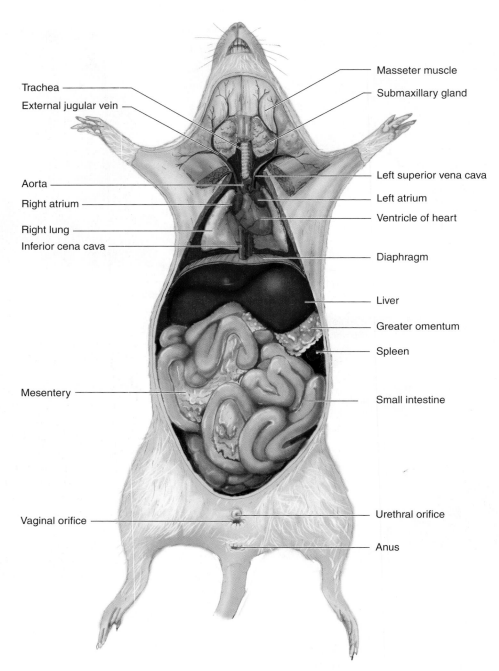

Trachea

External jugular vein

Masseter muscle

Submaxillary gland

Aorta

Right atrium

Right lung

Inferior cena cava

Left superior vena cava

Left atrium

Ventricle of heart

Diaphragm

Liver

Greater omentum

Spleen

Mesentery

Small intestine

Vaginal orifice

Urethral orifice

Anus

5. Using Figure 2.10 as a reference, identify all of the labeled viscera without moving the organs out of place. Note in particular the position and structure of the **diaphragm.**

Examination of the Thoracic Cavity

1. Using your scissors, cut along the left side of the rib cage as shown in Figure 2.11. Cut through all of the ribs and connective tissue. Then, cut along the right side of the rib cage in a similar manner.

2. Grasp the tip of the sternum with forceps as shown in Figure 2.12 and cut the diaphragm away from the rib cage with your scissors. Now you can lift up the rib cage and look into the thoracic cavity. See Figure 2.13.

3. With your scissors, complete the removal of the rib cage by cutting any remaining attachment tissue.

4. Now, examine the structures that are exposed in the thoracic cavity. Refer to Figure 2.10 and identify all the structures that are labeled.

Figure 2.11 Rib cage is severed on each side with scissors.

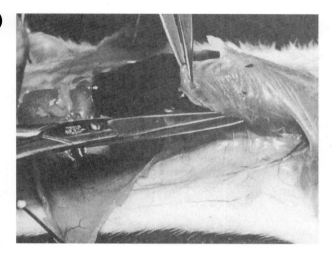

Figure 2.12 Diaphragm is cut free from edge of rib cage.

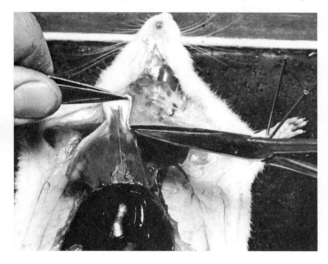

Figure 2.13 Thoracic organs are exposed as rib cage is lifted out.

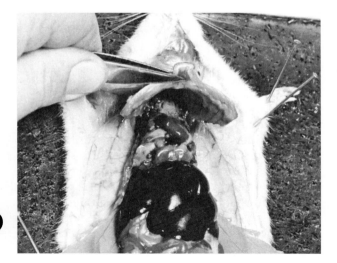

Figure 2.14 Specimen with heart, lungs, and thymus gland removed.

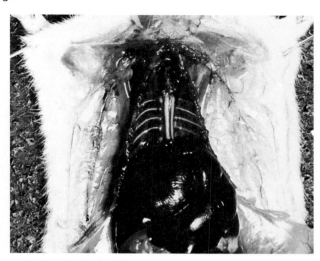

5. Note the pale **thymus gland,** which is located just above the heart. Remove this gland.
6. Carefully remove the thin **pericardial membrane** that encloses the **heart.**
7. Remove the heart by cutting through the major blood vessels attached to it. Gently sponge away pools of blood with Kimwipes or other soft tissues. See Figure 2.14.
8. Locate the **trachea** in the neck region and the **larynx** (voice box) at the anterior end of the trachea. Trace the trachea posteriorly to where it divides into two **bronchi** that enter the **lungs.** Squeeze the lungs with your fingers, noting how elastic they are. Remove the lungs.
9. Probe under the trachea to locate the soft tubular **esophagus** that runs from the oral cavity to the stomach. Excise a section of the trachea to reveal the esophagus as illustrated in Figure 2.15.

Deeper Examination of Abdominal Organs

1. Lift up the lobes of the reddish brown liver and examine them. Note that rats lack a **gallbladder.** *Carefully excise the liver* and wash out the abdominal cavity. The stomach and intestines are now clearly visible.
2. Lift out a portion of the intestines and identify the membranous **mesentery,** which holds the intestines in place. It contains blood vessels and nerves that supply the digestive tract. If your specimen is a mature healthy animal, the mesenteries will contain considerable fat.
3. Now, lift the intestines out of the abdominal cavity, cutting the mesenteries, as necessary,

Figure 2.15 Viscera of a male rat with heart, lungs, and thymus gland removed.

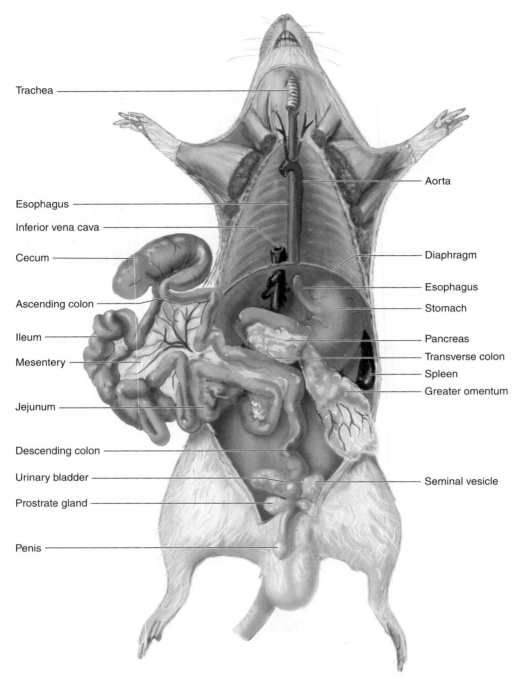

Trachea

Esophagus

Inferior vena cava

Cecum

Ascending colon

Ileum

Mesentery

Jejunum

Descending colon

Urinary bladder

Prostrate gland

Penis

Aorta

Diaphragm

Esophagus

Stomach

Pancreas

Transverse colon

Spleen

Greater omentum

Seminal vesicle

for a better view of the organs. Note the great length of the small intestine. Its name refers to its diameter, not its length. The first portion of the small intestine, which is connected to the stomach, is called the **duodenum.** At its distal end, the small intestine is connected to a large saclike structure, the **cecum.** The *appendix* in humans is a *vestigial* portion of the cecum. The cecum communicates with the **large**

intestine. This latter structure consists of the **ascending, transverse, descending,** and **sigmoid** segments. The last of these segments empties into the **rectum.**

4. Try to locate the **pancreas,** which is embedded in the mesentery alongside the duodenum. It is often difficult to see. Pancreatic enzymes enter the duodenum via the **pancreatic duct.** See if you can locate this minute tube.

Figure 2.16 Abdominopelvic cavity of a female rat with intestines and liver removed.

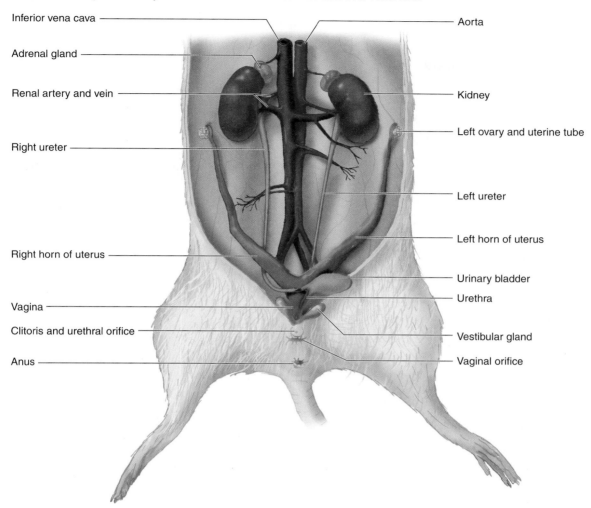

Inferior vena cava

Adrenal gland

Renal artery and vein

Right ureter

Right horn of uterus

Vagina

Clitoris and urethral orifice

Anus

Aorta

Kidney

Left ovary and uterine tube

Left ureter

Left horn of uterus

Urinary bladder

Urethra

Vestibular gland

Vaginal orifice

5. Locate the **spleen,** which is situated on the left side of the abdomen near the stomach. It is reddish brown and held in place with mesentery.
6. Remove the digestive tract by cutting through the esophagus next to the stomach and through the sigmoid colon. You can now see the descending **aorta** and the **inferior vena cava.** The aorta carries blood posteriorly to the body tissues. The inferior vena cava returns blood from the posterior regions to the heart.
7. Peel away the peritoneum and fat from the posterior wall of the abdominal cavity. *Removal of the fat will require special care to avoid damaging important structures.* The kidneys, blood vessels, and reproductive structures will now be more visible. Locate the two **kidneys** and **urinary bladder.** Trace the two **ureters,** which extend from the kidneys to the bladder. Examine the anterior surfaces of the kidneys and locate the **adrenal glands,** which

are important components of the endocrine gland system.
8. **Female.** If your specimen is a female, compare it with Figure 2.16. Locate the two **ovaries,** which lie lateral to the kidneys. From each ovary a **uterine tube** leads posteriorly to join the **uterus.** Note that the uterus is a Y-shaped structure joined to the **vagina.** If your specimen appears to be pregnant, open up the uterus and examine the developing embryos. Note how they are attached to the uterine wall.
9. **Male.** If your specimen is a male, compare it with Figure 2.17. The **urethra** is located in the **penis.** Apply pressure to one of the **testes** through the wall of the **scrotum** to see if it can be forced up into the **inguinal canal.** Carefully dissect out a testis, **epididymis,** and **vas deferens** from one side of the scrotum and, if possible, trace the vas deferens over the urinary bladder to where it penetrates the **prostate gland** to join the urethra.

Figure 2.17 Abdominopelvic cavity of a male rat with intestines and liver removed.

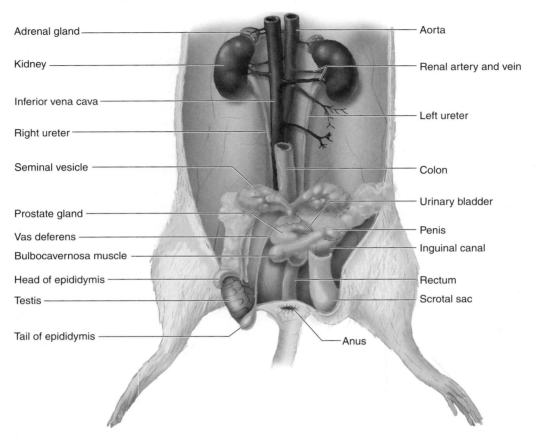

Adrenal gland

Kidney

Inferior vena cava

Right ureter

Seminal vesicle

Prostate gland

Vas deferens

Bulbocavernosa muscle

Head of epididymis

Testis

Tail of epididymis

Aorta

Renal artery and vein

Left ureter

Colon

Urinary bladder

Penis

Inguinal canal

Rectum

Scrotal sac

Anus

Conclusion

In this cursory dissection you have become ac-
quainted with the respiratory, circulatory, digestive,
urinary, and reproductive systems. Portions of the
endocrine system have also been observed. If you
have done a careful and thoughtful rat dissection,
you should have a good general understanding of
the basic structural organization of the human body.

Much that we see in rat anatomy has its human
counterpart.

Cleanup   Dispose of the specimen as directed by
your instructor. Scrub your instruments with soap
and water, rinse and dry them. Wash your hands with
soap; rinse, and dry thoroughly.

THE MICROSCOPE

After completing this exercise, you should be able to:

1. Identify the parts of a compound microscope and describe the function of each.
2. Describe and demonstrate the correct way to (a) carry a microscope, (b) clean the lenses, (c) focus with each objective, and (d) calculate the total magnification.
3. Define all terms in bold print.

Materials

    Compound microscope
    Prepared slide of the letter *e*
    Microscope slides
    Cover glasses
    Water in dropping bottle
    Scissors

The effective use of a **compound microscope** is essential for the study of cells and tissues. This exercise will introduce you to the parts, care, and use of a compound microscope.

Parts of the Microscope

Figure 3.1 shows the major parts of a compound microscope. Refer to it as you study this section. Your microscope may be somewhat different from the one illustrated, but you will be able to relate the illustration and the discussion to your microscope without difficulty.

The **base** is the bottom portion of the microscope. It contains the **lamp** and the on/off switch. Ideally, the lamp should have a **voltage control** to vary the intensity of light. The microscope shown in Figure 3.1 has a rotatable wheel on the right side of the base to regulate voltage. The lowest possible voltage that provides a good image should be used to extend the life of the lamp.

The **arm** rises from the base and supports the rest of the microscope. It serves as a convenient "handle" when carrying a microscope. The **body tube** has an **ocular lens** at the upper end and a **revolving nosepiece**, to which the **objective lenses** are attached, at the lower end. The nosepiece may be rotated to move an objective into viewing position.

The microscope in Figure 3.1 is binocular, and the head (and oculars) is rotatable so that the viewing position can be changed. However, many microscopes are monocular. The magnification of the ocular is usually 10×. Student microscopes usually have three objectives attached to the revolving nosepiece. The shortest objective is the **scanning objective** which has a magnification of 4×. It is used to scan the slide and locate the area of interest for viewing with more powerful objectives. The intermediate-length objective is the **low power objective,** which has a magnification of 10×. The longest objective is the **high-dry objective,** which has a magnification of 40× or 45× depending on the model of the microscope. Some student microscopes have a fourth objective that is slightly longer than the high-dry objective as shown in Figure 3.1. This is the **oil immersion objective,** and it has a magnification of 100×. You probably will not use the oil immersion objective in this course. The magnifications of the oil immersion and high-dry objectives may vary slightly in different models of microscopes.

The **stage** is the platform on which a microscope slide is placed for viewing. The opening in the center of the stage is the **stage aperture.** The **mechanical stage** is the device on the stage surface that holds the slide and enables precise movement of the slide. It is operated by the **mechanical stage control knobs.**

Below the stage is the **condenser,** which concentrates the light on the microscope slide. It may be raised or lowered by the **condenser control knob** located under the stage. Usually, it should be raised to its highest position.

An **iris diaphragm** within the condenser regulates the amount of light that reaches the slide. On the microscope shown in Figure 3.1, the diaphragm is adjusted by turning a knurled ring. Some microscopes have a diaphragm control lever.

Two focusing knobs are used to bring an object into clear focus. The **coarse focusing knob** has a larger diameter and is used to bring objects into rough focus. The **fine focusing knob** has a smaller diameter and is used to bring objects that are in rough focus into sharp focus.

Figure 3.1 A compound microscope.

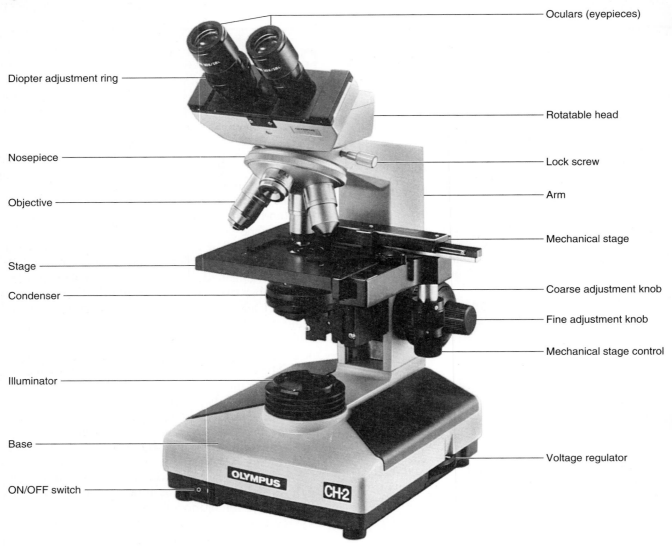

Oculars (eyepieces)

Diopter adjustment ring

Rotatable head

Nosepiece

Lock screw

Objective

Arm

Mechanical stage

Stage

Condenser

Coarse adjustment knob

Fine adjustment knob

Mechanical stage control

Illuminator

Base

Voltage regulator

ON/OFF switch

Courtesy of the Olympus Corporation, Lake Success, N.Y.

Magnification

The microscope lens system magnifies specimens so that very small structures can be distinguished. The **total magnification** is determined by the power of the ocular and objective being used. It is calculated by multiplying the power of the ocular by the power of the objective. For example, a $10\times$ ocular and a $45\times$ objective yield a total magnification of $450\times$.

Resolution

Magnification without resolution is of little value. **Resolution** is the ability to distinguish tiny adjacent objects as two distinct objects. **Resolving power** is a function of the wavelength of light and the design of the microscope lenses. The shortest wavelengths

(blue) of visible light provide maximum resolution. This is why microscopes have a blue light filter over the lamp. Use of the oil immersion objective is required for maximum resolution, and on the best light microscopes it will enable the distinction of microscopic objects that are 0.2 μm apart. If they are closer together, they will be seen as one object because of a fusion of the images.

Focusing

A microscope is focused by changing the distance between the object on the microscope slide and the objective lens. This change is accomplished by using the coarse and fine focusing knobs. The focusing knobs raise and lower either the stage *or* the body tube depending on the type of microscope. Both types of focusing procedures are described below. Determine which method is to be used for your microscope.

As a general rule, you should always start focusing with the low power (10×) objective. The coarse focusing knob is used *only* with the low power objective. With the high-dry and oil immersion objectives use *only* the fine focusing knob.

Focusing with a Movable Stage

1. Place the 10× objective into viewing position by rotating the nosepiece.
2. While looking from the side (not through the ocular), raise the stage with the coarse focusing knob until either it stops or the slide is about 3 mm from the objective.
3. While looking through the ocular, slowly lower the stage with the coarse focusing knob until the object comes into view.
4. Use the fine focusing knob to bring the object into sharp focus.

Focusing with a Movable Body Tube

1. Place the 10× objective in viewing position by rotating the nosepiece.
2. While looking from the side (not through the ocular), lower the body tube with the coarse focusing knob until either it stops or the slide is about 3 mm from the objective.
3. While looking through the ocular, slowly raise the body tube with the coarse focusing knob until the object comes into view.
4. Use the fine focusing knob to bring the object into sharp focus.

Switching Objectives

Modern microscopes are usually **parcentric** and **parfocal.** This means that when an object is centered in the field and in sharp focus with one objective, it will be centered and in focus when another objective is rotated into the viewing position. However, it may be necessary to make slight adjustments to re-center the object with the mechanical stage or bring it into sharper focus with the fine focusing knob.

Start your observations with the low power objective, even though you may want to observe the object with the high-dry or oil immersion objective. Once the object is centered and in focus with the low power objective, it is easy to switch to the high-dry objective simply by rotating the nosepiece. Note that the **working distance,** the distance between the objective and the slide, decreases as the power of the objective increases.

The amount of light entering the objective decreases as the power of the objective increases. Thus,

you will need to increase the voltage or open the diaphragm a bit to provide a light intensity that yields the sharpest image with the high-dry objective.

The easiest and safest way to bring the oil immersion objective into position is to progress from low power to high-dry to oil immersion. However, you may switch directly from the low power objective to the oil immersion objective. Before rotating the oil immersion objective into the viewing position, place a drop of immersion oil on the slide. Also, open the diaphragm to its maximum aperture to increase the amount of light. A *slight adjustment of the fine focusing knob* is all that will be required to bring the object into focus.

Care of the Microscope

Your instructor will assign a specific microscope to you for your use throughout the course. You will be responsible for its care. Do not use a different microscope without permission. Report any problems or malfunctions to your instructor immediately.

Transport and Placement

Care must be used in transporting the microscope from the storage cabinet to your workstation. Figure 3.2

Figure 3.2 The microscope should be held firmly with both hands while being carried.

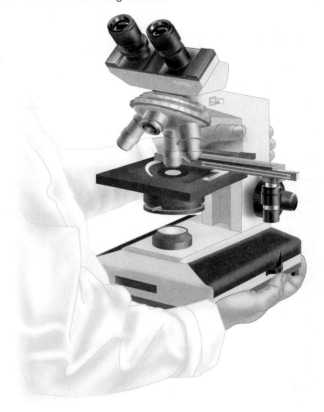

Figure 3.3 The microscope position on the right has the advantage of stage accessibility.

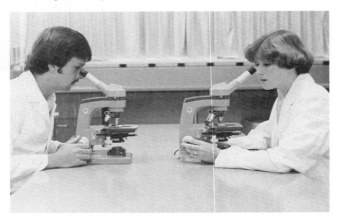

Figure 3.4 When oculars are removed for cleaning, cover the ocular opening with lens tissue. A blast from an air syringe or gas canister removes dust and lint.

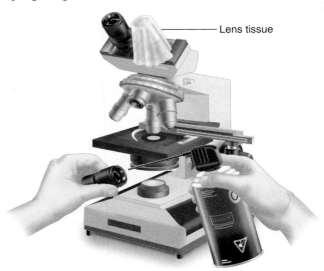

Lens tissue

shows the correct way to carry a microscope. Note that it is carried in front of you, with one hand supporting the base while the other grasps the arm. *Never* carry it with one hand at your side because a microscope in this position is apt to collide with furniture in the lab.

Most modern microscopes have an inclined body tube and a rotatable head that is secured by a set screw. You may use the microscope in either of the positions shown in Figure 3.3. The conventional position is with the arm facing the student. However, when the stage is facing the student, it is more easily accessible. Use the position recommended by your instructor.

Lens Care

Develop the habit of cleaning the lenses with **lens paper** before using the microscope. *Use only lens paper for cleaning the lenses.* If liquid gets on the objectives or stage of the microscope, wipe it off immediately with lens paper. If simply wiping the lenses with lens paper doesn't get them clean, it may be necessary to clean them with lens paper moistened with green soap or xylene. *Use such a procedure only as directed by your instructor. Never try to disassemble any part of the microscope.*

The best way to determine if the ocular is clean is to rotate it between your thumb and forefinger while looking through the microscope. A rotating pattern is evidence of a dirty lens. If cleaning the top lens fails to remove the debris, *inform your instructor.*

If the ocular is removed to clean the lower lens, it is imperative that a piece of lens paper be placed over the open end of the body tube as shown in Figure 3.4.

An image that appears cloudy or blurred indicates a dirty objective. If cleaning with lens paper moistened with water fails to clear the image, *consult your instructor* about cleaning the lens with xylene.

Routinely wipe off the upper surface of the condenser lens with lens paper to remove any accumulated dust.

Cleanup and Storage

When you have finished using the microscope, complete the following steps before returning it to the cabinet.

1. Remove the slide from the stage.
2. Clean the ocular, objective lenses, and the stage.
3. Rotate the nosepiece to place the lowest power objective in the viewing position.
4. Depending on the type of microscope, lower the body tube to its lowest position *or* raise the stage to its highest position.
5. Raise the condenser to its highest position.
6. Adjust the mechanical stage so that it projects minimally from each side of the stage.
7. Place the dustcover over the microscope.

ASSIGNMENT

Complete Sections A through C of the laboratory report.

Using the Microscope

Obtain the microscope assigned to you and carry it to your station in the manner just described. Locate on your microscope the parts labeled in Figure 3.1. Manipulate the control knobs and levers to see how they operate. Set up your microscope for viewing in one of the positions shown in Figure 3.3 as directed by your instructor.

Plug in the light cord, turn on the lamp, and look through the ocular. The circle of light that you see is called the **field of view** or just the **field.** If your microscope is binocular, adjust the oculars to match the distance between your eyes.

Clean the lenses with lens paper. Raise the condenser to its highest position. Once you feel comfortable with the microscope controls, proceed with the assignment below.

In this course, you will examine specimens on prepared slides and on liquid-mount slides. A *prepared slide* has a cover slip permanently mounted on the slide over the specimen. A *wet-mount slide* has a cover slip temporarily placed over the specimen, which is mounted in liquid.

ASSIGNMENT

1. Obtain a prepared slide of the letter *e,* handling it by the edges. *Never place your thumb or finger on the cover glass of a prepared slide.* Following the procedures shown in Figure 3.5, place the slide in the mechanical stage and center the letter *e* over the stage aperture.
2. With the low power objective (10✕) in the viewing position, use the focusing technique described earlier for your microscope to focus on the letter *e.* Adjust the voltage control to the lowest level that provides a sharp image. Practice moving the slide with the mechanical

Figure 3.5 The slide must be properly positioned as the retainer lever is moved to the right.

stage until you can center the *e* easily. What is different about the orientation of the *e* when viewed with the naked eye and through the microscope?
3. Center the *e* in the field. While viewing through the ocular, increase and decrease the light by adjusting the iris diaphragm or the voltage rheostat, noting how the sharpness of the image changes. Make drawings of your observations in Section D, number 1 of the laboratory report.
4. With the *e* centered, rotate the high-dry objective into position. Adjust the fine focus and the light to obtain a sharp image. Is the *e* centered and in focus? Do you need to make a large or small adjustment to see a sharp image? What does the *e* look like? At this magnification, you are seeing only ink blotches of a small part of the letter. Note how the **diameter of field** decreases as magnification increases. Complete Section D, number 1 of the laboratory report.
5. Practice steps 1–4 until you can easily place the slide in the mechanical stage and quickly focus on the *e* with the low power objective and then with the high-dry objective.

6. Obtain a clean microscope slide, cover glass, scissors, and dropping bottle of water. Make a wet-mount slide of crossed human hairs as shown in Figure 3.6.

7. Use the low power objective to examine the crossed hairs. Focus on the point where the hairs cross. Which hair is on top? Are both hairs in sharp focus? The vertical distance within which an object is in sharp focus is known as the **depth of field.** At this magnification, the depth of field is slightly less than the thickness of the crossed hairs.

8. Rotate the high-dry objective into viewing position and examine the crossed hairs. Are both hairs in sharp focus? At this magnification, only a portion of one hair is in sharp focus at a time.

9. Carefully focus through the thickness of the hair. In a surface view, you will see the margins of dead cells that form the hair, and the edges of the hair will be out of focus. When you focus through the midsection of the hair, the central core and edges of the hair will be in sharp focus. The depth of field is less than the thickness of a hair. Complete Section D, number 2 of the laboratory report.

ASSIGNMENT

Complete the laboratory report.

Figure 3.6 Making a wet-mount slide of crossed human hairs.

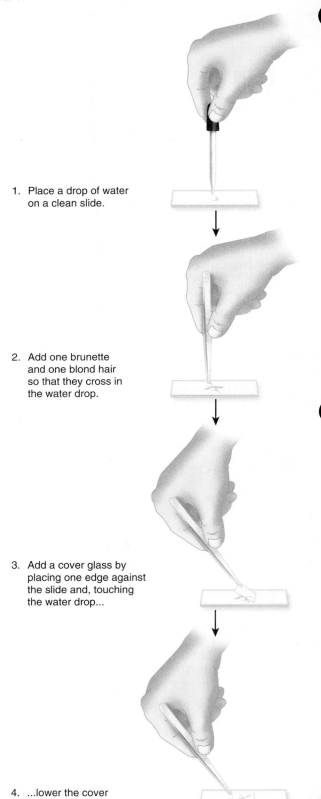

1. Place a drop of water on a clean slide.

2. Add one brunette and one blond hair so that they cross in the water drop.

3. Add a cover glass by placing one edge against the slide and, touching the water drop...

4. ...lower the cover glass.

Exercise 4

CELL ANATOMY

Materials

    Model of a generalized cell
    Biohazard container
    Compound microscope
    Microscope slides and cover glasses
    Medicine droppers, toothpicks
    Methylene blue, 0.01%, in dropping bottles
    Culture of *Amoeba*
    Prepared slides of:
        human sperm
        human blood
    Pond water

The **cell** is the structural and functional unit of all organisms, including humans. The processes of life take place within cells. An understanding of cell anatomy is essential to understanding the physiological processes that occur within cells.

All cells have many anatomical and functional characteristics in common, although differences in size, shape, and internal composition may result from specialization for particular functions. The human body is composed of trillions of cells. Most of these cells are specialized to perform specific functions. In this exercise, you will (1) study the anatomical features common to human cells and (2) examine the structure of selected cells with a compound microscope.

Basic Structure

When viewed with your compound microscope, three major components of a cell are visible. The **plasma membrane** forms the outer boundary of a cell, the **nucleus** appears as a dense spherical body within the cell, and the **cytoplasm** is the relatively clear area between the nucleus and the plasma membrane. However, cell structure is much more complex.

Electron microscopy has shown the cell to be a complex organization of components called **organelles.** Organelles are tiny organs that are specialized for specific functions. Because specific cellular functions are compartmentalized within specific organelles, many cellular functions can take place at the same time. Figure 4.1 shows the ultrastructure of a pancreatic cell, a rather unspecialized cell, as it might appear when viewed with an electron microscope. It is not possible to see the structure of organelles with your microscope. As you read the discussion of cell anatomy that follows, locate the structures in Figure 4.1.

The Plasma Membrane

Cell contents are separated from the surrounding environment by the **plasma (cell) membrane.** Like all cell membranes, the plasma membrane consists of two layers of phospholipid molecules (label 6), arranged back to back, in which protein molecules (label 5) are embedded. The membrane is flexible, with the consistency of a liquid film. See Figure 4.1.

The **selective permeability** of the plasma membrane allows some molecules to pass through the membrane while preventing the passage of others. In this way, the movement of materials into and out of the cell is controlled by the plasma membrane in both active and passive transport.

Several different types of proteins occur in the plasma membrane where they perform important and specific functions. Some (e.g., channel and carrier proteins) penetrate through the plasma membrane to form passageways for substances to enter or leave a cell. Membrane proteins with other functions (e.g., receptors) occur primarily on the intracellular or extracellular surface of the plasma membrane. Table 4.1 lists the major types of membrane proteins and their functions.

Figure 4.1 The microstructure of a cell.

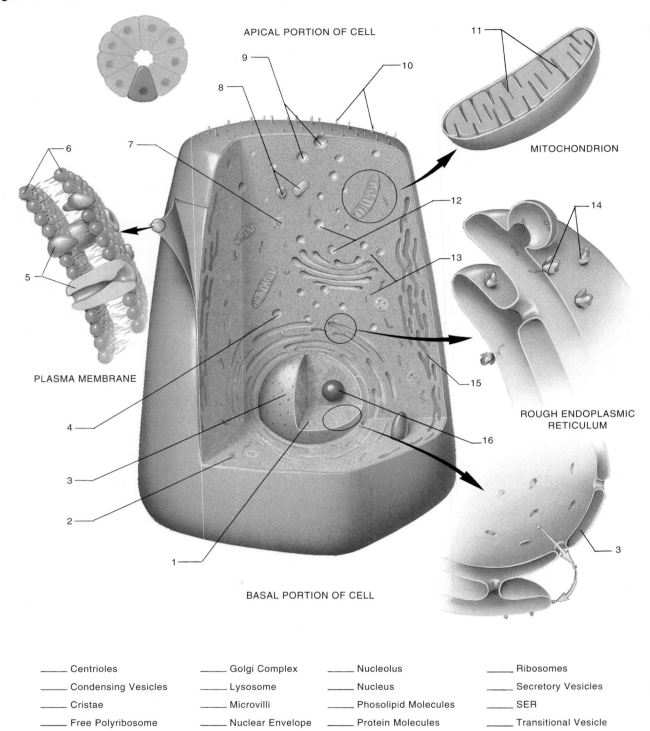

APICAL PORTION OF CELL

MITOCHONDRION

PLASMA MEMBRANE

ROUGH ENDOPLASMIC
RETICULUM

BASAL PORTION OF CELL

_____ Centrioles	_____ Golgi Complex	_____ Nucleolus	_____ Ribosomes
_____ Condensing Vesicles	_____ Lysosome	_____ Nucleus	_____ Secretory Vesicles
_____ Cristae	_____ Microvilli	_____ Phosolipid Molecules	_____ SER
_____ Free Polyribosome	_____ Nuclear Envelope	_____ Protein Molecules	_____ Transitional Vesicle

The Nucleus

The **nucleus** is a large, spherical organelle surrounded by a double-layered **nuclear envelope.** Unlike other cell membranes, the nuclear envelope contains pores that facilitate movement of materials between the nucleus and the cytoplasm. Note that a portion of the nuclear envelope is enlarged in Figure 4.1.

The nucleus is the control center of the cell because it contains the threadlike **chromosomes** composed of **deoxyribonucleic acid (DNA)** and protein. The genetic information that controls cellular functions is encoded in the DNA molecules. In nondividing cells, the uncoiled and dispersed chromosomes appear as **chromatin granules** (not shown). During cell division, they coil tightly to form the familiar rod-like shape of chromosomes.

A single spherical **nucleolus** is shown within the nucleus in Figure 4.1, although some types of cells may contain two or more nucleoli. Nucleoli are composed of **ribonucleic acid (RNA)** and proteins; they are assembly sites for these molecules, which will move into the cytoplasm to form ribosomes.

The Cytoplasm

The **cytoplasm** lies between the nucleus and the plasma membrane. It contains the fluid portion of the cell, the **cytosol,** which bathes the specialized organelles. The cytosol, or **intracellular fluid (ICF),** is composed of an aqueous mixture of many different kinds of chemicals. Cellular functions are primarily performed by the organelles, as described below.

Endoplasmic Reticulum (ER)

The endoplasmic reticulum (ER) is a network of membranes that permeates the cytoplasm between the nuclear envelope and plasma membrane and provides channels for the movement of molecules. **Rough endoplasmic reticulum (RER)** has ribosomes on its surface and provides storage and transport for proteins formed by the ribosomes. The RER usually surrounds the nucleus. In contrast, the **smooth endoplasmic reticulum (SER)** (label 15) lacks ribosomes. Its function may vary in different types of body cells. For example, it is involved in the synthesis of steroids in ovaries and testes, detoxification of drugs in the liver and kidneys, and storage and release of calcium in muscle cells to trigger muscle contraction.

Ribosomes

Ribosomes are tiny organelles composed of RNA and protein and serve as sites of protein synthesis. Ribosomes (label 14) associated with the RER produce proteins for export from the cell. The clusters or chains of **free polyribosomes** (label 2) scattered in the cytoplasm form proteins for intracellular use.

Golgi Complex

The Golgi complex is similar in structure to the ER. It consists of a stack of membranes adjacent to the RER. Proteins formed by ribosomes of the RER are packaged by the RER into transitional vesicles (label 4) that move to the Golgi complex. The role of the Golgi complex is to complete the processing and repackaging of such products for export from the cell. It places products for export into condensing vesicles (label 12), where final processing occurs as the vesicles move to the cell surface. Near the cell surface, condensing vesicles become secretory vesicles (label 9) that release their products outside the cell by *exocytosis*. The reverse process, endocytosis, moves substances into the cell within vesicles formed by the plasma membrane. The term **vesicle** refers to any small membranous sac of material within a cell.

Mitochondria

Mitochondria are ovoid or elongated organelles that have an outer membrane enclosing an inner membrane that has numerous partitionlike folds called **cristae** (label 11). The folds increase the surface area of the inner membrane to facilitate chemical reactions. Mitochondria can move about in the cytoplasm and can replicate themselves. They contain small amounts of RNA and DNA. They are sometimes called the "powerhouses" of the cell because they are sites of ATP synthesis by cellular respiration.

Lysosomes

Lysosomes are tiny oval sacs (label 7) that contain digestive enzymes and probably are formed by the Golgi complex. They release their enzymes into **vacuoles,** fluid-filled, membrane-enclosed sacs in the cytoplasm, to digest engulfed foreign particles or worn-out organelles. In fatally damaged cells, the lysosome membrane ruptures, and the released enzymes digest the entire cell.

Cytoskeletal Elements

Very thin protein fibers extend throughout the cytoplasm, providing support, contractility, and the movement of organelles within the cell. Two types of protein fibers compose these cytoskeletal elements: microtubules and microfilaments. They are not shown in Figure 4.1.

The cylindrical **microtubules** are the larger of the two. They serve as pathways for the movement of vesicles and other organelles within the cytoplasm and provide support for the cell. The thinner **microfilaments** are solid rather than tubular, and they are responsible for cell movements because of their contractility (shortening). Microfilaments are best developed in muscle cells.

Centrioles

Cells capable of cell division contain a pair of **centrioles** (label 8) located close together and at right angles to each other. Together they constitute the **centrosome.** Each centriole is a short hollow tube whose wall is formed by nine microtubules. Centrioles are centers of microtubule production, especially those that form the astral fibers and spindle fibers during cell division. You will observe this relationship in Exercise 5. In cells viewed with a light microscope, centrioles appear as a pair of dots near the nucleus.

Microvilli

Microvilli are tiny cytoplasmic projections on the free surfaces of certain cells. They greatly increase the surface areas of the cells. Microvilli (label 10) constitute the brush border on cells involved in the absorption of substances.

Cilia and Flagella

Both cilia and flagella (not shown in Figure 4.1) are composed of microtubules derived from centrioles or their products. **Cilia** are numerous, short, hairlike projections on the surface of certain cells that are used to move particles along the cell surface. They occur on cells lining the uterine tubes and respiratory passages. Spermatozoa are the only human cells with a **flagellum.** Each sperm has a single flagellum whose whiplike movements propel the sperm cell.

ASSIGNMENT

1. Label Figure 4.1 and record the labels on laboratory report 4.
2. Locate the parts of a cell on the cell model.
3. Complete Section B of the laboratory report.

Microscopic Study

The following microscopic observations of cells will aid your understanding of cell structure. Refer to Section C of the laboratory report as you make your observations.

The *Amoeba*

The *Amoeba* is a common freshwater protozoan that is large enough for you to observe the nucleus, cytoplasm, vacuoles, and cytoplasmic granules. A sol-gel reversibility of the cytoplasm enables the flowing **amoeboid movement** and the capture of minute organisms by **phagocytosis** (engulfment of particles into vacuoles). The organisms are digested within the food vacuoles. In a similar way, your white blood cells can move among body tissues, and some of them engulf and digest bacteria as part of your defense against disease organisms.

Contractile (excretory) vacuoles maintain the intracellular water balance by collecting and pumping out excessive water that accumulates because water continuously diffuses into the *Amoeba.* As you observe an *Amoeba,* these vacuoles will appear as clear bubbles in the cytoplasm that disappear periodically as the water is forced out of the cell.

1. Place a drop from the <u>bottom</u> of the *Amoeba* culture on a clean slide. Do not add a cover glass. Using the low power objective, locate an *Amoeba* and observe it. Locate the structures labeled in Figure 4.2. The cytosol in the interior of the cell is the *plasmasol,* which is in a fluid state. The *plasmagel* adjacent to the plasma membrane has the consistency of gelatin. Is the plasmasol granular or clear? Observe the amoeboid movement. Is the plasma membrane rigid or flexible?
2. Gently add a cover glass and examine the structures labeled in Figure 4.2 using the high-dry objective.

Figure 4.2 *Amoeba proteus.*

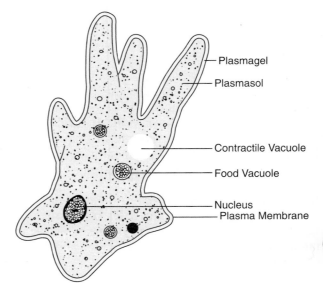

Plasmagel
Plasmasol
Contractile Vacuole
Food Vacuole
Nucleus
Plasma Membrane

Epithelial Cells

The epithelial cells lining the inside of your cheek are easily obtained for study. The cells that you will remove are dead and are continuously sloughed off normally.

1. Place a drop of methylene blue in the center of a clean slide.
2. *Gently* scrape the inside of your cheek with a clean toothpick, then swirl the end in the drop of methylene blue solution to dislodge the cells. *Immediately place the toothpick in the biohazard container. Do not place it on the tabletop.*
3. Add a cover glass and locate the cells with the low power objective.
4. Center the cells and rotate the high-dry objective into position for careful examination. Locate the nucleus, cytoplasm, and plasma membrane.
5. *When you are finished examining your slide, place the slide with the cover glass in the bio-hazard container.*

Human Sperm

Examine a prepared slide of human sperm set up under the oil immersion objective of a demonstration microscope. Note their small size. The head consists primarily of the sperm nucleus. Undulations of the flagellum enable forward movement by a sperm.

Human Blood Cells

Sharpen your microscopy skills by examining a pre-pared slide of human blood with the low power and high-dry objectives. The blood cells have been stained to enable recognition of cell types. Most of the cells that you see are red blood cells. They are stained pink and lack a nucleus. There are five types of white blood cells. A white blood cell is easy to recognize because its nucleus is stained purple. Some white blood cells have cytoplasmic granules stained lavender, red, or blue, but they are difficult to see at 400×.

Locate a white blood cell with the low power objective, then switch to the high-dry objective for your observations. Locate as many different types of blood cells as you can.

Microorganisms

Prepare a wet-mount slide of pond water. Place a drop from the bottom of the culture jar on a clean microscope slide and add a cover glass.

Examine the organisms, usually protozoans and algae, with both low and high-dry objectives. Look for cellular structures, such as nuclei, vacuoles, cilia, and flagella. Algae appear green because they possess the green pigment chlorophyll within organelles called chloroplasts. If you wish to identify the microorganisms, refer to Appendix D, which illustrates many different protozoans and algae. B)

pp. 341 -342

ASSIGNMENT

Complete the laboratory report.

Exercise 5

MITOTIC CELL DIVISION

Materials

Prepared slide of:

Whitefish mitosis

Mitotic cell division is the process by which a parent cell divides to form two daughter cells that have the same genetic composition as the parent cell. All new body cells are formed this way, and millions of new cells are formed each day. These new cells enable either growth or replacement of damaged and worn-out cells.

Hereditary information (directions controlling cellular functions) is encoded in the DNA molecule forming the core of each chromosome. Human body cells contain 23 pairs of chromosomes—a total of 46 chromosomes. Each pair of chromosomes contains its own unique encoded information. In cell division, it is vital that each chromosome is precisely replicated and that the replicated chromosomes are equally distributed to the new daughter cells. This is accomplished by mitotic cell division.

The Cell Cycle

The life cycle of a cell consists of three distinct stages: interphase, mitosis, and cytokinesis.

Interphase

Interphase is the stage of the life cycle when a cell is carrying out its normal metabolic functions and is not undergoing cell division. It is the stage between the end of one cell division and the start of the next cell division. During interphase, chromosomes are

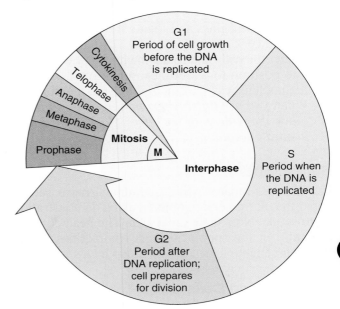

Figure 5.1 The cell cycle.

long, thin strands, which are loosely packed in the nucleus. When viewed microscopically, they appear as tiny dots and lines called **chromatin.**

As shown in Figure 5.1, interphase consists of three periods: G_1, S, and G_2. G_1 (Gap 1) is a period of growth in a newly formed cell. Cells that will not divide again remain in this stage and carry on their normal functions.

In the S (Synthesis) stage, chromosome replication occurs. The precise replication of the DNA molecules is the key component of the replication process.

In G_2, the cell makes final preparations, including completion of centriole replication, for the impending cell division.

Mitosis

Mitosis is the orderly separation and equal distribution of the replicated chromosomes to form two daughter nuclei with identical chromosomes and, therefore, the same genetic composition. Mitosis accounts for only 5% to 10% of the cell cycle. Once it begins, it is a continuous process, but it is subdivided into four phases for ease of understanding: prophase, metaphase,

Figure 5.2 Mitotic cell division.

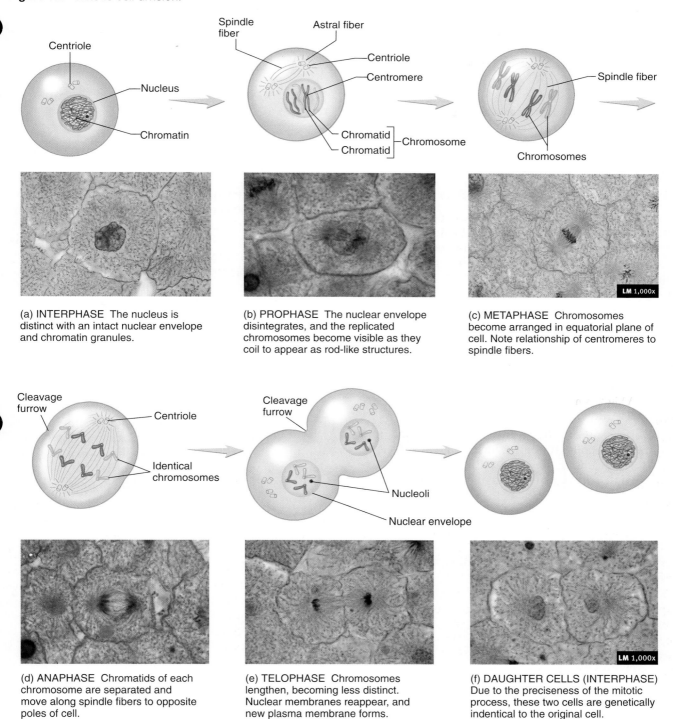

(a) INTERPHASE The nucleus is distinct with an intact nuclear envelope and chromatin granules.

(b) PROPHASE The nuclear envelope disintegrates, and the replicated chromosomes become visible as they coil to appear as rod-like structures.

(c) METAPHASE Chromosomes become arranged in equatorial plane of cell. Note relationship of centromeres to spindle fibers.

(d) ANAPHASE Chromatids of each chromosome are separated and move along spindle fibers to opposite poles of cell.

(e) TELOPHASE Chromosomes lengthen, becoming less distinct. Nuclear membranes reappear, and new plasma membrane forms.

(f) DAUGHTER CELLS (INTERPHASE) Due to the preciseness of the mitotic process, these two cells are genetically indentical to the original cell.

anaphase, and telophase. Examine Figure 5.2 as you study these phases.

Prophase

During **prophase,** the nuclear envelope disintegrates. The replicated chromosomes coil tightly and become visible as rod-shaped structures. Each replicated chromosome consists of two **sister chromatids** joined by the paired **centromeres.** The two centrosomes move to opposite ends of the cell. By the end of prophase, each pair of centrioles in each centrosome has formed **spindle fibers** (microtubules), which extend to the equator of the cell, and **astral fibers** (microtubules), which radiate to the plasma membrane. Each centrosome forms one pole (end) of the spindle and is called an **aster** because of the radiating astral fibers. The replicated chromosomes begin their migration to the equator of the spindle.

Metaphase

During the brief **metaphase,** the replicated chromosomes line up on the equatorial plane of the spindle. The centromeres of sister chromatids are attached to spindle fibers extending to opposite poles of the spindle.

Anaphase

In anaphase, the paired centromeres separate, separating the sister chromatids, which migrate along spindle fibers to opposite poles of the cell. They appear to be drawn to the poles by the spindle fibers. The separated sister chromatids are now called **daughter chromosomes.**

Telophase

During telophase, the daughter chromosomes extend and become less distinct. A new nuclear envelope forms around each set of chromosomes, forming the daughter nuclei. The spindle fibers gradually disappear.

Cytokinesis

Cytokinesis is the division of the parent cell into two daughter cells. It begins during late anaphase or early telophase with the formation of a **cleavage furrow,** which forms as the plasma membrane constricts at the equator of the cell. The cleavage furrow gradually constricts and finally separates the parent cell into two daughter cells. The daughter cells then enter interphase of the cell cycle.

ASSIGNMENT

Complete Sections A and B of the laboratory report.

Microscopic Study

In this section, you will examine mitotic cell division in cells of whitefish blastula, an early embryonic stage. Use the low power objective to locate cells in interphase and the stages of mitosis. Then switch to the high-dry objective for careful examination. Your study will be simplified by studying mitotic stages as you encounter them on your slides rather than by trying to locate them as they occur in the mitotic sequence. There are several clusters of cells on your slides, and you may need to examine all of them, or even additional slides, to locate cells in all of the mitotic stages.

ASSIGNMENT

Examine a prepared slide of whitefish mitosis and locate cells in interphase and in each mitotic stage. Whitefish cells contain many chromosomes, but you can identify the various stages by the chromosome distribution. Diagram and label these cells in Section C of the laboratory report.

Exercise 6

DIFFUSION AND OSMOSIS

Materials

>   Compound microscopes
>   Hot plates
>   Beakers, 250 ml
>   Biohazard container
>   Depression slides and cover glasses
>   Forceps
>   Mechanical pipetting devices
>   Petri dishes of agar-agar
>   Serological pipettes
>   Serological test tubes and test tube racks
>   Syringes and needles
>   Thermometers, Celsius
>   Thistle-tube osmometers
>   Fresh blood, mammalian
>   Glucose solutions, 2.0%, 5.0%
>   Ice
>   India ink
>   Methylene blue granules
>   NaCl solutions, 0.3%, 0.9%, and 2.0%
>   Potassium permanganate granules
>   Sucrose solutions, 5% and 15%, for osmometers

Before You Proceed

Consult with your instructor about using protective disposable gloves when performing portions of this exercise.

Materials are constantly exchanged between body cells and the tissue fluid that bathes them. These substances, including organic nutrients, inorganic ions, or water, pass into and out of cells through the plasma membrane. Some molecules, usually small ones, move passively through the membrane by diffusion along a concentration gradient. Others move either with or against a concentration gradient by active transport. In either case, the selective permeability of the plasma membrane plays an important role.

In this exercise you will investigate the passive transport of molecules.

Brownian Movement

Molecules in a liquid or gaseous state are in constant, random motion. Each molecule moves in a straight line until it bumps into another molecule, and then it bounces off in another straight-line path. This molecular motion cannot be observed directly, but it may be observed indirectly by noting the random motion of microscopic particles suspended in water. The random bombardment of these particles by water molecules causes their random, vibratory movement, which is called **Brownian movement** after Robert Brown, a Scottish botanist who first described it in 1827.

Diffusion

The net movement of the same kind of molecules from an area of their higher concentration to an area of their lower concentration is called **diffusion.** Molecules move away from the area of higher concentration because the area of lower concentration has fewer molecules to obstruct straight-line movement of the molecules. For example, if a soluble substance such as a lump of sugar is placed in water, it will gradually diffuse throughout the liquid until it is equally distributed.

Diffusion occurs in liquids and gases and is an important means of distributing materials within cells and of moving some materials into and out of cells. It is a passive process that results from the normal random motion of molecules and does not require an expenditure of energy by the cells. However, the rate of diffusion is affected by both temperature and molecular size.

Osmosis

Water is the most abundant substance in living cells. It is the **solvent** of living systems—the liquid medium in which the chemical reactions of life occur. Organic and inorganic materials constitute the **solutes.** Water molecules are small and move freely through cell membranes. Like other molecules, water molecules diffuse from an area of their higher concentration to an area of their lower concentration. The diffusion of water through a semipermeable, or selectively permeable, membrane is called **osmosis.**

The direction of a net movement of water across a membrane is dependent upon the concentration of water and thus the concentration of solutes on each side of the membrane. If pure water is separated from a 10% sucrose solution by a semipermeable membrane, there will be a net movement of water into the sucrose solution. The force required to prevent this movement is called **osmotic pressure,** and its value is assigned to the sucrose solution. The greater the concentration of solute particles (molecules or ions), the greater is the osmotic pressure of the solution. Thus, the degree of ionization of solute molecules markedly affects the osmotic pressure of a solution.

If two solutions with the same osmotic pressure are separated by a semipermeable membrane, there will be no net movement of water from one solution into the other. Such solutions are said to be **isotonic,** and they are at **osmotic equilibrium.** The integrity of cell membranes is dependent upon osmotic equilibrium between the cells and the fluid that bathes them.

Now consider hypothetical sucrose solutions A and B that are separated by a semipermeable membrane. Solution A has a lower solute concentration but a greater water concentration than B. Thus, A has a lower osmotic pressure than B, and the net movement of water will be from A to B. Solution A is said to be **hypotonic** to B, because it has a lower solute concentration and a lower osmotic pressure. Solution B is **hypertonic** to A, because it has a higher solute concentration and a higher osmotic pressure. Whenever aqueous solutions with unequal osmotic pressures are separated by a semipermeable membrane, the net movement of water is always from the hypotonic solution into the hypertonic solution.

ASSIGNMENT

Complete Section A of the laboratory report.

Demonstrations and Experiments

The demonstrations and experiments in this section will aid your understanding of diffusion and osmosis. Record your observations and results on the laboratory report.

Brownian Movement

A hanging drop slide of diluted India ink has been set up under a demonstration microscope using the oil immersion objective. Observe the movement of the ink particles. What causes their movement?

Diffusion and Temperature

1. Fill three 250-ml beakers about two-thirds full with tap water at different temperatures.

Beaker A	Ice water
Beaker B	Room temperature
Beaker C	60° to 70° C

2. Measure and record the temperature of the water in each beaker.
3. Place the beakers on your table where they will not be disturbed and can remain motionless.
4. Drop a small granule of potassium permanganate into each beaker without disturbing the water.
5. Observe the beakers at 5-minute intervals and record the time required for the molecules of potassium permanganate to diffuse throughout the water in each beaker.

Diffusion and Molecular Weight

1. Obtain a Petri dish containing about 12 ml of 1.5% agar-agar. The agar gel is 98.5% water, and water-soluble molecules readily diffuse through it.
2. Use forceps to place approximately equal-sized granules of potassium permanganate and methylene blue on the agar about 5 cm apart as shown in Figure 6.1. Set the dish aside where it will not be disturbed.
3. After 1 hour, measure and record the diameter (in mm) of the colored area around each granule to determine any difference in the rate of diffusion. The molecular weight of potassium permanganate is 158; the molecular weight of methylene blue is 320.

Osmosis

1. Examine the two thistle-tube osmometers, like the one shown in Figure 6.2, that were set up at the start of the laboratory session. Each osmometer was prepared by securing a

Figure 6.1 Determining the effect of molecular weight on the rate of diffusion.

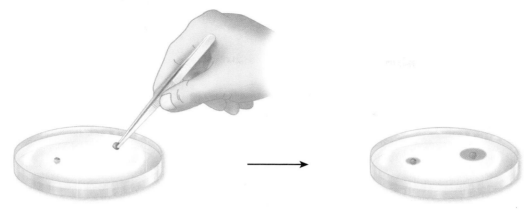

Equal-sized granules of methylene blue and potassium permanganate placed on agar medium.

Diffusion distances measured after one hour.

semipermeable membrane over the bulb of a thistle-tube and filling the bulb with a sucrose solution. One tube received a 5% solution, and the other received a 15% solution. Then, the bulb of each thistle-tube was immersed in a beaker of distilled water as shown in Figure 6.2.

2. Measure and record the height (mm) of the water-sucrose column in osmometers A and B at the start of the experiment and 30 minutes later.

Osmosis and Cell Membrane Integrity

In this experiment, you will determine the effect of the osmotic pressure of five solutions on red blood

cells. Figure 6.3 shows the protocol for the experiment. Recall that osmotic equilibrium between cells and the fluid bathing them is required for the normal functioning of cells. Figure 6.4 shows the effect of hypertonic, isotonic, and hypotonic solutions on red blood cells. Hypertonic solutions cause water to leave

Figure 6.3 Routine for making cell suspensions.

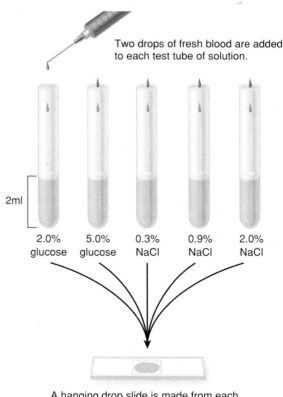

Two drops of fresh blood are added to each test tube of solution.

2ml

| 2.0% glucose | 5.0% glucose | 0.3% NaCl | 0.9% NaCl | 2.0% NaCl |

A hanging drop slide is made from each test tube and examined microscopically.

Figure 6.2 Osmosis setup.

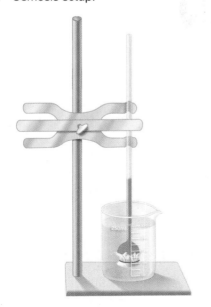

Figure 6.4 The effects of hypertonic, isotonic, and hypotonic solutions on red blood cells.

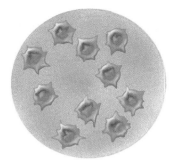
Hypertonic
(Crenation)

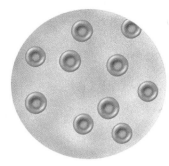
Isotonic
(Osmotic Equilibrium)

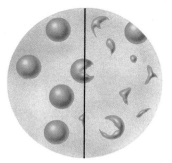
Hypotonic
(Hemolysis)

the cells, and they shrivel (a process called crenation). Isotonic solutions cause no net movement of water. Hypotonic solutions cause water to enter the cells, and they ultimately burst (lyse), because of the excessive internal pressure that is generated. The rupture of red blood cells is called **hemolysis.**

Your instructor has prepared a demonstration of osmotic pressure on red blood cells as described next. Your instructor will place *all items that come in contact with blood in a biohazard container immediately after use.*

1. Label five clean serological test tubes 1 to 5 and place them in a test tube rack.

Figure 6.5 Use a mechanical pipetting device for all pipetting.

2. Use a 5-ml pipette to place 2 ml of solution in each tube:

   Tube 1—2.0% glucose
   Tube 2—5.0% glucose
   Tube 3—0.3% NaCl
   Tube 4—0.9% NaCl
   Tube 5—2.0% NaCl

   Use the mechanical pipetting device as shown in Figure 6.5 to dispense the solutions. Rinse the pipette with distilled water after the delivery of each solution.

3. Use a syringe to add 2 drops of blood to each tube. Shake the tubes from side to side to mix thoroughly and return them to the test tube rack. The test tube rack and test tubes are on the demonstration table for your observation. *Do not handle the tubes.*

4. After 5 minutes, examine the relative transparency of the liquid in the tubes by holding a page of this text behind the test tube rack and viewing the text through the tubes. Interpret the clarity of the solutions as follows:

   Hypotonic solution: transparent due to hemolysis.
   Isotonic solution: cloudy due to the intact red blood cells.
   Hypertonic solution: clarity intermediate between transparent and cloudy due to the crenation of the cells.

5. Examine the demonstration microscope setups using hanging drop slides and oil immersion objectives, which show the effect of these solutions on the shape of red blood cells.

 ASSIGNMENT

Complete the laboratory report.

Exercise 7

EPITHELIAL AND CONNECTIVE TISSUES

OBJECTIVES

After completing this exercise, you should be able to:

1. List the types of epithelial and connective tissues and, for each, describe its structure, functions, and locations.
2. Recognize each type of epithelial and connective tissue when viewed with a microscope.
3. Define all terms in bold print.

Materials

Prepared slides of epithelial tissues:
    columnar, simple ciliated
    columnar, simple nonciliated
    columnar, stratified
    columnar, pseudostratified ciliated
    columnar, pseudostratified nonciliated
    cuboidal, simple
    cuboidal, stratified
    squamous, simple
    squamous, stratified
    transitional
Prepared slides of connective tissues:
    adipose tissue
    bone, compact, l.s., x.s.
    dense fibrous tissue
    elastic cartilage
    fibrocartilage
    hyaline cartilage
    loose fibrous
    reticular tissue

Although body cells share many common structures and functions, they may differ considerably in size, shape, and structure in accordance with their specialized functions. Cells of a similar type usually occur in groups. An aggregation of cells that are similar in structure and function is a **tissue.** The scientific study of tissues is **histology.**

There are four basic types of tissues in the body: epithelial, connective, muscle, and nerve. In this exercise, you will study only epithelial and connective tissues. Muscle and nerve tissues will be studied in Exercise 18.

Epithelial Tissues

Epithelial tissues cover surfaces of the body and internal organs, line cavities, and form glands. They may be protective, absorptive, secretory, or excretory in function. They are characterized by (1) closely packed cells without intercellular substances or blood vessels and (2) the presence of a noncellular **basement membrane** that attaches the tissue to underlying connective tissue. Most epithelial tissues have a layer of epithelial cells with a **free surface** (i.e., cells with one surface that is not in contact with other cells). Epithelial cells have a high rate of cell division to replace cells that are worn out or sloughed off by normal processes.

Epithelial tissues are categorized on the basis of (1) single or multiple cell layers, (2) cell shape, and (3) presence or absence of cilia. The three basic categories are simple, stratified, and pseudostratified. Figure 7.1 depicts the morphological classification of epithelial tissues.

Figure 7.1 A morphological classification of epithelial types.

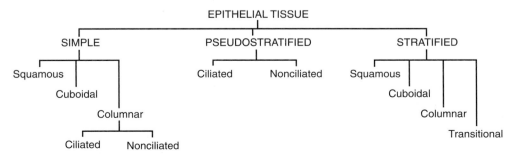

Simple Epithelia

Simple epithelial tissues consist of a single layer of cells. Figure 7.2 illustrates the basic types.

Simple Squamous

Simple squamous cells are thin and flat and appear as irregular polygons in a surface view. The peritoneum, pleurae, air sacs of lungs, and inner linings of arteries, veins, capillaries, and kidney tubules are formed of simple squamous epithelium. This epithelium is involved in filtering fluids and in the exchange of materials.

Simple Cuboidal

The cells of simple cuboidal epithelium are block-like in cross section and hexagonal in a surface view. Secretion and absorption are its chief functions. Cuboidal epithelium occurs in kidney tubules and in both endocrine and exocrine glands. For example, it is found in the thyroid gland, pancreas, ovaries, sweat glands, and salivary glands.

Simple Columnar

The cells of simple columnar epithelium are elongated and appear polygonal on their free surfaces. Scattered among the columnar cells are special **goblet cells** that secrete **mucus,** which flows over the surfaces of adjacent cells and protects the tissue.

There are two types of simple columnar epithelium. **Ciliated columnar epithelium** lines the uterine tubes and the small bronchi of the lungs. **Nonciliated (plain) columnar epithelium** lines the digestive tract from stomach to anus. These intestinal cells have **microvilli** on their free surface, which greatly increase the absorptive surface area of the cells.

Stratified Epithelia

Stratified epithelial tissues consist of two or more layers of cells. Three types of stratified epithelia are shown in Figure 7.3. The tissues are classified according to the shape of the surface cells.

Stratified Squamous

The superficial cells of stratified squamous epithelium are distinctly squamous in shape, but the deepest cells are cuboidal or columnar. Cells change in shape as they migrate to the surface of the tissue. Protection is the chief function of this tissue, found in regions where abrasions or friction occur, such as the outer layer of the skin, oral cavity, vagina, cornea of the eye, and esophagus.

Stratified Cuboidal

The shape of the surface cells is cubelike or round. Stratified cuboidal epithelium is found in ducts of sweat glands, in sperm-forming tubules of the testes, and in egg-producing follicles of the ovaries. Secretion is a major function.

Figure 7.2 Simple epithelia.

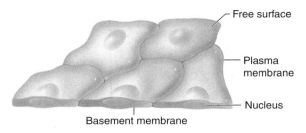

A. Simple Squamous

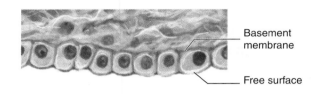

B. Simple Cuboidal

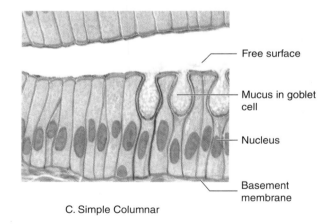

C. Simple Columnar

Stratified Columnar

The columnar cells at the surface of stratified columnar epithelium are formed from the deeper rounded cells as they migrate upward. This tissue is not shown here because it occurs in only a few areas of the body, such as the conjunctiva, parts of the pharynx and epiglottis, and the anal canal. Protection and secretion are its primary functions.

Transitional

Transitional epithelium occurs in organs of the body where stretching of the tissue may occur, such as the urinary bladder and ureters. Elasticity is a key characteristic. The deepest cells are columnar

Figure 7.3 Stratified epithelia.

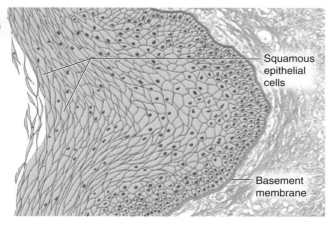

A. Stratified Squamous

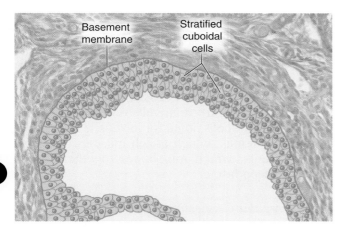

B. Stratified Cuboidal

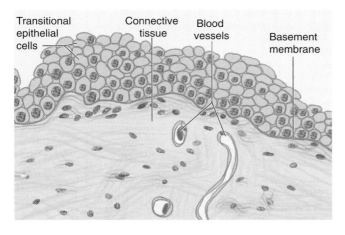

C. Transitional Epithelium

and loosely held together, whereas the surface cells may be dome-shaped and cuboidal or squamous, depending upon the degree of stretching.

Figure 7.4 Pseudostratified ciliated columnar epithelium.

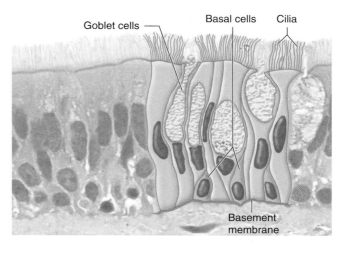

Pseudostratified Epithelia

Pseudostratified epithelial tissues consist of a single layer of columnar cells but appear to be composed of more than one layer (Figure 7.4). This illusion occurs because (1) each cell is in contact with the basement lamina, but not all cells reach the free surface of the tissue; and (2) the nuclei occur at different levels in the cells. Ciliated forms of this tissue line the nasal cavity, trachea, primary bronchi, and auditory tubes. Nonciliated types are found in parts of the male urethra and in large ducts of the parotid salivary gland. Mucus-secreting goblet cells occur in both ciliated and nonciliated types.

ASSIGNMENT

Complete Section A of the laboratory report.

Connective Tissues

Connective tissues provide support and attachment for various organs and fill spaces in the body. They are characterized by an abundance of a nonliving intercellular substance, the **matrix,** and few cells. Connective tissue may be classified in several ways. The system used here is based on the nature of the matrix. See Figure 7.5.

Connective Tissue Proper

Connective tissue proper consists of those ordinary connective tissues that are not highly specialized. They are pliable and have a matrix that consists of a semifluid or gelatinous **ground substance** and usually protein fibers produced by **fibroblasts.** Three types of fibers are produced by fibroblasts.

Figure 7.5 Types of connective tissue.

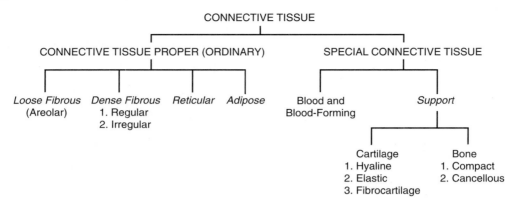

The strong, non-elastic **collagenous fibers** provide great tensile strength. **Elastic fibers,** formed of the protein elastin, provide elasticity. They are often branched and not as strong as collagenous fibers. The thin, branching **reticular fibers** form a network of very fine collagenous threads.

Loose Fibrous (Areolar) Connective Tissue

Loose fibrous connective tissue is widespread in the body. It has all three types of fibers scattered throughout the loosely organized matrix. Fibroblasts are the most common cells, although mast cells and macrophages are present. See Figure 7.6. Loose fibrous connective tissue provides a flexible supporting framework within and between organs, contains blood vessels that nourish surrounding tissues, and is a site of immune reactions. It occurs beneath epithelial tissues, around and within muscles and nerves, and in serous membranes, and it composes the superficial fascia.

Adipose Tissue

Small groups of fat or adipose cells (adipocytes) are scattered throughout the body. Accumulations of fat cells form adipose tissue that serves as a protective cushion and insulation for the body. The vacuoles of fat cells are filled with lipids and usually compress the cytoplasm and nucleus to the edge of the cell. The lipids stored in fat cells provide an important energy reserve for the body.

Reticular Tissue

Reticular tissue is characterized by a network of fine reticular fibers plus an abundance of leukocytes (white blood cells), which are active in providing immunity. It provides a supporting framework for many vascular organs such as the liver, hemopoietic tissue, and lymph nodes. See Figure 7.6.

Dense Fibrous Connective Tissue

Dense fibrous connective tissue differs from loose connective tissue in that its matrix consists of a predominance of densely packed fibers and relatively few cells, mostly fibroblasts. There are two subcategories of dense connective tissue based on the arrangement of the fibers. See Figure 7.7.

Dense Regular Connective Tissue This tissue consists of closely packed bundles of collagenous fibers that are roughly parallel to each other and a few elastic fibers. The few cells present are nearly all fibroblasts. There are relatively few blood vessels present. This tissue is noted for its tensile strength. It composes tendons, which attach muscles to bones, and ligaments, which join bones together, and it forms the deep fascia.

Dense Irregular Connective Tissue This tissue consists of bundles of collagenous fibers and a few elastic fibers that are not parallel but are interwoven in a random fashion. This tissue forms much of the dermis of the skin, the sheaths around nerves and tendons, and the outer layer of the periosteum.

Special Connective Tissue

Special connective tissue is subdivided into blood and hemopoietic (blood-forming) tissue and the specialized support tissues, cartilage and bone. See Figure 7.5. Hemopoietic tissue produces blood cells and platelets. It will not be studied here, but blood will be considered in Exercise 27. Blood is an atypical special connective tissue consisting of blood cells and platelets transported by a fluid matrix (plasma).

Cartilage and bone possess (1) a matrix that is solid and strong, (2) cells located in **lacunae,** tiny cavities in the matrix, and (3) an external membrane capable of producing new tissue.

Cartilage

All three types of cartilage are surrounded by a fibrous membrane, the **perichondrium,** and have cartilage cells, **chondrocytes,** in the lacunae of the matrix.

Figure 7.6 Loose fibrous (areolar), reticular, and adipose connective tissues.

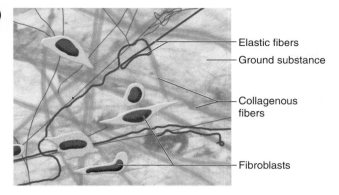

A. Loose Fibrous Tissue

- Elastic fibers
- Ground substance
- Collagenous fibers
- Fibroblasts

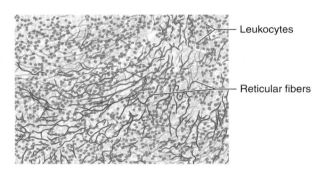

B. Reticular Tissue

- Leukocytes
- Reticular fibers

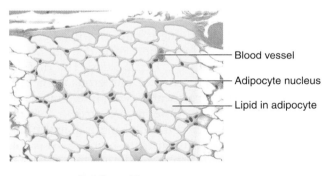

C. Adipose Tissue

- Blood vessel
- Adipocyte nucleus
- Lipid in adipocyte

Figure 7.7 Dense fibrous connective tissues.

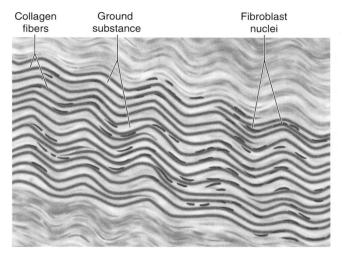

A. Dense Regular Tissue

- Collagen fibers
- Ground substance
- Fibroblast nuclei

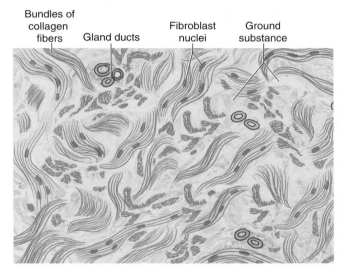

B. Dense Irregular Tissue

- Bundles of collagen fibers
- Gland ducts
- Fibroblast nuclei
- Ground substance

Unlike other connective tissue, cartilage lacks blood vessels, so its cells derive nourishment by diffusion from capillaries in the perichondrium. See Figure 7.8.

Hyaline Cartilage   This is the most common type of cartilage. It has a white, glassy appearance due to the absence of visible fibers in the matrix. Hyaline cartilage composes much of the fetal skeleton before ossification and covers the ends of long bones in adults. It also forms the cartilaginous portion of the nose and the cartilaginous supports of the respiratory passages.

Elastic Cartilage   The presence of an abundance of elastic fibers makes this cartilage more flexible and elastic than hyaline cartilage. It is located in such places as the external ear and the epiglottis.

Fibrocartilage   This very tough cartilage contains densely packed bundles of collagenous fibers that separate short rows of chondrocytes. It forms the intervertebral disks and parts of the knee joint and pelvic girdle, where it serves as a cushioning shock absorber.

Bone
Bone, the most rigid connective tissue, is hard, strong, and lightweight. It protects vital organs, provides an internal support for the body, and serves as a site for the attachment of muscles. It contains **red marrow,** which forms blood cells.

Figure 7.8 Types of cartilage.

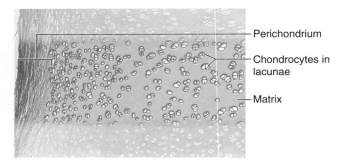

A. Hyaline Cartilage

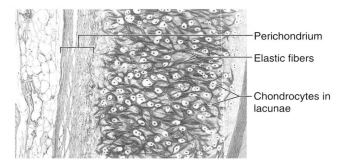

B. Elastic Cartilage

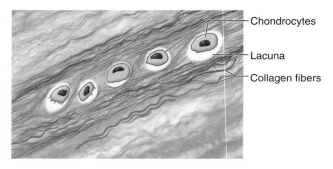

C. Fibrocartilage

In embryonic and fetal development, bones are preformed in either hyaline cartilage or fibrous connective tissue. These tissues provide a framework for the subsequent deposition of calcium salts, which form the bone matrix. Like other connective tissues, bone is continually modified and reconstructed.

There are two basic types of bone tissue, as shown in Figure 7.9. Solid, dense **compact bone** (label 1) forms the outer portion of the bone, and porous **spongy** or **cancellous bone** (label 2) forms the interior framework in the ends of long bones and the interior of all other bones.

Compact Bone The structural and functional unit of compact bone is the **osteon** (Haversian system),

label 10 in Figure 7.9. Each osteon is an elongated hollow cylinder formed of concentric rings, called **lamellae,** of bone matrix. Running through the center of each osteon is a **central** (Haversian) **canal** that contains capillaries (tiny blood vessels) and nerves serving bone cells in the osteon.

The lamellae are secreted by **osteoblasts** that have numerous extensions radiating from the cell. Once osteoblasts are trapped in lacunae (label 5) by their own depositions, they are called **osteocytes** (bone cells). The radiating microcanals between lacunae are the **canaliculi** (label 12). They contain the cellular extensions of the osteocytes and serve as passageways for materials moving between blood vessels in the central canals and the lacunae.

Adjacent osteons are fused together at their outer lamellae and by bone deposits between them to yield a solid, hard matrix of calcium salts. Central canals of adjacent osteons are connected by **perforating canals** (label 4) that penetrate osteons at right angles to the osteon axis. These canals carry capillaries and nerves from the periosteum to the central canals.

The outer surface of compact bone is covered by the **periosteum.** It consists of (1) an outer, fibrous membrane and (2) an inner layer that is a source of new bone-forming cells and capillaries. The periosteum is firmly attached to the compact bone by **collagenous fibers** that penetrate the outer lamellae.

Spongy Bone This tissue is interior to and continuous with compact bone. There is no distinct line of demarcation between the two tissues. The distinctive feature is the presence of numerous branching bony plates, called **trabeculae,** with interconnecting spaces between them. Osteons are absent. Instead, osteocytes are scattered through the trabeculae and are nourished by diffusion from blood vessels that meander through the spaces. The spaces are filled with red marrow and serve to decrease the weight of the bone.

 ASSIGNMENT

1. Label Figure 7.9.
2. Complete Sections B, C, and D of the laboratory report.

Microscopic Study

Now that you know the basic characteristics of epithelial and connective tissues, examine prepared slides of these tissues so that you can learn to recognize

Figure 7.9 Bone tissue.

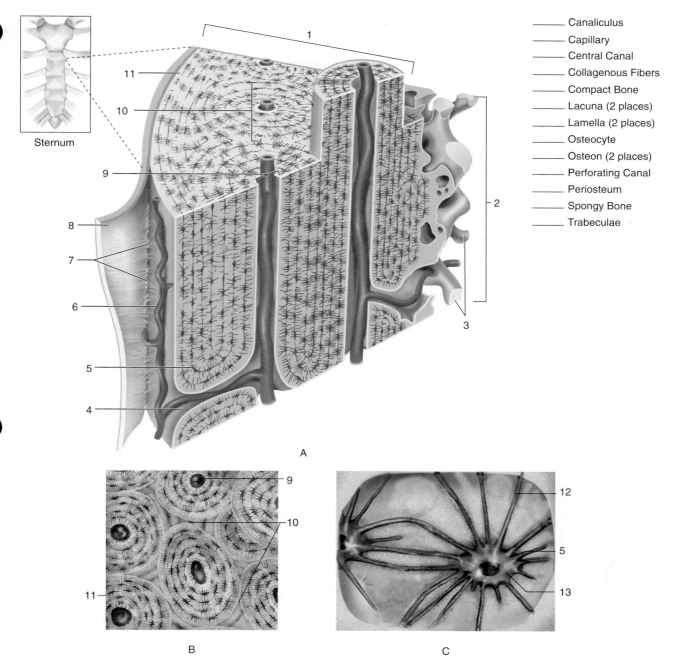

Sternum

	Canaliculus
	Capillary
	Central Canal
	Collagenous Fibers
	Compact Bone
	Lacuna (2 places)
	Lamella (2 places)
	Osteocyte
	Osteon (2 places)
	Perforating Canal
	Periosteum
	Spongy Bone
	Trabeculae

A

B

C

them when viewed microscopically. Compare your slides with appropriate figures in this exercise and also with the photomicrographs in Figures HA-1 through HA-8 of the Histology Atlas, which starts on page 219.

ASSIGNMENT

1. Examine prepared slides of epithelial and connective tissues.
2. Draw a small portion of each tissue, using Section E of the laboratory report; label pertinent parts. These drawings will be valuable when you are studying for the lab practicum.

THE INTEGUMENT

After completing this exercise, you should be able to:

1. Describe the structure and function of the integument.
2. Identify the components of the skin when viewed microscopically.
3. Define all terms in bold print.

Materials

Dividers (or pointed scissors)
Metric ruler
Model of human skin
Prepared slides of:
    human skin showing hair follicles, glands, and receptors
    human skin from the sole of a foot

The **integument,** or **skin,** consists of epithelial and connective tissues with associated nerve and muscle tissues. The functions of the skin include (1) protection of underlying tissues from abrasion, dehydration, microorganisms, and ultraviolet radiation; (2) excretion; (3) maintenance of body temperature; and (4) detection of certain stimuli. The skin consists of two layers: an outer, multilayered **epidermis** and an underlying **dermis.**

The integument is attached to underlying tissues by the **hypodermis,** or **subcutaneous layer.** The hypodermis is sometimes called the **superficial fascia.** The hypodermis consists primarily of loose fibrous connective tissue and adipose tissue, and it contains abundant blood vessels and nerves that service the integument. The adipose tissue provides cushioning and insulation for underlying tissues and serves as a storehouse for energy reserves.

The Epidermis

The epidermis, label 1 in Figure 8.1, consists of stratified squamous epithelium. The epidermis is normally thin in hairy skin, but it can be thick and tough in hairless skin on the soles of the feet. Prepared slides of thick, hairless skin reveal five distinct layers: an outer **stratum corneum,** a thin translucent **stratum lucidum,** a darkly stained **stratum granulosum,** a multilayered **stratum spinosum,** and an innermost **stratum basale,** a single layer of columnar cells. Cells of all five layers originate from cells produced by the stratum basale. As the cells are pushed to the surface by new cells formed below them, they change in shape, structure, and composition.

The brown-black skin pigment **melanin** is produced by **melanocytes** (not shown) in the stratum basale and is incorporated into the cells of this layer and the cells formed by this layer. Inherited differences in skin color are due to the amount of melanin produced rather than the number of melanocytes. Melanin protects underlying tissues from harmful ultraviolet radiation. In light-skinned persons, exposure to ultraviolet radiation increases the production of melanin, resulting in a "tan." Clustering of melanocytes results in freckles.

The granules in cells of the stratum granulosum are precursors of **keratin,** a water-repellent protein. As the cells are pushed toward the surface, they become the keratinized, flattened, dead cells of the stratum corneum. Surface cells of this outer layer are constantly sloughed off and replaced by underlying cells. The thin stratum lucidum is apparent only in very thick skin.

The Dermis

Often called the true skin, the dermis varies in thickness from less than 1 mm to over 6 mm. It is composed of fibrous connective tissue. It contains numerous capillaries that provide nourishment not only for the dermis but also for the epidermis. Nerve fibers and sensory receptors are also present in the dermis.

The inner surface of the epidermis is attached to the outer surface of the dermis. The dermis has numerous projections called **dermal papillae** (label 6) that extend into the epidermis. These papillae are most abundant and pronounced in skin of the fingertips, palms of the hands, and soles of the feet, where they produce epidermal ridges that improve the frictional characteristics of these areas. The pattern of these epidermal ridges is unique in each person,

Figure 8.1 Skin structure. (a) Basic components; (b) epidermal layers.

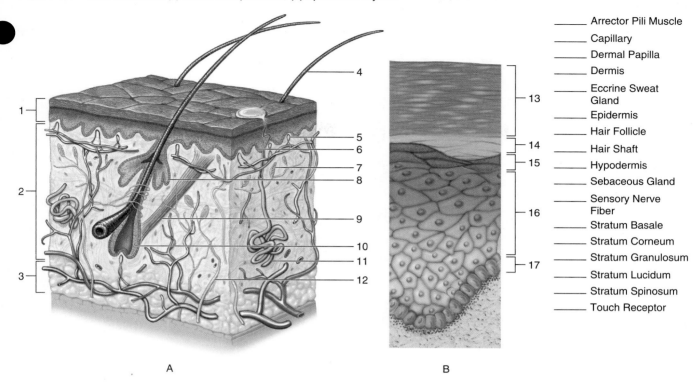

Arrector Pili Muscle
_____ Capillary
_____ Dermal Papilla
_____ Dermis
_____ Eccrine Sweat Gland
_____ Epidermis
_____ Hair Follicle
_____ Hair Shaft
_____ Hypodermis
_____ Sebaceous Gland
_____ Sensory Nerve Fiber
_____ Stratum Basale
_____ Stratum Corneum
_____ Stratum Granulosum
_____ Stratum Lucidum
_____ Stratum Spinosum
_____ Touch Receptor

A   B

for example, they are responsible for each person's unique fingerprint.

Abundant collagenous and elastic fibers are found in the dermis. They are responsible for the strength and elasticity of the skin.

Hair

A hair is formed of dead, heavily keratinized epidermal cells. The **hair shaft** extends above the skin, and the **hair root** is located below the skin surface. The root is enveloped by **internal** and **external root sheaths,** formed of epidermal cells, that compose the **hair follicle** (Figure 8.2).

The hair follicle extends inward into the dermis and sometimes into the subcutaneous layer. The tip of the hair root forms an onion-shaped enlargement called the **bulb.** The bulb is composed of stratum basale cells that produce the cells forming the hair. The bulb encompasses a dermal papilla known as a **follicular** (hair) **papilla.** It contains capillaries that nourish the bulb and follicle.

An **arrector pili muscle,** formed of smooth muscle fibers, extends diagonally from the hair follicle upward to the epidermis. Contraction of this muscle raises the hair into a more upright position and produces "goose bumps."

Glands

Two kinds of glands are present in the skin: sebaceous glands and sweat glands.

Sebaceous Glands

Sebaceous glands (label 7 in Figure 8.2) are located within the epithelial tissue of the hair follicle. They secrete an oily secretion called **sebum** into the follicle, which serves to keep the hair soft and pliable and to kill bacteria on the skin. Sebum also helps to waterproof the skin.

Sweat Glands

There are two types of sweat glands: eccrine and apocrine. The smaller **eccrine (merocrine) sweat glands** (Figures 8.1 and 8.2) carry their secretions directly to the surface of the skin. These are simple tubular glands that have a coiled basilar portion located in the dermis. They are widely distributed over the body and are abundant on the head, neck, back, palms of the hands, and soles of the feet. They are absent on the lips, glans penis, and clitoris. Eccrine glands located in the axillae and palms may be stimulated by psychic factors such as fear and stress, but most eccrine glands are activated primarily by thermal stimuli. The watery perspiration secreted by these glands serves to remove waste

Figure 8.2 Hair follicle and associated structures.

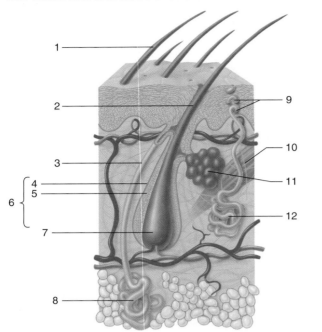

_____ Apocrine sweat gland
_____ Arrector pili muscle
_____ Duct of apocrine sweat gland
_____ Eccrine sweat gland
_____ Duct of eccrine sweat gland
_____ External root sheath
_____ Hair bulb
_____ Hair follicle
_____ Hair root
_____ Hair shaft
_____ Internal root sheath
_____ Sebaceous gland

materials from the blood and cools the body surface by evaporation.

The larger **apocrine sweat glands** (Figure 8.2) empty their secretions into a hair follicle. The coiled basilar portion is located in the hypodermis. These glands are essentially scent glands producing a white or yellow odiferous secretion that contributes to body odors. They are most abundant in the axillae, scrotum of the male, and perigenital region of the female. Apocrine secretions are stimulated more by psychic factors than by thermal factors.

Sensory Receptors

The skin plays an important role in sensory perception because it contains numerous sensory receptors for touch, pressure, heat, cold, and pain stimuli. **Touch receptors (Meissner's corpuscles)** are located in the dermis near the epidermis (label 7 in Figure 8.1), often within dermal papillae. They are especially important receptors in hairless skin. Free nerve endings wrapped around hair follicles (Figure 8.2) also serve as touch receptors. **Pressure receptors (Pacinian corpuscles)** (not shown) are spherical or ovoid, multilayered structures that lie deep in the dermis.

ASSIGNMENT

1. Label Figures 8.1 and 8.2.
2. Complete Sections A and B of the laboratory report.
3. Examine the model of human skin. Locate the parts described in this exercise.

Microscopic Study

Examine prepared slides of human skin, including hair follicles, glands, and receptors, and human skin from the sole of a foot. Compare your observations with Figures 8.1 and 8.2 and with Figure HA-9 in the Histology Atlas.

Distribution of Touch Receptors

In this experiment, you will determine the relative density of Meissner's corpuscles in skin at four locations. In order for you to perceive two simultaneous touch stimuli as two touch sensations, the stimuli must be far enough apart to stimulate two Meissner's corpuscles that are separated by an unstimulated touch receptor. Perform the experiment with your laboratory partner as follows.

1. The subject's eyes are to be closed during the experiment. Avoiding any hair that may be present, lightly touch the subject's skin with one or two points of the dividers. The subject responds either "one" or "two" to indicate either a one-point sensation or a two-point sensation.
2. Start with the points of the dividers close together and gradually increase the distance between the points until the subject reports a two-point sensation about 75% of the time.

Measure and record this distance between the tips of the dividers as the minimum distance producing a two-point sensation.
3. Using this procedure, determine and record the minimum distance producing a two-point sensation on (1) the back of the neck, (2) the inner forearm, (3) the palm of the hand, and (4) the tip of the index finger.

 ASSIGNMENT

Complete the laboratory report.

Exercise 9

THE SKELETAL PLAN

OBJECTIVES

After completing this exercise, you should be able to:

1. Identify the parts of a longitudinally sectioned long bone.
2. Identify the major types of bone fractures.
3. Identify the major bones of the body and their relationships.
4. Identify the surface features of bones.
5. Define all terms in bold print.

Materials

Fresh beef bones sawed longitudinally
Human femur sawed longitudinally
Articulated human skeleton
Dissection instruments

In this exercise, you will study the structure of a typical long bone, certain types of bone fractures, and the skeleton as a whole.

Long Bone Structure

Figure 9.1 depicts a tibia that has been sectioned to reveal its internal structure. The bone consists of an elongated shaft, the **diaphysis,** and expanded terminal portions, the **epiphyses.** During the growing years, a plate of hyaline cartilage, the **epiphyseal disk** (or **metaphysis**), is present at the junction of the epiphyses and the diaphysis. The bone grows in length as new cartilage is formed on the epiphyseal side of the disk and cartilage is replaced by bone on the diaphyseal side. Linear growth ceases when the cartilage ceases to reproduce and is completely replaced by bone. The lines of fusion that form between the epiphyses and the diaphysis are called **epiphyseal lines.**

The interior of the epiphyses and ends of the diaphysis consist of **cancellous bone** (spongy bone). **Compact bone** forms the rigid tube of the diaphysis and is reduced in thickness where it covers the underlying cancellous bone.

The hollow chamber in the diaphysis is the **medullary cavity.** It is lined with a thin epithelial

Figure 9.1 Anatomy of a long bone, the tibia.

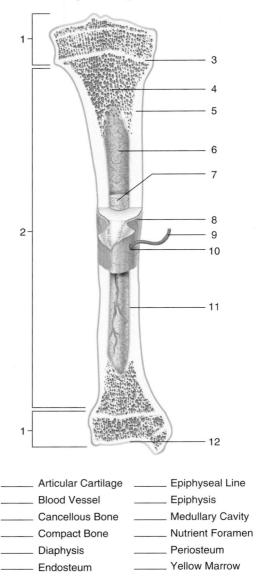

_____ Articular Cartilage	_____ Epiphyseal Line
_____ Blood Vessel	_____ Epiphysis
_____ Cancellous Bone	_____ Medullary Cavity
_____ Compact Bone	_____ Nutrient Foramen
_____ Diaphysis	_____ Periosteum
_____ Endosteum	_____ Yellow Marrow

membrane, the **endosteum,** that extends into the spaces in cancellous bone and into the central canals of osteons. The medullary cavity and the adjoining spaces in cancellous bone are filled with fatty **yellow** marrow. **Red marrow,** which forms blood cells, occurs in cancellous bone in the epiphyses of the femur and humerus, but the epiphyses of other long bones contain yellow marrow. Most red marrow in adults is found in cancellous bone of the ribs, sternum, pelvic girdle, and vertebrae.

Except for the articular surfaces, the bone is covered with the **periosteum,** a tough, fibrous membrane that tightly adheres to the bone surface. The periosteum is richly supplied with blood vessels and nerves. Nutrient blood vessels enter the bone via small channels called **nutrient foramina.** The articular surfaces are covered with a smooth **articular cartilage** (hyaline type) that reduces friction and protects the ends of the bone.

ASSIGNMENT

1. Label Figure 9.1.
2. Examine the cut surface of a split beef bone. Locate the parts shown in Figure 9.1.

   a. Remove a little yellow marrow from the medullary cavity. Note its texture.
   b. Determine the distribution of compact and cancellous bone.
   c. Locate the red marrow.
   d. Use a scalpel to cut away a small portion of the periosteum. Determine if the periosteum is elastic or nonelastic and strong or weak.
   e. Muscles are attached to bones by tendons, and bones are bound to other bones by ligaments. Examine the attachment of a tendon or ligament to the bone. Determine how the periosteum is involved in the attachment.
   f. Examine the articular cartilages to determine how they aid movement at a joint.

3. Examine the split human femur. Compare its epiphyseal lines with the epiphyseal cartilages in the beef bone. Note the distribution of compact and cancellous bone.
4. Complete Section B of the laboratory report.

Parts of the Skeleton

The adult skeleton consists of 206 named bones and several small unnamed ones. Bones vary considerably in size and shape and are categorized into two main groups: (1) bones of the **axial skeleton** and (2) bones of the **appendicular skeleton.** Only the major bones of the skeleton will be considered in this section. Refer to Figure 9.2.

The Axial Skeleton

The axial skeleton consists of the **skull, vertebral column** (spine), and **thoracic cage.** The skull is formed of twenty-one fused bones plus the movable **mandible** (lower jaw). The vertebral column includes twenty-four somewhat movable vertebrae plus the **sacrum** and **coccyx,** or tail bone. The thoracic cage is formed by 12 pairs of **ribs,** the **sternum,** or breastbone, and the **costal cartilages** (label 9), which join the ribs to the sternum. Not shown in Figure 9.2 are the **hyoid bone,** a U-shaped bone located under the lower jaw and associated with the larynx, and the **auditory ossicles,** which play a major role in hearing.

The Appendicular Skeleton

The appendicular skeleton is formed by the bones of the pectoral (shoulder) and pelvic girdles and the upper and lower limbs.

The **pectoral** (shoulder) **girdle** consists of the **clavicles** (collarbones) and **scapulae** (shoulder blades), which attach the upper limbs to the trunk. An upper limb consists of the **humerus** (upper arm bone), **radius** (label 10) and **ulna** (lower arm bones), **carpals** (wrist bones), **metacarpals** (bones of the palm), and **phalanges** (finger bones).

The **pelvic girdle** (label 11) is formed by two **coxal bones** (hipbones) that unite with the **sacrum** of the vertebral column, forming the **sacroiliac joint,** and with each other anteriorly at the **symphysis pubis** joint. A lower limb consists of the **femur** (thighbone), **tibia** (shinbone), **fibula** (smaller bone of the lower leg), **patella** (kneecap), **tarsals** (ankle bones), **metatarsals** (bones of the instep), and **phalanges** (toe bones). Note that both finger and toe bones are called phalanges.

Surface Features of Bones

As you examine the bones of the skeleton, you will notice a variety of surface features, such as grooves and ridges. Study the major surface features of bones in Table 9.1. Surface features of bones will be emphasized in subsequent exercises on the skeletal system, so learn to recognize them as soon as possible.

ASSIGNMENT

1. Label Figure 9.2.
2. Identify the major bones on an articulated skeleton.
3. Locate as many of the surface features of bones listed in Table 9.1 as you can on bones of the articulated skeleton.
4. Complete Sections C and D of the laboratory report.

Figure 9.2   Major bones of the body. Axial skeleton in yellow, appendicular skeleton in gray.

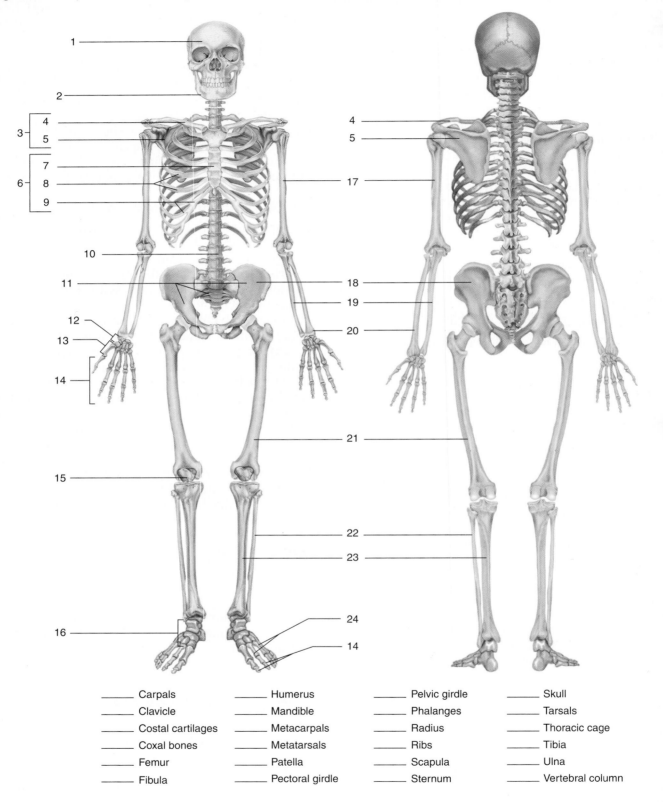

_____ Carpals	_____ Humerus	_____ Pelvic girdle	_____ Skull
_____ Clavicle	_____ Mandible	_____ Phalanges	_____ Tarsals
_____ Costal cartilages	_____ Metacarpals	_____ Radius	_____ Thoracic cage
_____ Coxal bones	_____ Metatarsals	_____ Ribs	_____ Tibia
_____ Femur	_____ Patella	_____ Scapula	_____ Ulna
_____ Fibula	_____ Pectoral girdle	_____ Sternum	_____ Vertebral column

TABLE 9.1

Surface Features of Bones

Term	Description
Cavities or Depressions	
Fissure	Narrow slit
Foramen	Opening that is a passageway for blood vessels and nerves
Fossa	Shallow depression
Meatus	Large canal or passageway
Sinus	Air-filled cavity
Sulcus	Groove or furrow
Processes	
Condyle	Rounded knucklelike process that articulates with another bone
Crest	Narrow ridge
Head	Enlarged end of a bone supported by a narrow neck
Spine	Sharp, slender process
Trochanter	Very large process
Tubercle	Small rounded process
Tuberosity	Low, roughened process serving as an attachment site for a muscle

Bone Fractures

Figure 9.3 shows some common types of fractures. If a broken bone pierces the skin or mucous membrane, it is a **compound fracture;** if not, it is a **simple fracture. Complete fractures** are those in which the bone is broken completely through. In **incomplete fractures,** the bone is only partially broken or splintered.

There are several types of complete fractures. If the break is at right angles to the long axis of the bone, it is a **transverse fracture.** Fractures at some other angle are **oblique fractures.** A **spiral fracture** results from excessive twisting of the bone. Several pieces are broken out of the bone in a **comminuted fracture.** If only one piece is broken out of the bone, it is a **segmental fracture.** When bone fragments have been moved out of alignment as in illustration G, the fracture may also be said to be **displaced.**

Figure 9.3 Types of bone fractures.

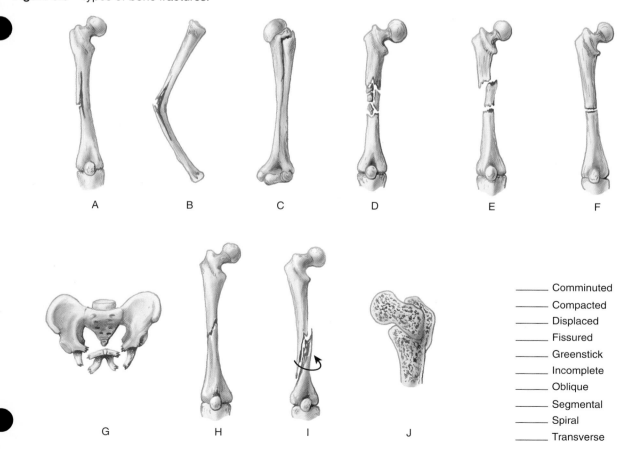

A   B   C   D   E   F

G   H   I   J

_____ Comminuted
_____ Compacted
_____ Displaced
_____ Fissured
_____ Greenstick
_____ Incomplete
_____ Oblique
_____ Segmental
_____ Spiral
_____ Transverse

There are two types of incomplete fractures. A **greenstick fracture** occurs on only one side of a bone when the bone is bent. A **fissured fracture** is a lengthwise split in a long bone.

When one piece of the bone is forced into another part of the same bone, a **compacted fracture** (illustration J) results. **Compression fractures** (not shown) of the vertebrae (crushed vertebrae) sometimes occur because of excessive vertical forces.

ASSIGNMENT

1. Label Figure 9.3.
2. Complete the laboratory report.

THE SKULL

OBJECTIVES

After completing this exercise, you should be able to:

1. Identify the skull bones and their distinctive features, sutures, paranasal sinuses, and major foramina.
2. Describe the relationships of the bones forming the skull.
3. Identify the fontanels and unossified sutures of the fetal skull.
4. Define all terms in bold print.

Materials

Human skulls with removable top of cranium
Fetal skulls
Pipe cleaners

The adult skull consists of twenty-one bones, which are firmly joined together by immovable joints called **sutures,** plus the movable lower jaw—twenty-two bones in all. Eight of the interlocked bones form the **cranium** encasing the brain, and thirteen compose the face. Bones of the skull contain numerous **foramina,** passageways for cranial nerves and blood vessels. Only a few of the larger foramina will be considered here.

While studying the laboratory skulls, be very careful not to damage them. Some parts are delicate and easily broken. *Never use a pencil or pen as a pointer or probe.* Use a pipe cleaner instead.

The Cranium

The cranium is formed of the following bones: one **frontal,** two **parietals,** one **occipital,** two **temporals,** one **sphenoid,** and one **ethmoid.** Locate the bones and their distinctive features in Figures 10.1 through 10.4 as you study this section.

Frontal Bone

The frontal bone forms the anterior superior portion of the cranium, including the forehead, upper parts of the eye orbits, and the roof of the nasal cavity. Above each eye orbit is a **supraorbital foramen** that is reduced in some skulls to a **supraorbital notch.**

Parietal Bones

The parietal bones form the roof and sides of the cranium posterior to the frontal bone, to which they are joined by **coronal sutures.** The parietal bones are joined to each other at the superior midline by the **sagittal suture.**

Occipital Bone

The posterior inferior portion of the skull consists of the occipital bone. It joins to the parietal bones by the **lambdoidal suture.** As shown in Figure 10.2, the floor of the occipital bone contains the large **foramen magnum,** which surrounds the brain stem. On each side of this foramen are the **occipital condyles.** These knucklelike processes articulate with the atlas, the first vertebra of the vertebral column.

Temporal Bones

The temporal bones are located just below the parietal bones on each side of the skull. Each joins to the parietal bone above it by a **squamosal suture** and articulates posteriorly with the occipital bone by the **lambdoidal suture.** The ear components are located within the temporal bones. The **external auditory** (or acoustic) **meatus,** which leads inward toward the middle ear, is located near the inferior margin. Just anterior to the auditory meatus is the **mandibular fossa** (depression). The mandibular condyle fits into this fossa to form the **temporomandibular joint.**

There are three major processes on each temporal bone. The **zygomatic process** is an anterior extension that articulates with the cheekbone (zygomatic). The **styloid process** is a slender spinelike process that extends downward below the auditory meatus. It serves as an attachment site for some tongue and pharyngeal muscles. The **mastoid process** is a rounded eminence inferior

Figure 10.1 Lateral view of the skull.

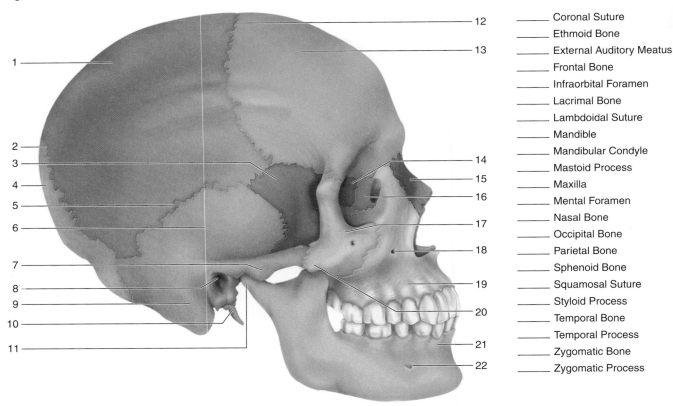

12	_____ Coronal Suture
13	_____ Ethmoid Bone
	_____ External Auditory Meatus
	_____ Frontal Bone
	_____ Infraorbital Foramen
	_____ Lacrimal Bone
	_____ Lambdoidal Suture
	_____ Mandible
14	_____ Mandibular Condyle
15	_____ Mastoid Process
16	_____ Maxilla
	_____ Mental Foramen
17	_____ Nasal Bone
	_____ Occipital Bone
18	_____ Parietal Bone
	_____ Sphenoid Bone
19	_____ Squamosal Suture
	_____ Styloid Process
20	_____ Temporal Bone
	_____ Temporal Process
21	_____ Zygomatic Bone
22	_____ Zygomatic Process

Figure 10.2 Inferior view of the skull.

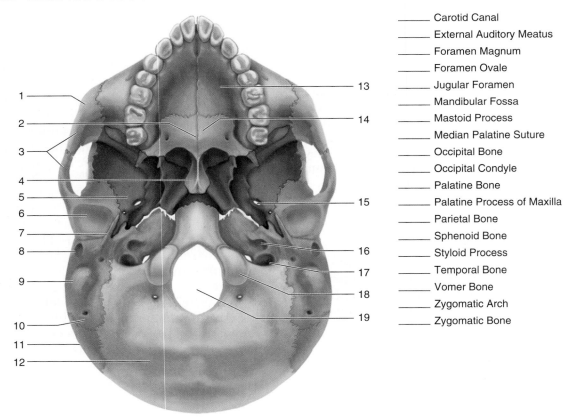

	_____ Carotid Canal
	_____ External Auditory Meatus
	_____ Foramen Magnum
	_____ Foramen Ovale
13	_____ Jugular Foramen
	_____ Mandibular Fossa
14	_____ Mastoid Process
	_____ Median Palatine Suture
	_____ Occipital Bone
	_____ Occipital Condyle
	_____ Palatine Bone
15	_____ Palatine Process of Maxilla
	_____ Parietal Bone
	_____ Sphenoid Bone
16	_____ Styloid Process
17	_____ Temporal Bone
18	_____ Vomer Bone
19	_____ Zygomatic Arch
	_____ Zygomatic Bone

Figure 10.3 Internal view of the cranial floor.

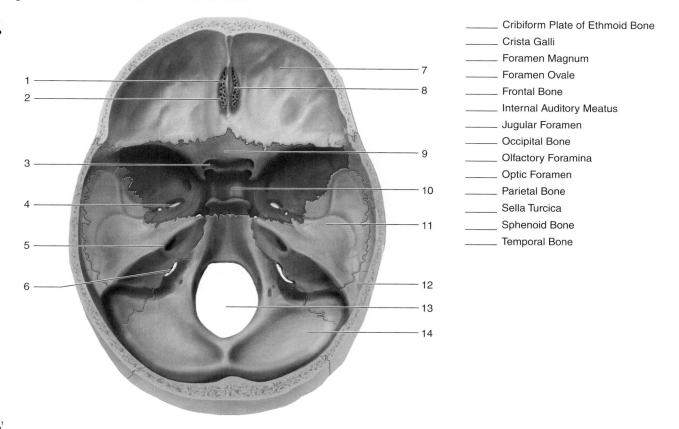

1	
2	
3	
4	
5	
6	

_____	Cribiform Plate of Ethmoid Bone
_____	Crista Galli
_____	Foramen Magnum
_____	Foramen Ovale
_____	Frontal Bone
_____	Internal Auditory Meatus
_____	Jugular Foramen
_____	Occipital Bone
_____	Olfactory Foramina
_____	Optic Foramen
_____	Parietal Bone
_____	Sella Turcica
_____	Sphenoid Bone
_____	Temporal Bone

7, 8, 9, 10, 11, 12, 13, 14

and posterior to the auditory canal. It is an attachment site for certain neck muscles.

Figure 10.2 depicts two important paired foramina of the temporal bones. The narrow **jugular foramina** (label 17) are located just medial to the styloid processes at the junction of the temporal and occipital bones. They allow passage of the jugular veins from the brain to the neck. Just anterior to the jugular foramina are the **carotid canals,** through which the carotid arteries pass to the brain.

In Figure 10.3, the **internal auditory** (or acoustic) **meatus** (label 5) is the opening on the thick, sloping portion of the temporal bone that contains the inner ear. Facial and acoustic nerves pass through this foramen. A jugular foramen is just posterior to each internal auditory meatus at the junction of the temporal and occipital bones. Insert a pipe cleaner through a jugular foramen and carotid canal to clarify their external and internal openings.

Sphenoid Bone

The sphenoid is colored blue in the figures. It consists of a *central body* and two winglike structures that extend laterally. The sphenoid forms part of the floor of the cranium, the sides of the cranium in the "temple" areas, and the posterior walls of the eye orbits. A ventral view of the skull (Figure 10.2) shows an oval opening, the **foramen ovale** (label 15), on each side of the sphenoid bone. The mandibular nerve passes through this opening.

As shown in the view of the cranial floor (Figure 10.3), the central body forms a saddle-shaped structure called the **sella turcica** (Turk's saddle). The deep **hypophyseal fossa** within the sella turcica is occupied by the hypophysis (pituitary gland), which hangs downward from the brain. The **optic foramina** (label 3), passageways for the optic nerves, are located on each side in the anterior portion of the sella turcica. The foramina

Figure 10.5 The mandible.

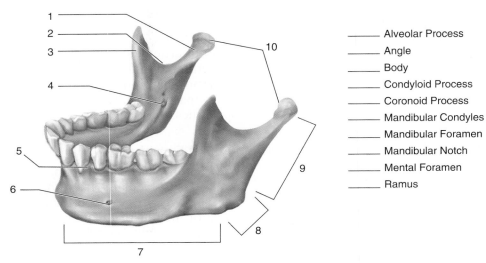

_____	Alveolar Process
_____	Angle
_____	Body
_____	Condyloid Process
_____	Coronoid Process
_____	Mandibular Condyles
_____	Mandibular Foramen
_____	Mandibular Notch
_____	Mental Foramen
_____	Ramus

Figure 10.6 The paranasal sinuses.

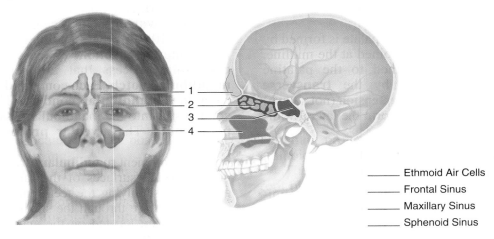

_____	Ethmoid Air Cells
_____	Frontal Sinus
_____	Maxillary Sinus
_____	Sphenoid Sinus

lined with a mucous membrane and have passageways leading into the nasal cavity. The sinuses are named after the bones in which they are located and are shown in Figure 10.6. The **frontal sinuses** occur in the forehead above the eyes. The large **maxillary sinuses** are below the eyes. The **sphenoidal sinus** (label 3 in Figure 10.7) is centrally located under the sella turcica. The **ethmoidal sinuses** consist of a number of small, air-filled spaces.

ASSIGNMENT

1. Label Figures 10.6 and 10.7.
2. Complete Sections A (except for Figure 10.9), B, and C of the laboratory report.
3. Examine a human skull and locate the parts illustrated and described. *Remember to use only pipe cleaners as pointers.*

Figure 10.7 Sagittal view of skull.

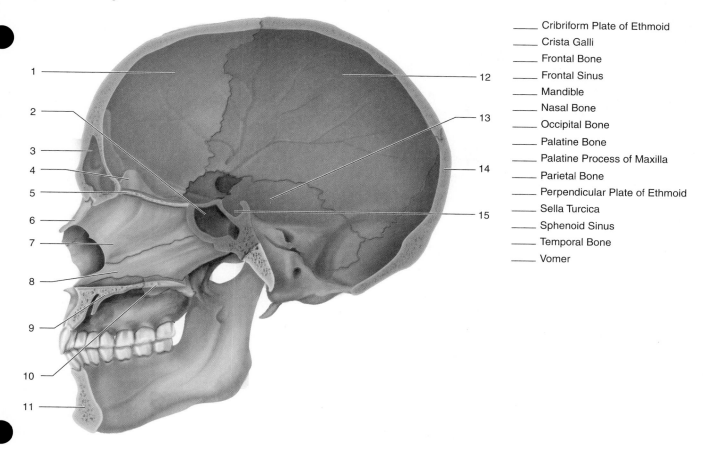

_____ Cribriform Plate of Ethmoid
_____ Crista Galli
_____ Frontal Bone
_____ Frontal Sinus
_____ Mandible
_____ Nasal Bone
_____ Occipital Bone
_____ Palatine Bone
_____ Palatine Process of Maxilla
_____ Parietal Bone
_____ Perpendicular Plate of Ethmoid
_____ Sella Turcica
_____ Sphenoid Sinus
_____ Temporal Bone
_____ Vomer

Bones Associated with the Skull

Seven additional bones are associated with the skull, although they are not considered a part of it. There are three **auditory ossicles** located in each middle-ear cavity, where they play a vital role in hearing. They will be considered in Exercise 26, The Ear. The **hyoid bone** is a small U-shaped bone located below the mandible on the superior surface of the larynx. See Figure 10.8. It does not articulate with another bone. Instead, it is suspended from the styloid processes by slender stylohyoid muscles and ligaments. The hyoid is an attachment site for certain muscles controlling the tongue, larynx, and mandible.

Figure 10.8 The hyoid bone.

Styloid process

Stylohyoid muscle

Hyoid

Larynx

Greater cornu
Lesser cornu
Body

Figure 10.9 The fetal skull. (a) Lateral view; (b) superior view.

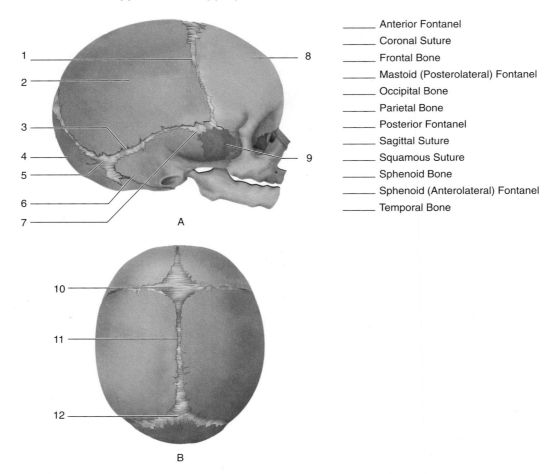

_____ Anterior Fontanel
_____ Coronal Suture
_____ Frontal Bone
_____ Mastoid (Posterolateral) Fontanel
_____ Occipital Bone
_____ Parietal Bone
_____ Posterior Fontanel
_____ Sagittal Suture
_____ Squamous Suture
_____ Sphenoid Bone
_____ Sphenoid (Anterolateral) Fontanel
_____ Temporal Bone

The Fetal Skull

The incompletely ossified fetal skull is shown in Figure 10.9. The unossified sutures and **fontanels,** membranous areas at the junction of several cranial bones, allow the compression of the skull during childbirth and growth of the brain after birth.

There are six fontanels. The large **anterior fontanel** lies on the midline at the junction of the frontal and parietal bones. The smaller **posterior fontanel** is on the midline at the junction of the parietal and occipital bones. A **sphenoid (anterolateral) fontanel** occurs on each side of the skull at the junction of the frontal, parietal, zygomatic, sphenoid,

and temporal bones. A **mastoid (posterolateral) fontanel** is on each side of the skull at the junction of the temporal, parietal, and occipital bones.

ASSIGNMENT

1. Label Figure 10.9 and complete Section A of the laboratory report.
2. Complete Section D of the laboratory report.
3. Examine a fetal skull. Locate the skull bones and fontanels shown in Figure 10.9

THE VERTEBRAL COLUMN AND THORAX

OBJECTIVES

After completing this exercise, you should be able to:

1. Identify the bones, and their significant parts, that form the vertebral column and thorax.
2. Define all terms in bold print.

Materials

Skeletons, articulated and disarticulated
Vertebral column, mounted
Pipe cleaners

The Vertebral Column

The vertebral column forms a flexible but sturdy vertical axis extending from the skull to the pelvis. It is composed of twenty-four movable vertebrae, the sacrum, and the coccyx. They are all bound together by ligaments and muscles to form a unified structure. Refer back to Figure 9.2 and note the relationship between the vertebral column and the rest of the skeletal system. The vertebral column exhibits four defined **spinal curvatures** that are shown in Figure 11.1. From top to bottom, they are the **cervical, thoracic, lumbar,** and **pelvic curvatures.**

Vertebrae share many common features, although they vary in size and shape in different regions of the vertebral column. Each has an anterior structural mass, the **body,** that is the major load-bearing contact between adjacent vertebrae. Between the bodies of adjacent vertebrae are fibrocartilaginous **intervertebral disks** that serve as protective cushions.

Projecting posteriorly from each side of the vertebral body are two processes, the **pedicles** (label 25), that form the anterolateral portion of the **neural arch.** The posterolateral portion of the neural arch is formed by the **laminae** (label 24), which join to form the posteriorly projecting **spinous process.**

The **spinal (vertebral) foramen** is located in the center of the neural arch. The spinal cord descends from the brain through the spinal foramina, where it is protected by the surrounding neural arches of the vertebrae. Spinal nerves emerging from the spinal cord exit through the **intervertebral foramina** (label 29), small openings between pedicles of adjacent vertebrae.

In addition to the bodies, vertebrae contact each other by two pairs of articular surfaces located on the **transverse processes** (label 20). The **superior articular surfaces** (label 21) of one vertebra articulates with the **inferior articular surfaces** (label 26) of the adjacent vertebra above (superior to) it.

Cervical Vertebrae

The first seven vertebrae are the **cervical vertebrae** of the neck. The first vertebra is the **atlas.** Illustration A shows its superior surface. Its vertebral foramen is larger than that of other vertebrae to accommodate the brain stem. The superior articular surfaces articulate with the occipital condyles of the skull. A small **transverse foramen** occurs in each transverse process. These foramina are found only in cervical vertebrae and serve as passageways for the vertebral arteries and veins.

The second cervical vertebra, the **axis,** is shown in illustration B. It is unique in having an upward-projecting **odontoid process** (dens) arising from the body, which serves as a pivot for the rotation of the atlas. When the head is turned, the atlas rotates on the axis. This pivotal movement is enabled because the atlas lacks a body.

Illustration C shows the superior surface of a nonspecialized cervical vertebra. Note the Y-shaped spinous process.

Thoracic Vertebrae

Below the cervical vertebrae are the twelve **thoracic vertebrae.** A superior view of one is shown in illustration D. These vertebrae are distinguished in having ribs attached to them. They have **articular facets for ribs** (label 23) on their transverse processes and on the sides of their body. Also, their spinous processes project downward.

Lumbar Vertebrae

The next five vertebrae are the **lumbar vertebrae** (illustration E). They have a larger body than other vertebrae because of the greater stress that occurs in this region of the vertebral column.

Figure 11.1 The vertebral column.

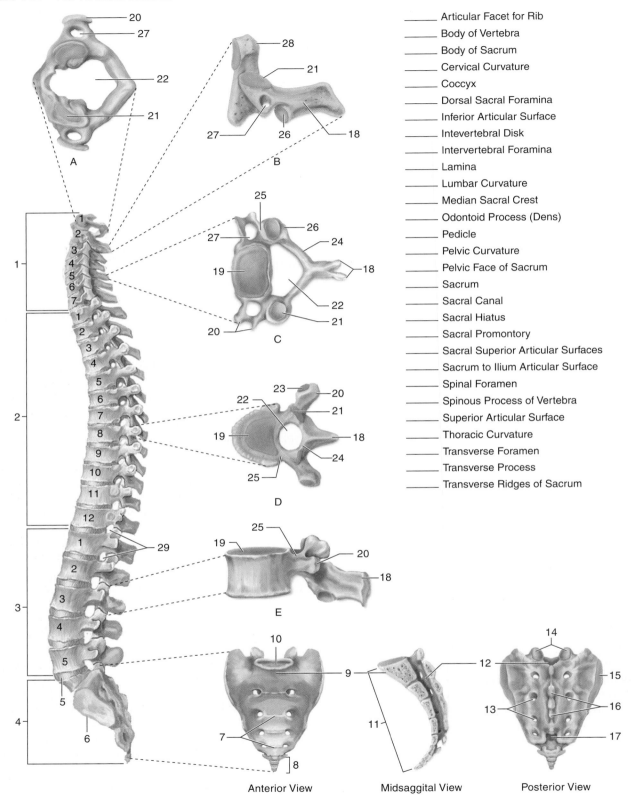

_____	Articular Facet for Rib
_____	Body of Vertebra
_____	Body of Sacrum
_____	Cervical Curvature
_____	Coccyx
_____	Dorsal Sacral Foramina
_____	Inferior Articular Surface
_____	Intevertebral Disk
_____	Intervertebral Foramina
_____	Lamina
_____	Lumbar Curvature
_____	Median Sacral Crest
_____	Odontoid Process (Dens)
_____	Pedicle
_____	Pelvic Curvature
_____	Pelvic Face of Sacrum
_____	Sacrum
_____	Sacral Canal
_____	Sacral Hiatus
_____	Sacral Promontory
_____	Sacral Superior Articular Surfaces
_____	Sacrum to Ilium Articular Surface
_____	Spinal Foramen
_____	Spinous Process of Vertebra
_____	Superior Articular Surface
_____	Thoracic Curvature
_____	Transverse Foramen
_____	Transverse Process
_____	Transverse Ridges of Sacrum

Anterior View

Midsaggital View

Posterior View

Figure 11.2 The thoracic cage.

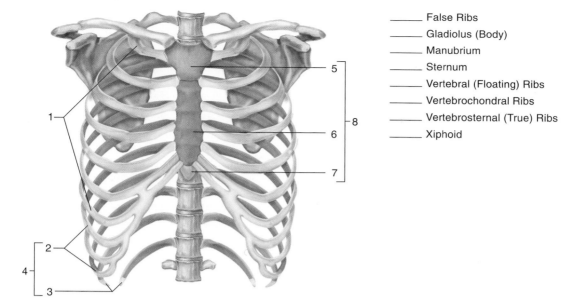

_____ False Ribs
_____ Gladiolus (Body)
_____ Manubrium
_____ Sternum
_____ Vertebral (Floating) Ribs
_____ Vertebrochondral Ribs
_____ Vertebrosternal (True) Ribs
_____ Xiphoid

The Sacrum

Inferior to the fifth lumbar vertebra is the **sacrum,** which composes the posterior wall of the pelvic cavity. It consists of five fused vertebrae. The **transverse ridges** where the vertebrae have fused are visible. The **pelvic face of the sacrum** curves posteriorly, causing an anterior protrusion of the body of the first sacral vertebra. This protrusion is the **sacral promontory.**

On the posterior surface, the spinous processes have fused to form the tubercles of the **median sacral crest.** On each side are rows of **dorsal sacral foramina** through which nerves and blood vessels pass. The neural arches of the fused vertebrae form the **sacral canal,** which continues to an inferior opening, the **sacral hiatus.**

The **sacral superior articular surfaces** (label 14) articulate with the inferior articular surfaces of the fifth lumbar vertebra. On each lateral surface is a **sacrum to ilium articular surface** (label 15).

The Coccyx

Four or five rudimentary vertebrae fuse together to form the **coccyx,** or tailbone. The triangular coccyx is joined to the sacrum by ligaments.

The Thoracic Cage

The thoracic vertebrae, ribs, costal cartilages, and sternum form the skeleton (rib cage) of the thorax. It protects the heart and lungs and provides a support for the pectoral (shoulder) girdles. See Figure 11.2.

Sternum

The **sternum,** or breastbone, is notched along the sides where the costal cartilages attach. It consists of three parts: an upper **manubrium,** a middle **body** or **gladiolus,** and a lower **xiphoid process.**

Ribs

There are twelve pairs of **ribs** attached to thoracic vertebrae. The first seven pairs of ribs attach directly to the sternum by **costal cartilages.** They are the **vertebrosternal,** or **true, ribs.** The remaining five pairs are **false ribs.** The cartilages of the first three pairs of false ribs are fused to the costal cartilages of ribs above them. These are the **vertebrochondral ribs.** The last two pairs of false ribs are not attached anteriorly and are called **vertebral,** or **floating, ribs.**

Figure 11.3 shows the structure of a thoracic vertebra, articulation of a rib with a thoracic vertebra, and the structure of a rib. The lateral view of the thoracic vertebra in illustration A shows, in addition to structures previously studied, two **articular facets** for attachment of a rib: one facet is near the body and the other facet is on the transverse process.

As shown in illustration C, a rib consists of head, neck, tubercle, and body. The **head** (label 13) articulates with the vertebral column via two **articular facets.** The **tubercle** (label 14) consists of an **articular portion** and a **nonarticular portion** (label 8). The **neck** is the short segment between the head and tubercle. The **body** is the flattened, curved remainder of the rib.

Figure 11.3 Rib anatomy and its articulation.

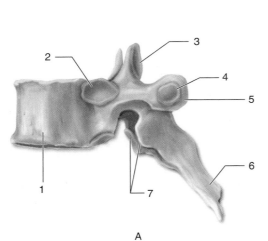

A
THORACIC VERTEBRA
(Lateral View)

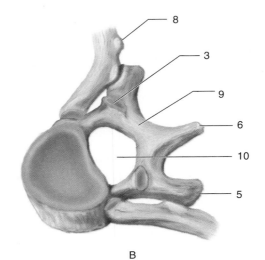

B
RIB-TO-VERTEBRA ARTICULATION
(Superior Anteroventral View)

_____ Articular Facet of Rib (2 places)
_____ Articular Portion of Tubercle
_____ Body of Rib
_____ Body of Vertebra
_____ Head of Rib
_____ Inferior Articular Processes
_____ Lamina of Vertebra
_____ Neck of Rib
_____ Nonarticular Portion of Tubercle
_____ Rib Facet (near body)
_____ Rib Facet (on transverse process)
_____ Spinal Foramen
_____ Spinous Process (2 places)
_____ Superior Articular Process (2 places)
_____ Transverse Process (2 places)
_____ Tubercle of Rib

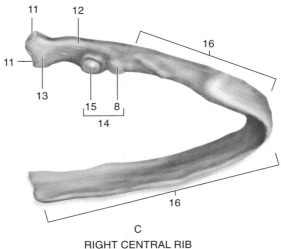

C
RIGHT CENTRAL RIB
(Posterior Ventral View)

Illustration B shows a rib-vertebra articulation. The head of the rib articulates with the articular surface near the body of its own vertebra and also with the body of the adjacent vertebra (not shown) above it. The articular portion of the tubercle articulates with a facet on the vertebra's transverse process.

ASSIGNMENT

1. Label Figures 11.1, 11.2, and 11.3.
2. Complete the laboratory report.
3. Examine the skeletal material available in the laboratory and locate the parts discussed. *Use only pipe cleaners as pointers.*

THE APPENDICULAR SKELETON

OBJECTIVES

After completing this exercise, you should be able to:

1. Identify the bones, including their significant parts, that compose the appendicular skeleton.
2. Describe the relationships of the bones involved.
3. Define all terms in bold print.

Materials

Skeletons, articulated and disarticulated
Pelvic girdles, male and female
Pipe cleaners

The **appendicular skeleton** consists of the pectoral girdle and bones of the upper limbs and the pelvic girdle and bones of the lower limbs. Refer to the figures to locate the bones and their parts as you study this exercise.

The Pectoral Girdle

The **pectoral,** or **shoulder, girdle** is formed by a clavicle (collarbone) and a scapula (shoulder blade) on each side of the body. It supports the upper limbs and permits great freedom of movement.

Each **clavicle** is a slender S-shaped bone. Its lateral end articulates with the acromion process of a shoulder blade, and its medial end articulates with the superiolateral margin of the sternum. Thus, the sternum closes the pectoral girdle anteriorly, but the girdle is open posteriorly.

The **scapula,** or shoulder blade, is a flat, triangular bone that does not articulate directly with the axial skeleton. Instead, it is held in place by muscles, thus giving greater mobility to the shoulder. On its **axillary margin** is the shallow **glenoid cavity,** which articulates with the head of the humerus. Above the glenoid cavity are two large processes. The **coracoid process** (label 8 in Figure 12.1) projects anteriorly under the clavicle. The **acromion process** projects posteriorly and articulates with the clavicle. On the scapula's posterior surface, the **scapular spine** is a slender process that runs diagonally from the medial **vertebral margin** to the acromion process. A **scapular notch** occurs on the **superior margin** at the base of the coracoid.

The Upper Limb

The **upper limb** consists of the upper arm, forearm, and hand.

Figure 12.1 The scapula.

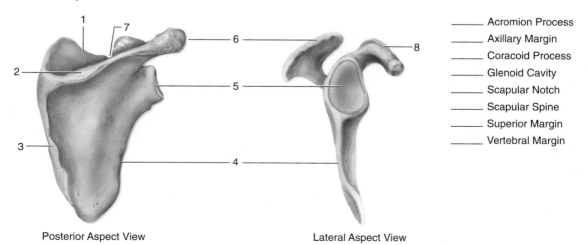

Acromion Process
Axillary Margin
Coracoid Process
Glenoid Cavity
Scapular Notch
Scapular Spine
Superior Margin
Vertebral Margin

Posterior Aspect View   Lateral Aspect View

Figure 12.2 The arm and pectoral girdle.

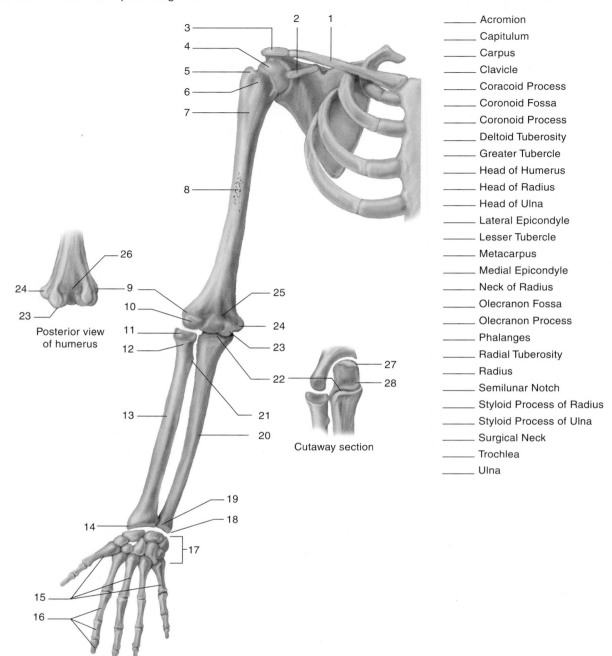

Posterior view of humerus

Cutaway section

_____	Acromion
_____	Capitulum
_____	Carpus
_____	Clavicle
_____	Coracoid Process
_____	Coronoid Fossa
_____	Coronoid Process
_____	Deltoid Tuberosity
_____	Greater Tubercle
_____	Head of Humerus
_____	Head of Radius
_____	Head of Ulna
_____	Lateral Epicondyle
_____	Lesser Tubercle
_____	Metacarpus
_____	Medial Epicondyle
_____	Neck of Radius
_____	Olecranon Fossa
_____	Olecranon Process
_____	Phalanges
_____	Radial Tuberosity
_____	Radius
_____	Semilunar Notch
_____	Styloid Process of Radius
_____	Styloid Process of Ulna
_____	Surgical Neck
_____	Trochlea
_____	Ulna

The Upper Arm

The **humerus** is the bone of the upper arm. The rounded **head** of the humerus fits into the glenoid cavity of the scapula. Just inferior to the head are two processes: a **greater tubercle** on the lateral surface and a **lesser tubercle** on the anterior surface. The **surgical neck** is inferior to the tubercles and is so named because of the frequency of fractures in this area. The **deltoid tuberosity** is a rough, raised area near the midpoint of the lateral surface of the shaft to which the _deltoid_ muscle attaches.

The distal end of the humerus has two condyles. The **capitulum** is the lateral condyle articulating with the radius. The **trochlea** is the medial condyle, which articulates with the ulna. Superior to these condyles are the **lateral** and **medial epicondyles.** The depression on the anterior surface just superior to the trochlea is the **coronoid fossa.** The **olecranon fossa** (label 26 in Figure 12.2) is in a similar location on the posterior surface of the humerus.

Figure 12.3 The coxal bone.

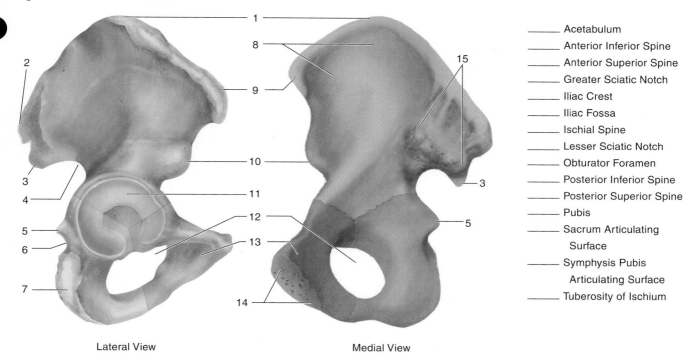

Lateral View

Medial View

_____ Acetabulum
_____ Anterior Inferior Spine
_____ Anterior Superior Spine
_____ Greater Sciatic Notch
_____ Iliac Crest
_____ Iliac Fossa
_____ Ischial Spine
_____ Lesser Sciatic Notch
_____ Obturator Foramen
_____ Posterior Inferior Spine
_____ Posterior Superior Spine
_____ Pubis
_____ Sacrum Articulating
        Surface
_____ Symphysis Pubis
        Articulating Surface
_____ Tuberosity of Ischium

The Forearm

Two bones occur in the forearm: a lateral **radius** and a medial **ulna.** The disklike **head** of the radius articulates with the capitulum of the humerus and enables the head to rotate when the position of the hand is changed from palm up to palm down. A short distance below the head is the **radial tuberosity** which is the attachment site for the _biceps brachii_ muscle, a major flexor of the arm. The **neck** (label 12) is located between the head and the tuberosity. At the distal end of the radius, a lateral **styloid process** is present at the articulation with the hand.

The proximal posterior prominence of the ulna, the **olecranon process** (label 27 in Figure 12.2), forms the point of the elbow and fits into the olecranon fossa when the arm is extended. It has a depression, the **semilunar notch,** that articulates with the trochlea of the humerus. A small eminence at the anterior margin of the trochlear notch is the **coronoid process** (label 22). At the distal end of the ulna, the knoblike **head** articulates with the radius and a fibrocartilaginous disk that separates it from the hand. The **styloid process** is the distal medial prominence.

The Hand

The skeleton of each hand consists of carpus, metacarpus, and phalanges. Eight small bones form the **carpus,** or wrist. The **metacarpus,** or palm, consists of five metacarpal bones numbered 1–5 starting with the thumb. The **phalanges** are the bones of the fingers: two in the thumb and three in each finger.

The Pelvic Girdle

The **pelvic girdle,** or **pelvis,** is formed by two **coxal bones** (hipbones), the sacrum, and the coccyx. Each coxal bone is formed of three fused bones as shown in Figure 12.3. The **ilium** is the broad upper bone whose superior margin forms the **iliac crest,** the ridgelike prominence of the hip. A broad, shallow concavity, the **iliac fossa,** occupies most of the medial surface of each ilium. The **ischium** (pink in Figure 12.3) is the lower posterior portion. The **pubis** is the lower anterior part. Note the **symphysis pubis articulating surface** on the pubis and the **sacrum articulating surface** on the ilium. A large cup-shaped fossa, the **acetabulum,** is located at the junction of the ilium, ischium, and pubis on the lateral surface of each coxal bone. The acetabulum receives the head of the femur. The coxal bones articulate with the sacrum posteriorly at the **sacroiliac joints** and with each other anteriorly at the **symphysis pubis.** See Figure 12.4.

The iliac **crest** extends between the **anterior superior spine** (label 9 in Figure 12.3) and the **posterior superior spine** (label 2). Just inferior to each of these prominences are smaller iliac spines:

Figure 12.4   The leg and pelvic girdle.

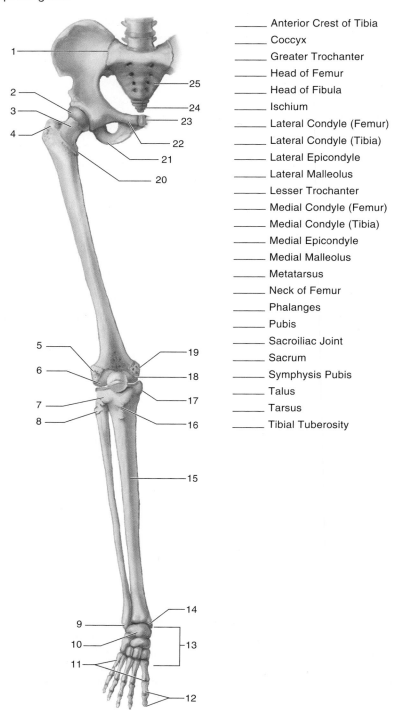

_____ Anterior Crest of Tibia
_____ Coccyx
_____ Greater Trochanter
_____ Head of Femur
_____ Head of Fibula
_____ Ischium
_____ Lateral Condyle (Femur)
_____ Lateral Condyle (Tibia)
_____ Lateral Epicondyle
_____ Lateral Malleolus
_____ Lesser Trochanter
_____ Medial Condyle (Femur)
_____ Medial Condyle (Tibia)
_____ Medial Epicondyle
_____ Medial Malleolus
_____ Metatarsus
_____ Neck of Femur
_____ Phalanges
_____ Pubis
_____ Sacroiliac Joint
_____ Sacrum
_____ Symphysis Pubis
_____ Talus
_____ Tarsus
_____ Tibial Tuberosity

the **anterior inferior spine** and the **posterior inferior spine.** Between the posterior inferior spine and the **ischial spine** (label 5) is the **greater sciatic notch.** The **lesser sciatic notch** is just inferior to the ischial spine. The **tuberosity of the ischium** (label 7) is located at the posterior inferior angle of the ischium. The large opening surrounded by the ischium and pubis is the **obturator foramen.**

The Lower Limb

The **lower limb** consists of the thigh, leg, and foot.

The Thigh

The **femur** is the bone of the upper leg. The rounded **head** of the femur fits into the acetabulum of a coxal bone. Two large processes occur at the base of the

neck (label 3 in Figure 12.4): the lateral **greater trochanter** and the medial **lesser trochanter.** The enlarged lower end of the femur terminates with the **lateral** and **medial condyles,** which have **lateral** and **medial epicondyles** just above them.

The Leg

The bones of the leg are the **tibia** (shinbone) and the smaller **fibula.** The upper end of the tibia consists of **lateral** and **medial condyles** that articulate with the corresponding condyles of the femur. The **tibial tuberosity** is located on the anterior surface just inferior to the condyles. A sharp **anterior crest** is evident on the shaft. At the distal end a process called the **medial malleolus** forms the medial prominence of the ankle.

The superior end, or **head,** of the fibula articulates with the tibia but does not form part of the knee joint. The neck lies just below the head and an anterior crest is present. At the inferior end, the **lateral malleolus** forms the lateral prominence of the ankle.

The Foot

The skeleton of each foot consists of tarsus, metatarsus, and phalanges. There are seven tarsal bones forming the **tarsus,** or ankle. The most prominent tarsal bones are the **talus,** which articulates with the tibia and fibula, and the **calcaneus,** or heelbone (not shown). The instep, or **metatarsus,** is formed by five metatarsal bones numbered 1–5 starting on the medial side. The bones of the toes are the **phalanges:** two in the great toe and three in each of the other toes.

ASSIGNMENT

1. Label Figures 12.1 through 12.4.
2. Complete Section A of the laboratory report.
3. Examine the skeletal material and locate the parts discussed.
4. Compare the differences in male and female pelves in accordance with Section B of the laboratory report.
5. Complete the laboratory report.

ARTICULATIONS

OBJECTIVES

After completing this exercise, you should be able to:

1. State the basic functional types of joints and describe their characteristics.
2. Classify the joints of the body as to functional type.
3. Identify the components of diarthrotic joints.
4. Define all terms in bold print.

Materials

Skeleton, articulated
Fresh knee joint of cow or lamb, sectioned longitudinally
Models of various joints

Before You Proceed

Consult with your instructor about using protective disposable gloves when performing portions of this exercise.

The Basic Functional Types

All joints of the body are classified according to three functional categories as shown in Figure 13.1.

Immovable Joints

Synarthrotic joints are immovable because the bones forming the joint are tightly bonded to each other by fibrous connective tissue or cartilage. There are two types of immovable joints: sutures and synchondroses.

Figure 13.1 Types of articulations.

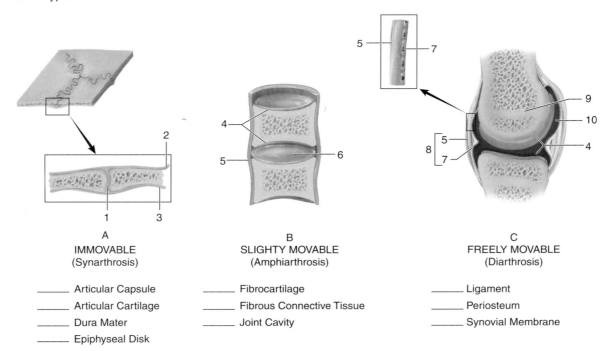

A	B	C
IMMOVABLE (Synarthrosis)	SLIGHTY MOVABLE (Amphiarthrosis)	FREELY MOVABLE (Diarthrosis)

_____ Articular Capsule   _____ Fibrocartilage   _____ Ligament
_____ Articular Cartilage   _____ Fibrous Connective Tissue _____ Periosteum
_____ Dura Mater   _____ Joint Cavity   _____ Synovial Membrane
_____ Epiphyseal Disk

Figure 13.2 The shoulder joint.

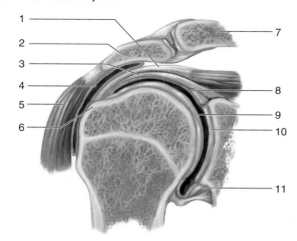

_____	Acromion Process
_____	Articular Capsule
_____	Articular Cartilage of the Glenoid Cavity
_____	Articular Cartilage of the Humerus
_____	Clavicle
_____	Deltoid Muscle
_____	Glenoid Labrum
_____	Greater Tubercle of the Humerus
_____	Subdeltoid Bursa
_____	Supraspinatus Tendon
_____	Synovial Membrane

Sutures are irregular joints between the immovable bones of the skull (illustration A in Figure 13.1). The bones are bonded together by **fibrous connective tissue** that is continuous with the **periosteum** (label 2) on the outer surface of the bones and with the **dura mater** on the inner surface of the bones.

Synchondroses have cartilage as the bonding tissue. An example is the bonding of the epiphyses to the diaphysis by the epiphyseal cartilages in the long bones of children.

Slightly Movable Joints

The bones of **amphiarthrotic** (slightly movable) **joints** are also bound by fibrous connective tissue or cartilage but not as tightly as in immovable joints. The two types of slightly movable joints are symphyses and syndesmoses.

Symphyses have a cushioning pad of **fibrocartilage** between the bones. The articulations of the bodies of vertebrae and the joining of coxal bones at the symphysis pubis are examples. Adjacent to the cartilagenous pads, the articular surfaces of the bones are covered with **articular cartilages** (label 4) that reduce friction in the joint. The joint is wrapped in a fibrous capsule formed of **ligaments** (label 5).

Syndesmoses lack fibrocartilage, but the bones are held together by fibrous connective tissue forming interosseous ligaments. The attachment of the fibula to the tibia, as shown in Figure 13.3, is an example.

Freely Movable Joints

Bones forming **diarthrotic** (freely movable) **joints** are bound together by a fibrous **articular capsule** formed of ligaments. A **synovial membrane** lines the inside of the capsule and secretes **synovial fluid** that lubricates the joint. The articular surfaces of the bones are covered by protective, friction-reducing **articular cartilages.** These joints may also contain **bursae,** sacs of synovial fluid that reduce friction. There are six kinds of diarthrotic joints.

Gliding joints occur between small bones with flat or slightly convex surfaces, such as the carpal and tarsal bones.

Hinge joints allow movement in only one plane, such as in the elbow and knee.

Condyloid joints allow movement in two planes. They are formed by a rounded condyle articulating with an elliptical depression, such as between the carpus and the radius.

Saddle joints occur where the ends of both bones are saddle-shaped, convex in one direction and concave in the other. An example is the joint between the trapezium (a carpal bone) and the metacarpal bone of the thumb, which permits a variety of movements.

Pivot joints allow rotational movement around a pivot point, such as the atlas around the odontoid process of the axis.

Ball-and-socket joints allow angular movement in all directions, such as in the shoulder and hip joints.

ASSIGNMENT

Label Figure 13.1.

The Shoulder Joint

The shoulder joint (Figure 13.2) is the most freely movable joint in the body, due to the shallowness

Figure 13.3   The knee joint. (a) Anterior view; (b) posterior view

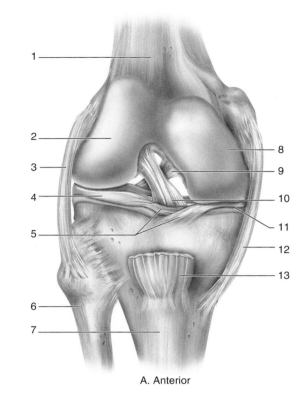

A. Anterior

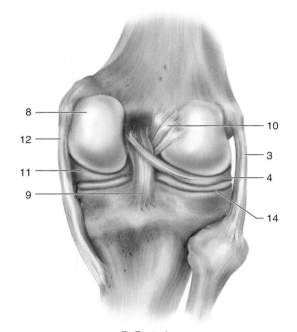

B. Posterior

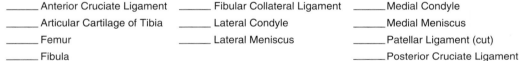

_____ Anterior Cruciate Ligament	_____ Fibular Collateral Ligament
_____ Articular Cartilage of Tibia	_____ Lateral Condyle
_____ Femur	_____ Lateral Meniscus
_____ Fibula	

_____ Medial Condyle	_____ Tibia
_____ Medial Meniscus	_____ Tibial Collateral Ligament
_____ Patellar Ligament (cut)	_____ Transverse Ligament
_____ Posterior Cruciate Ligament	

of the glenoid cavity and the looseness of the joint capsule. A fibrocartilage ring, the **glenoid labrum** (label 11), around the edge of the glenoid cavity slightly increases the depth of the cavity. Ligaments of the articular capsule and rotator cuff tendons of associated muscles support the joint. The **acromion process** and the lateral end of the **clavicle** are shown above the joint. The tendon of **deltoid muscle** is attached to the acromion process, and the **tendon of the supraspinatus muscle** inserts on the **greater tubercle of the humerus.** Note the **subdeltoid bursa** under the deltoid muscle and tendon. A portion of the **articular capsule** (label 3), lined with **synovial membrane,** is shown just inferior to the supraspinatus tendon. Both the head of the humerus and the glenoid cavity are coated with **articular cartilages.**

ASSIGNMENT

Label Figure 13.2.

The Knee Joint

The structure of the knee joint, a complex hinge joint, is shown in Figure 13.3 in anterior and posterior views. The knee is probably the most highly stressed joint in the body. Some of this stress is absorbed by two lateral and medial fibrocartilage pads, the **menisci** (labels 4 and 11). They are thicker at the periphery and thinner at the center of the joint, providing a recess for the condyles. The menisci are wrapped together peripherally by a **transverse ligament** (label 5). **Articular cartilages** cover the articular surfaces of the femur (not labeled) and the tibia (label 14).

The joint is enveloped by a fibrous capsule composed of a complex of several layers of ligaments and muscle tendons. Innermost are the **anterior cruciate ligament** (label 10) and the **posterior cruciate ligament** (label 9). They are named for their anterior or posterior attachment to the tibia and because they cross to form an X. The **fibular collateral ligament** is the lateral ligament extending from the fibula to the lateral epicondyle of the

Figure 13.4 A sagittal section of the knee joint.

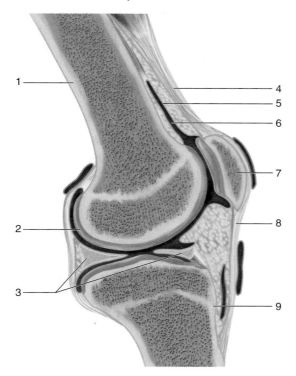

_____ Articular Cartilage
_____ Femur
_____ Menisci
_____ Patella
_____ Patellar Ligament
_____ Quadriceps Tendon
_____ Suprapatellar Bursa
_____ Synovial Membrane
_____ Tibia

femur. The **tibial collateral ligament** extends from the tibia to the medial epicondyle of the femur. In the anterior view, a stub of the **patellar ligament** (cut) is shown attached to the tibia. Other ligaments are not shown.

Figure 13.4 shows a sagittal section of the knee joint. Note the articular cartilages and menisci within the joint. The **tendon of the quadriceps muscle,** the large muscle on the anterior of the thigh, and the **patellar ligament,** which attaches to the tibia, hold the patella in place. A **suprapatellar bursa** lies posterior to the quadriceps tendon. There are a dozen or so bursae associated with muscle tendons of the knee joint but not shown here. **Synovial membranes** line the bursae and the articular capsule.

As you can see, there is no bony reinforcement to prevent dislocation of the knee in any direction. The knee is especially vulnerable to lateral and rotational forces, so it is often injured in vigorous sport activities. A torn ligament involves a collateral ligament but may also involve a cruciate ligament. Damage to a miniscus is called a torn cartilage.

ASSIGNMENT

1. Label Figures 13.3 and 13.4.
2. Complete Sections A through C of the laboratory report.
3. Examine the joint models and the major structures of each.
4. Examine an articulated skeleton and classify the joints in accordance with Section D of the laboratory report.
5. Examine the freely movable joints on your body and correlate the degree and type of movement with the type of freely movable joint.
6. Examine a fresh cow or sheep hinge joint that has been sectioned longitudinally. Compare it with Figures 13.3 and 13.4 and identify as many structures as possible.

MUSCLE ORGANIZATION AND BODY MOVEMENTS

OBJECTIVES

After completing this exercise, you should be able to:

1. Identify and describe the structure of a skeletal muscle.
2. Describe the arrangement of muscles in the body.
3. Identify the various types of body movements in diarthrotic joints.
4. Define all terms in bold print.

Movement of limbs and other body parts by the contraction (shortening) of skeletal muscles results from the contraction of muscle fibers (cells) composing the muscles involved. The type of movement is determined by the sites of muscle attachment and the type of articulation involved. Typically, a muscle is attached to the skeleton at each end and spans across a joint where the movement occurs. In this exercise, you will study the structure of skeletal muscles and types of body movements at freely movable joints.

Figure 14.1 shows the orientation of two muscles of the upper arm as examples. The **triceps brachii** extends the forearm, and the **biceps brachii** and **brachialis** flex the forearm.

Muscle Attachments

A muscle is attached to a bone by the fusion of its fibrous connective tissue with the periosteum. This attachment may be direct or indirect. In a **direct attachment,** the muscle's connective tissue does not extend beyond the muscle but fuses directly to the periosteum. In an **indirect attachment,** the muscle's connective tissue extends beyond the muscle as a narrow bandlike or ropelike **tendon,** or a broad sheetlike **aponeurosis** (not shown). The immovable end of a muscle is called the **origin,** and the movable end—where the action occurs—is called the **insertion.**

Figure 14.1A shows the orientation of three muscles of the upper arm that span the elbow joint: the **triceps brachii, brachialis,** and **biceps brachii.**

The triceps brachii has three origins or **heads** (label 1). Two origins fuse by direct attachment to the posterior surface of the humerus, and one origin is attached by a tendon to the scapula. Its insertion is on the olecranon process of the ulna. The origin of the brachialis is fused by direct attachment (label 5) to the lower half of the anterior surface of the humerus, and its insertion (cut) is on the coronoid process of the ulna. The biceps brachii has two origins (label 4) attached by tendons to the scapula, and its insertion is on the radial tuberosity. The insertions of all three muscles are attached to bones by tendons.

When a muscle contracts, it exerts a pull on the insertion-bone attachment. Contraction of the triceps brachii exerts a pull on the olecranon, which extends the forearm. Contraction of the brachialis and biceps brachii exert a pull on the ulna and radius, respectively, which flexes the forearm.

Microscopic Structure

Illustrations B, C, and D in Figure 14.1 show the detailed structure of a skeletal muscle. A muscle is composed of many **muscle fibers** that are bound together in small bundles called **fasciculi** (fasciculus is singular). Only one of three fasciculi is labeled (label 8) in illustration B, which shows the fusion of connective tissue to form a tendon.

Illustration C shows how adjacent fasciculi are separated from, and bound to, each other by connective tissue called **perimysium** (label 7). A coarser connective tissue, the **epimysium** (label 11), surrounds and binds together all of the fasciculi in a muscle. Exterior to the epimysium, the **deep fascia** also envelops the entire muscle.

Illustration D shows a single fasciculus. Note that each muscle fiber is separated from, and bound to, adjacent muscle fibers by a thin layer of connective tissue called **endomysium.** A single muscle fiber is extended for easy recognition.

ASSIGNMENT

Label Figure 14.1.

Figure 14.1 Muscle anatomy and attachments.

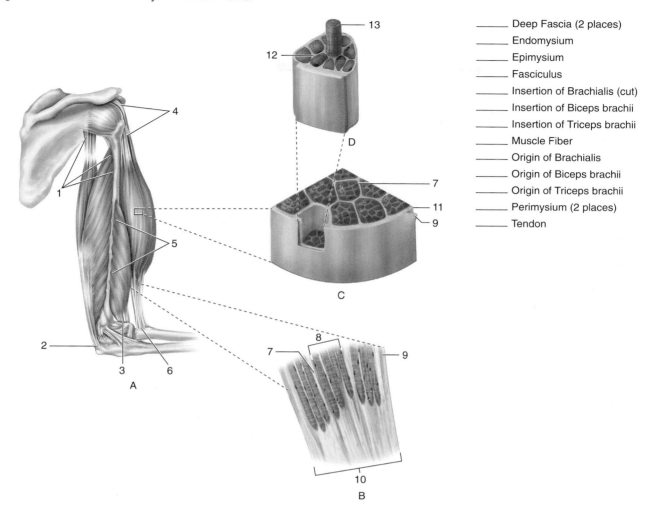

_____ Deep Fascia (2 places)
_____ Endomysium
_____ Epimysium
_____ Fasciculus
_____ Insertion of Brachialis (cut)
_____ Insertion of Biceps brachii
_____ Insertion of Triceps brachii
_____ Muscle Fiber
_____ Origin of Brachialis
_____ Origin of Biceps brachii
_____ Origin of Triceps brachii
_____ Perimysium (2 places)
_____ Tendon

Body Movements

The arrangement of muscles in the body allows them to produce specific body movements upon contraction. The common types of movements are noted below.

Flexion and Extension

Flexion is the decrease of the angle formed by the bones at a joint. **Extension** is the increase of this angle. Flexion of the arm occurs at the elbow when the hand is brought toward the shoulder. Straightening of the arm produces extension.

Specific terms are applied to the movement of the foot. Flexing the foot upward is called **dorsiflexion.** Moving the sole of the foot downward, or flexion of the toes, is called **plantar flexion.** Extension of the foot at the ankle is the same as plantar flexion. When extension is excessive, as when leaning backward, it is called **hyperextension.**

Abduction and Adduction

The movement of a limb away from the midline of the body is **abduction.** Movement toward the midline is **adduction.** These terms may be applied to parts of a limb, such as fingers and toes, by using the axis of the limb as the point of reference.

Pronation and Supination

The palm of the hand faces anteriorly in the anatomical position. Rotating the hand so the palm faces posteriorly is **pronation. Supination** rotates the palm to again face anteriorly. These terms also apply when the arm is not in the anatomical position. Turning the palm downward is pronation, and turning it upward is supination.

Inversion and Eversion

These terms apply to foot movements. **Inversion** is when the sole of the foot is turned inward toward the

Figure 14.2 Body movements.

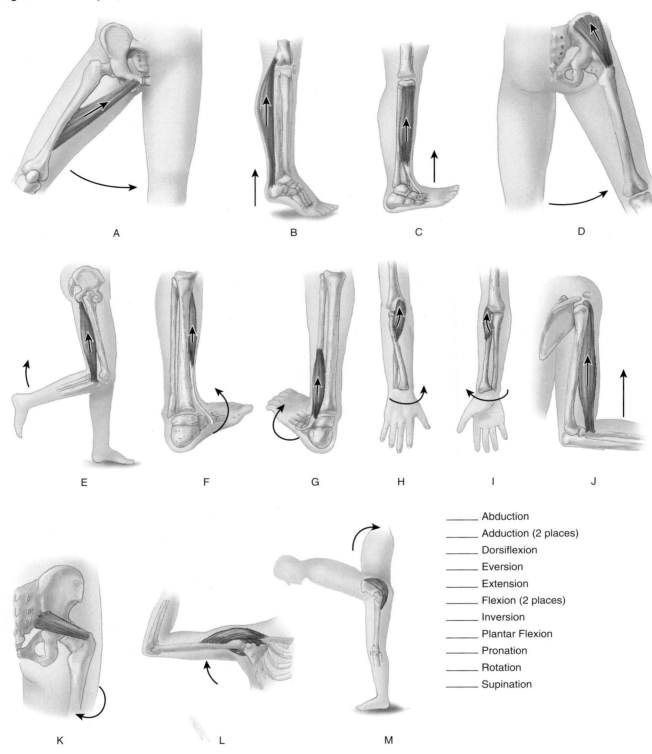

A   B   C   D

E   F   G   H   I   J

_____ Abduction

_____ Adduction (2 places)

_____ Dorsiflexion

_____ Eversion

_____ Extension

_____ Flexion (2 places)

_____ Inversion

_____ Plantar Flexion

_____ Pronation

_____ Rotation

_____ Supination

K   L   M

midline of the body. **Eversion** is turning the sole outward away from the midline.

Rotation and Circumduction

The movement of a body part around its longitudinal axis is **rotation.** If the body part can be rotated so that it describes a circle, the movement is called **circumduction.** For example, the head, arm, and leg can be rotated, but only the arm and leg can be circumducted.

Constriction and Dilation

Sphincter muscles are circular muscles that surround openings of the body. Contraction causes **constriction** of the opening; relaxation causes **dilation** of the opening.

Muscle Arrangements

Muscles usually function in groups to bring about body movements. The **prime mover** is the primary muscle in the group that produces the desired action. The assisting muscles, the **synergists,** impart steadiness and prevent unnecessary action. Other muscles that hold structures in position for action are called **fixators.**

Muscles are usually arranged in **antagonistic groups** so that the contraction of one muscle, the **agonist,** causes a specific action, and the contraction of the other muscle, the **antagonist,** causes the opposite action. Smooth movement requires that each muscle relax while the opposing muscle contracts.

ASSIGNMENT

1. Identify the types of movements in Figure 14.2. Some movements are illustrated more than once.
2. Complete the laboratory report.

HEAD AND TRUNK MUSCLES

OBJECTIVES

After completing this exercise, you should be able to:

1. Identify the major superficial muscles of the head and trunk on charts or manikin.
2. Describe the action of the muscles studied.
3. Define all terms in bold print.

Materials

Skeleton, articulated
String

Head Muscles

Refer to Figure 15.1 as you study this section.

Orbicularis Oris This sphincter muscle encircles the mouth. Its *origin* is on several facial muscles, maxilla, mandible, and nasal septum. The *insertion* is on the lips.

Action: Closes and puckers the lips.

Orbicularis Oculi This is the sphincter muscle encircling the eye. Its *origin* is on the frontal and maxilla bones. Its *insertion* is on the eyelids.

Action: Closes the eyelids.

Epicranius This muscle is composed of two muscular parts that lie over the frontal and occipital bones. They are joined by the **epicranial aponeurosis,** which covers the top of the skull. The **frontalis** has its *origin* on the aponeurosis and its *insertion* on the soft tissue under the eyebrows. The **occipitalis** has its *origin* on the mastoid process and occipital bone. Its *insertion* is on the aponeurosis.

Action: The frontalis elevates the eyebrows and wrinkles the forehead. The occipitalis pulls the scalp backward.

Zygomaticus These two muscles have their *origin* on the zygomatic arch and extend diagonally to their *insertion* on the orbicularis oris at the corner of the mouth.

Action: Draws the corner of the mouth upward and backward as in smiling and laughing.

Triangularis This muscle has its *origin* on the mandible and its *insertion* on the orbicularis oris at the corner of the mouth. It is an antagonist of the zygomaticus.

Action: Draws the corner of the mouth downward.

Platysma This is a broad sheetlike muscle that covers the front and side of the neck. Its *origin* is on the fascia of the upper chest and shoulder. The *insertion* is on the mandible and the muscles around the mouth. It is involved in many facial expressions.

Action: Pulls the angle of the mouth downward and backward and assists in opening the mouth.

Buccinator This horizontal muscle is located in the walls of the cheeks. Its *origin* is on the mandible and maxilla, and its *insertion* is on the orbicularis oris at the corner of the mouth.

Action: Compresses the cheek, holds food between teeth during chewing, and interacts with other muscles in facial expressions.

Masseter This is the primary chewing muscle. Its *origin* is on the zygomatic arch, and its *insertion* is on the lateral surface of the ramus and angle of the mandible.

Action: Raises the mandible.

Temporalis This is a large, fan-shaped muscle on the side of the head. Its *origin* is on the temporal, parietal, and frontal bones. The *insertion* is on the coronoid process of the mandible.

Action: Acts synergistically with the masseter to raise the mandible.

Sternocleidomastoid This muscle is located on the side of the neck and is partially covered by the

Figure 15.1 Head muscles.

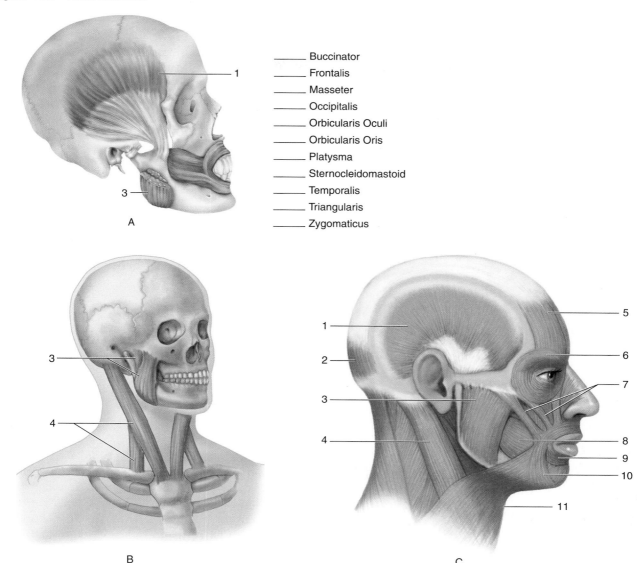

_____ Buccinator
_____ Frontalis
_____ Masseter
_____ Occipitalis
_____ Orbicularis Oculi
_____ Orbicularis Oris
_____ Platysma
_____ Sternocleidomastoid
_____ Temporalis
_____ Triangularis
_____ Zygomaticus

A

B

C

platysma. Its _origin_ is on the sternum and medial end of the clavicle, and its _insertion_ is on the mastoid process.

> _Action:_ Contraction of both muscles flexes the head down toward the chest. Contraction of one muscle only turns the face away from the side of the contracting muscle.

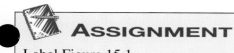

ASSIGNMENT

Label Figure 15.1.

Anterior Trunk Muscles

Refer to Figure 15.2 as you study this section.

**Pectoralis Major**   This fan-shaped muscle occupies the upper quadrant of the chest. Its _origin_ is on the clavicle, sternum, costal cartilages, and aponeurosis of the external oblique. Its _insertion_ is in a groove between the greater and lesser tubercles of the humerus.

> _Action:_ Adduction and medial rotation of the upper arm.

**Deltoid**   This primary muscle of the shoulder _originates_ on the lateral third of the clavicle and on the

Figure 15.2 Muscles of the anterior trunk.

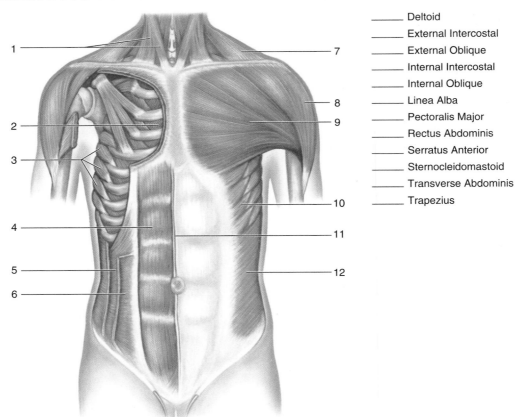

_____ Deltoid
_____ External Intercostal
_____ External Oblique
_____ Internal Intercostal
_____ Internal Oblique
_____ Linea Alba
_____ Pectoralis Major
_____ Rectus Abdominis
_____ Serratus Anterior
_____ Sternocleidomastoid
_____ Transverse Abdominis
_____ Trapezius

acromion and spine of the scapula. It *inserts* on the deltoid tuberosity of the humerus.

> *Action:* Abduction and extension of the upper arm.

Serratus Anterior   This muscle covers the upper lateral surface of the ribs. Its *origin* is on the upper eight or nine ribs, and its *insertion* is on the anterior surface of the scapula.

> *Action:* Pulls scapula downward and forward toward the chest.

External Intercostals   Intercostal muscles are found between the ribs. The external intercostals (label 10) *originate* on the lower border of the upper rib and *insert* on the upper edge of the lower rib. The fibers are directed obliquely forward.

> *Action:* Elevate rib cage and increase volume of the thorax, resulting in inspiration of air in breathing.

Internal Intercostals   These muscles (label 2) *originate* on the upper border of the lower rib and *insert* on the lower edge of the upper rib. The muscle fibers are arranged in a direction opposite to the external intercostals.

> *Action:* Lower rib cage and draw ribs closer together. This decreases volume of the thorax, causing expiration of air in breathing.

Abdominal Muscles

The abdominal wall consists of four pairs of thin muscles that have a common collective action. The right side of the abdominal wall has been removed in Figure 15.2 to show the three layers composed of the first three muscles described next, starting with the outermost muscle.

External Oblique   The *origin* of this superficial muscle is on the lower eight ribs. Its *insertion* is on the iliac crest and the **linea alba** (label 11), the white line at the midline of the abdomen where the aponeuroses of the right and left external obliques meet. The lower margin of each aponeurosis forms the **inguinal ligament,** which extends from the anterior superior spine of the ilium to the pubic tubercle. The muscle fibers run diagonally downward from the ribs toward the linea alba.

Internal Oblique   This muscle (label 5) lies under the external oblique. Its *origin* is on the iliac crest

Figure 15.3 Muscles of the posterior shoulder and trunk.

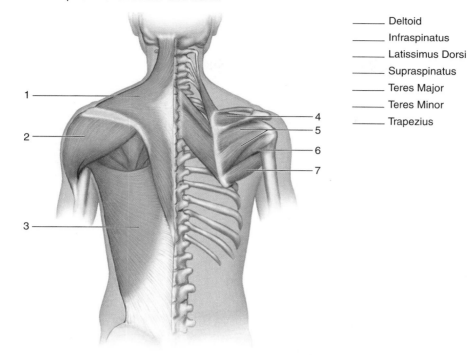

_____ Deltoid
_____ Infraspinatus
_____ Latissimus Dorsi
_____ Supraspinatus
_____ Teres Major
_____ Teres Minor
_____ Trapezius

and inguinal ligament. Its *insertion* is on the costal cartilages of the lower three ribs, the linea alba, and the pubic crest. The muscle fibers also run diagonally.

Transversus Abdominis This innermost muscle *originates* from the inguinal ligament, iliac crest, and the costal cartilages of the lower six ribs. It *inserts* on the linea alba and the pubic crest. The muscle fibers run horizontally across the abdomen.

Rectus Abdominis The right rectus abdominis is the narrow segmented muscle extending from the rib cage to the pubic bone in Figure 15.2. It is embedded within the aponeurosis formed by the preceding three muscles. Its *origin* is on the pubic bone, and its *insertion* is on the xiphoid process and the cartilages of the fifth, sixth, and seventh ribs. The rectus abdominis aids in flexion of the trunk in the lumbar region.

> *Collective Action:* These four muscles compress the abdominal organs and maintain or increase intra-abdominal pressure, aiding in urination, defecation, and childbirth. They are antagonists to the diaphragm and aid in forcing air from the lungs.

 ASSIGNMENT

Label Figure 15.2.

Posterior Trunk Muscles

Refer to Figure 15.3 as you study this section.

Trapezius This large triangular muscle of the upper back has its *origin* on the occipital bone and the spinous processes of the cervical and thoracic vertebrae. It *inserts* on the spine and acromion of the scapula and on the lateral third of the clavicle.

> *Action:* Adducts and elevates scapula; pulls head backward (hyperextension) if scapulae are fixed.

Latissimus Dorsi This large muscle covers the lower back. Its *origin* is a large aponeurosis attached to thoracic and lumbar vertebrae, spine of the sacrum, iliac crest, and lower ribs. Its *insertion* is on the intertubercular groove of the humerus.

> *Action:* Extends, adducts, and rotates the arm medially; pulls shoulder downward and backward.

Infraspinatus This muscle (label 5) *originates* on the inferior margin of the spine and posterior surface of the scapula. Its *insertion* is on the greater tubercle of the humerus. It is partially covered by the trapezius and deltoid in illustration A. The small muscle just below it, the **teres minor,** synergistically assists the infraspinatus.

> *Action:* Rotates upper arm laterally.

Supraspinatus This muscle (label 4) is covered by the trapezius and deltoid. It *originates* from the fossa above the scapular spine and *inserts* on the greater tubercle of the humerus.

Action: Assists deltoid in abducting upper arm.

Teres Major This muscle is the most inferior of the three muscles that *originate* on the posterior surface of the scapula. Its *insertion* is on the lesser tubercle of the humerus.

Action: Rotates upper arm medially.

ASSIGNMENT

1. Label Figure 15.3.
2. Complete the laboratory report.
3. Locate these muscles on a wall chart or manikin.
4. Study the action of these muscles by locating their origins and insertions on an articulated skeleton. Place one end of a piece of string on the origin and the other end on the insertion to help you visualize the effect of muscle contraction.

Exercise 16

MUSCLES OF THE UPPER LIMB

OBJECTIVES

After completing this exercise, you should be able to:

1. Identify the muscles studied on a chart or manikin.
2. Describe the action of the muscles studied.
3. Define all terms in bold print.

Materials

Skeleton, articulated
String

Upper Arm Movements

The major muscles moving the upper arm were studied in Exercise 15 except for the coracobrachialis. It is included here because it is primarily located on the upper arm instead of on the trunk. See illustration D, Figure 16.1.

Coracobrachialis This muscle covers the upper medial surface of the humerus. Its *origin* is on the coracoid process of the scapula, and its *insertion* is on the middle medial surface of the humerus.

 Action: Flexes and adducts upper arm.

Forearm Movements

The first four muscles described occur on the upper arm and are shown in Figure 16.1. The last three are forearm muscles and are depicted in illustrations A (posterior view) and B (anterior view) in Figure 16.2.

Biceps Brachii This large muscle on the anterior surface of the upper arm bulges when the forearm is flexed. It *originates* at two sites on the scapula: the coracoid process and the tubercle above the glenoid cavity. Its *insertion* is on the radial tuberosity.

 Action: Flexes forearm; rotates forearm laterally (supination).

Figure 16.1 Arm muscles.

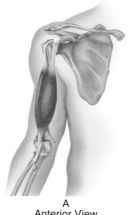

_____ Biceps Brachii
_____ Brachialis
_____ Brachioradialis
_____ Coracobrachialis
_____ Triceps Brachii

A
Anterior View

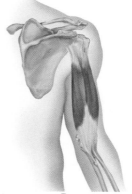

B
Posterior View

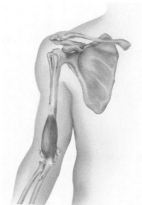

C
Anterior View

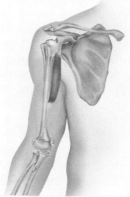

D
Anterior View

E
Anterior View

Figure 16.2 Forearm muscles.

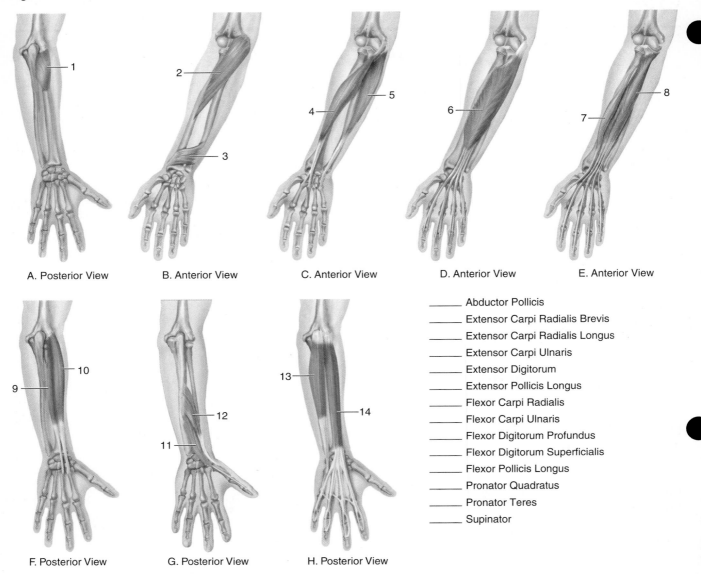

A. Posterior View   B. Anterior View   C. Anterior View   D. Anterior View   E. Anterior View

F. Posterior View   G. Posterior View   H. Posterior View

_____ Abductor Pollicis
_____ Extensor Carpi Radialis Brevis
_____ Extensor Carpi Radialis Longus
_____ Extensor Carpi Ulnaris
_____ Extensor Digitorum
_____ Extensor Pollicis Longus
_____ Flexor Carpi Radialis
_____ Flexor Carpi Ulnaris
_____ Flexor Digitorum Profundus
_____ Flexor Digitorum Superficialis
_____ Flexor Pollicis Longus
_____ Pronator Quadratus
_____ Pronator Teres
_____ Supinator

Brachialis   This muscle lies just under the biceps. Its *origin* is on the lower half of the humerus. Its *insertion* is on the coronoid process of the ulna.

   Action: Flexes the forearm.

Brachioradialis   This is the most superficial muscle on the lateral side of the forearm. Its *origin* is on the lateral epicondyle of the humerus. Its *insertion* is on the lateral surface of the radius just above the styloid process.

   Action: Flexes the forearm.

Triceps Brachii   This muscle covers the posterior surface of the upper arm and is the antagonist of the brachialis. Its *origin* occurs at three sites: the tubercle below the glenoid cavity, and the posterior and

medial surfaces of the humerus. It *inserts* on the olecranon process of the ulna.

   Action: Extends the forearm.

Supinator   This small muscle (label 1, Figure 16.2) near the elbow *originates* from the lateral epicondyle of the humerus and proximal crest of the ulna. It curves around the radius and *inserts* on the lateral margin of the radial tuberosity.

   Action: Rotates forearm laterally (supination).

Pronator Teres   This muscle *originates* from the medial epicondyle of the humerus and *inserts* on the proximal lateral surface of the radius.

   Action: Rotates forearm medially (pronation).

Pronator Quadratus   This small muscle located near the wrist is a synergist of the pronator teres. Its *origin* is on the distal part of the ulna, and its *insertion* is on the distal lateral part of the radius.

Action: Pronates forearm.

Hand Movements

Muscles moving the hand are shown in Figure 16.2. All illustrations are of the right arm. Anterior views are shown in illustrations B through E; posterior views are in illustrations A, F, G, and H.

Flexor Carpi Radialis   This muscle, shown in illustration C, *originates* from the medial epicondyle of the humerus and extends diagonally across the arm where its tendon divides to *insert* on the proximal ends of the second and third metacarpals.

Action: Flexes and abducts the hand.

Flexor Carpi Ulnaris   This muscle also *originates* on the medial epicondyle of the humerus, but it *inserts* on the base of the fifth metacarpal.

Action: Flexes and adducts the hand.

Flexor Digitorum Superficialis   This muscle (label 6) *originates* on the humerus, ulna, and radius. Its tendon divides to *insert* on the middle phalanges of fingers 2–5. It is partially covered by the flexor carpi radialis and flexor carpi ulnaris.

Action: Flexes middle phalanges of fingers 2–5.

Flexor Digitorum Profundus   This muscle (label 8) *originates* on the anterior surface of the ulna. Its tendon divides to *insert* on the distal phalanges of fingers 2–5.

Action: Flexes the distal phalanges of fingers 2–5.

Flexor Pollicis Longus   This smaller muscle in illustration E *originates* on the radius and ulna and *inserts* on the distal phalanx of the thumb.

Action: Flexes the thumb.

Extensor Carpi Radialis Longus and Brevis   These two muscles are shown in illustration F. Both *originate* on the lateral epicondyle of the humerus; the longus is attached in the more proximal position. The longus *inserts* on the base of the second metacarpal; the brevis *inserts* on the base of the third metacarpal.

Action: Both muscles extend and abduct the hand.

Extensor Carpi Ulnaris   This muscle lies on the medial side of the arm in illustration H. It *originates* on the lateral epicondyle of the humerus, crosses over, and *inserts* on the base of the fifth metacarpal.

Action: Extends and adducts the hand.

Extensor Digitorum   This muscle (label 14) also *originates* from the lateral epicondyle of the humerus. Its tendon divides to *insert* on the posterior surfaces of the distal phalanges of fingers 2–5.

Action: Extends fingers 2–5.

Extensor Pollicis Longus   This muscle (label 11) *originates* on the radius and ulna and *inserts* on the distal phalanx of the thumb.

Action: Extends the thumb.

Abductor Pollicis   This is the superior muscle in illustration G. It *originates* on the interosseus ligament between the radius and ulna and *inserts* on the lateral surface of the first metacarpal.

Action: Abducts the thumb.

ASSIGNMENT

1. Label Figures 16.1 and 16.2.
2. For each muscle in the figures, label the origin with an O and the insertion with an I.
3. Complete the laboratory report.
4. Locate the muscles on a wall chart or manikin.
5. Study the action of the muscles by locating their origins and insertions on an articulated skeleton. Place one end of a piece of string on the origin and the other end on the insertion to help you visualize the effect of muscle contraction.

MUSCLES OF THE LOWER LIMB

OBJECTIVES

After completing this exercise, you should be able to:

1. Identify the major leg muscles on a chart or manikin.
2. Describe the action of the major leg muscles.
3. Define all terms in bold print.

Materials
Skeleton, articulated
String

The major muscles of the leg are described in this exercise according to functional groups.

Thigh Movements

The muscles that move the thigh originate on the pelvis and insert on the femur. See Figure 17.1.

Gluteus Maximus This superficial buttocks muscle is covered by the deep fascia that invests the thigh muscles. A broad tendon, the **iliotibial tract** (label 2), extends downward from the deep fascia and attaches to the tibia. The gluteus maximus *originates* from the ilium, sacrum, and coccyx and *inserts* on the posterior surface of the femur and the iliotibial tract.

Action: Extends and rotates the thigh laterally.

Gluteus Medius and Gluteus Minimus These muscles lie under the gluteus maximus, with the minimus being

Figure 17.1 Muscles that move the thigh.

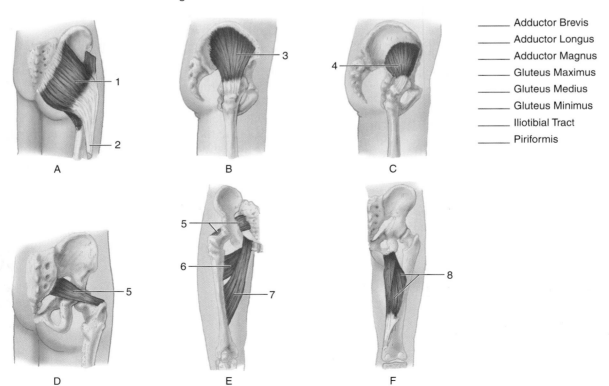

_____	Adductor Brevis
_____	Adductor Longus
_____	Adductor Magnus
_____	Gluteus Maximus
_____	Gluteus Medius
_____	Gluteus Minimus
_____	Iliotibial Tract
_____	Piriformis

A   B   C

D   E   F

the innermost. They *originate* from the ilium and *insert* on the greater trochanter of the femur.

Action: Abduct and rotate the thigh medially.

Piriformis   This small muscle (label 5) *originates* from the anterior surface of the sacrum and *inserts* on the greater trochanter of the femur.

Action: Abducts and rotates the thigh laterally.

Adductor Longus and Adductor Brevis   Both muscles *originate* from the pubis and *insert* on the posterior surface of the femur. The adductor brevis (label 6) is shorter and inserts superior to the adductor longus.

Action: Adduct, flex, and rotate the thigh laterally.

Adductor Magnus   This muscle, the strongest of the adductors, *originates* from the ischium and pubis and *inserts* on the posterior surface and the medial epicondyle of the femur.

Action: Adducts, flexes, and rotates the thigh laterally.

ASSIGNMENT

Label Figure 17.1.

Thigh and Leg Movements

The following muscles, shown in Figure 17.2, are primarily involved in the flexion and extension of the leg.

Hamstrings   The following three muscles, shown in illustration A, are collectively known as the hamstrings.

Biceps Femoris   This is the most lateral of the three hamstring muscles. It *originates* from the ischial tuberosity and the posterior surface of the femur and *inserts* on the head of the fibula and lateral epicondyle of the tibia.

Semimembranosus   This is the most medial muscle of the three hamstrings. Its *origin* is on the ischial tuberosity. *Insertion* is on the posterior surface of the medial epicondyle of the tibia.

Semitendinosus   Located between the other two hamstrings, this muscle *originates* on the ischial tuberosity and *inserts* on the upper medial surface of the tibia.

Collective Action: Flex lower leg; extend thigh; biceps rotates thigh laterally; the other two rotate the thigh medially.

Quadriceps Femoris   The quadriceps femoris is a large muscle that forms the anterior portion of the thigh. It is composed of the following four smaller muscles that unite to form a common tendon. The tendon envelopes the patella and continues to insert on the tibial tuberosity.

Rectus Femoris   This is the anterior central portion of the quadriceps. It *originates* from the inferior ischial spine and a site just above the acetabulum.

Figure 17.2   Muscles that move the leg.

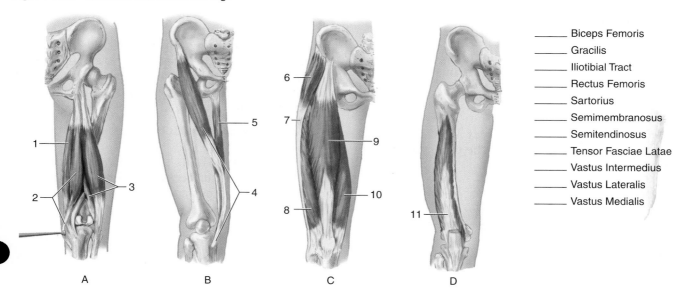

- _____ Biceps Femoris
- _____ Gracilis
- _____ Iliotibial Tract
- _____ Rectus Femoris
- _____ Sartorius
- _____ Semimembranosus
- _____ Semitendinosus
- _____ Tensor Fasciae Latae
- _____ Vastus Intermedius
- _____ Vastus Lateralis
- _____ Vastus Medialis

A   B   C   D

Figure 17.3 Leg muscles.

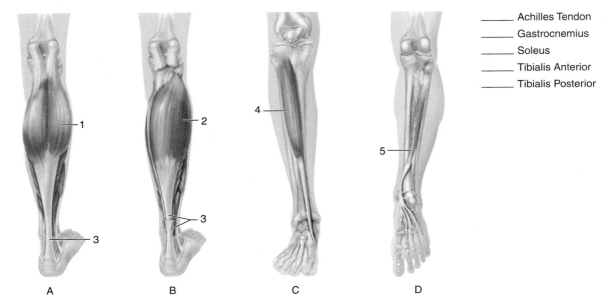

_____ Achilles Tendon
_____ Gastrocnemius
_____ Soleus
_____ Tibialis Anterior
_____ Tibialis Posterior

A   B   C   D

Vastus Lateralis This lateral part of the quadriceps *originates* from the greater trochanter and the posterior surface of the femur.

Vastus Medialis This medial portion *originates* from the posterior surface of the femur.

Vastus Intermedius This portion (label 11) is obscured by the other portions of the quadriceps. It *originates* from the anterior and lateral surfaces of the femur.

> *Collective Action:* Extend the lower leg; the rectus femoris also flexes the thigh.

Sartorius This long muscle *originates* from the anterior superior spine of the ilium and *inserts* on the proximal medial surface of the tibia.

> *Action:* Flexes the lower leg and the thigh; rotates the thigh laterally and the leg medially.

Tensor Fasciae Latae This muscle *originates* from the anterior iliac crest and *inserts* in the iliotibial tract (label 7).

> *Action:* Flexes and abducts thigh.

Gracilis This muscle *originates* from the lower edge of the pubis and *inserts* on the proximal medial surface of the tibia.

> *Action:* Adducts the thigh and flexes the leg.

ASSIGNMENT

Label Figure 17.2.

Leg and Foot Movements

Muscles that move the leg and foot are shown in Figures 17.3 and 17.4.

Gastrocnemius This large superficial muscle covers the calf of the leg. It *originates* from the lateral and medial condyles of the femur and *inserts* on the posterior surface of the calcaneous (heelbone).

> *Action:* Flexes leg; plantar flexes the foot.

Soleus This muscle lies under the gastrocnemius and *originates* from the head and posterior surface of the fibula and shaft of the tibia. Its tendon unites with the tendon of the gastrocnemius to form the **Achilles tendon,** which *inserts* on the calcaneus.

> *Action:* Plantar flexes the foot.

Tibialis Anterior This muscle *originates* from the anterior surface of the lateral condyle and upper two-thirds of the tibia. It crosses over at the ankle to *insert* on the inferior surface of a tarsal bone and the first metatarsus.

> *Action:* Produces dorsiflexion and inversion of the foot.

Tibialis Posterior This deep muscle *originates* from the posterior surfaces of the tibia and fibula and *inserts* on the inferior surfaces of several tarsal bones and metatarsal bones 2–4.

> *Action:* Plantar flexes and inverts the foot.

Peroneus Longus This muscle (label 1, Figure 17.4) *originates* from the head and proximal portion of the

Figure 17.4 Leg muscles.

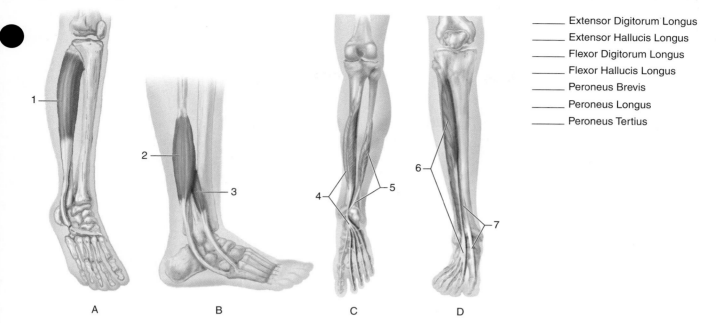

_____ Extensor Digitorum Longus
_____ Extensor Hallucis Longus
_____ Flexor Digitorum Longus
_____ Flexor Hallucis Longus
_____ Peroneus Brevis
_____ Peroneus Longus
_____ Peroneus Tertius

A   B   C   D

fibula. Its tendon passes under the arch of the foot to *insert* on the first metatarsal and a tarsal bone.

> *Action:* Plantar flexes and everts the foot; supports transverse arch.

Peroneus Brevis   This muscle (label 2) lies under the peroneus longus; it *originates* on the middle portion of the fibula and *inserts* on the lateral proximal margin of the fifth metatarsal.

> *Action:* Plantar flexes and everts the foot.

Peroneus Tertius   This is the smallest of the three peroneus muscles. It *originates* from the lower portion of the fibula and *inserts* on the medial proximal margin of the fifth metatarsal.

> *Action:* Dorsiflexes and everts the foot.

Flexor Hallucis Longus   This muscle *originates* from the lower two-thirds of the fibula. Its tendon runs diagonally under the foot to *insert* on the distal phalanx of the great toe.

> *Action:* Flexes the great toe.

Flexor Digitorum Longus   This muscle *originates* on the lower two-thirds of the medial posterior surface of the tibia. Its tendon passes along the bottom of the foot and divides to *insert* on the distal phalanges of toes 2–5.

> *Action:* Flexes toes 2–5; aids in plantar flexion of the foot.

Extensor Digitorum Longus   This muscle (label 6) *originates* on the lateral condyle of the tibia and the

anterior surface of the fibula. Its tendon divides to *insert* on the upper surfaces of the second and third phalanges of toes 2–5. It lies lateral and posterior to the tibialis anterior.

> *Action:* Extends toes 2–5; aids in dorsiflexion of the foot.

Extensor Hallucis Longus   This muscle (label 7) *originates* on the anterior surface of the fibula and *inserts* on the upper surface of the distal phalanx of the great toe.

> *Action:* Extends the great toe; aids in dorsiflexion of the foot.

ASSIGNMENT

1. Label Figures 17.3 and 17.4.
2. For each muscle in the figures, label the origin with an O and the insertion with an I.
3. Complete Sections A through D of the laboratory report.
4. Locate the muscles on a wall chart or manikin.
5. Study the actions of the muscles by locating their origins and insertions on an articulated skeleton. Place one end of a piece of string on the origin and the other end on the insertion to help you visualize the effect of muscle contraction.
6. Review your understanding of superficial muscles by labeling Figure 17.5.
7. Complete Section E of the laboratory report.

Figure 17.5 Major surface muscles of the body. (a) Anterior view; (b) posterior view.

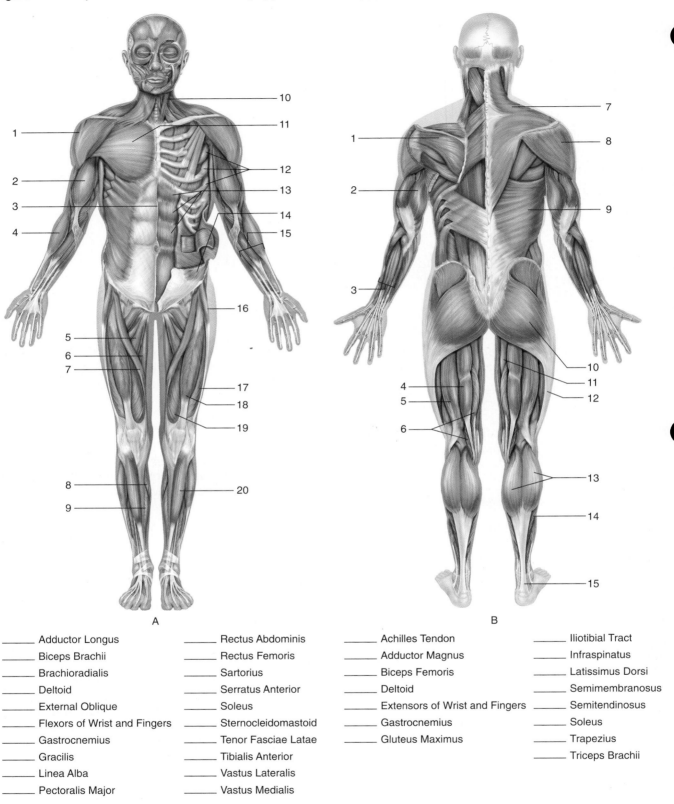

A   B

_____ Adductor Longus

_____ Biceps Brachii

_____ Brachioradialis

_____ Deltoid

_____ External Oblique

_____ Flexors of Wrist and Fingers

_____ Gastrocnemius

_____ Gracilis

_____ Linea Alba

_____ Pectoralis Major

_____ Rectus Abdominis

_____ Rectus Femoris

_____ Sartorius

_____ Serratus Anterior

_____ Soleus

_____ Sternocleidomastoid

_____ Tenor Fasciae Latae

_____ Tibialis Anterior

_____ Vastus Lateralis

_____ Vastus Medialis

_____ Achilles Tendon

_____ Adductor Magnus

_____ Biceps Femoris

_____ Deltoid

_____ Extensors of Wrist and Fingers

_____ Gastrocnemius

_____ Gluteus Maximus

_____ Iliotibial Tract

_____ Infraspinatus

_____ Latissimus Dorsi

_____ Semimembranosus

_____ Semitendinosus

_____ Soleus

_____ Trapezius

_____ Triceps Brachii

Exercise 18

MUSCLE AND NERVE TISSUES

OBJECTIVES

After completing this exercise, you should be able to:

1. Describe the characteristics of smooth, skeletal, and cardiac muscle tissue and identify each type when viewed microscopically.
2. Describe the characteristics of motor, sensory, and internuncial neurons and identify neurons when viewed microscopically.
3. Define all terms in bold print.

Materials

Prepared slides of:
cardiac muscle tissue
skeletal muscle tissue, teased and unteased
smooth muscle tissue, teased and unteased
giant multipolar neurons

Muscle and nerve cells exhibit **excitability,** the ability to transmit electrochemical impulses along their membranes. The adaptive use of excitability results in (1) the contraction of muscle cells to enable movement and (2) the transmission of impulses by nerve cells to coordinate body functions.

Muscle Tissues

Contractility, the ability to shorten or contract, is a major characteristic of muscle tissue. During contraction the elongate **muscle cells,** or **fibers,** shorten and exert a pull at their attached ends that causes movement of body parts. Muscle tissues may be classified according to location, structure, and function.

Skeletal Muscle

Skeletal muscle tissue occurs in muscles attached to bones or other muscles. The actions of skeletal muscles are consciously controlled. Thus, this tissue is classified functionally as **voluntary.**

Skeletal muscle fibers (cells) are elongate and cylindrical. Review in Exercise 14 and Figure 14.1 how they are grouped together in bundles called **fasciculi.** Each fasciculus is surrounded by fibrous connective tissue, the **perimysium,** and the individual fibers are separated by a thin layer of connective tissue, the **endomysium.** Each fiber is distinguished by numerous peripherally located nuclei and alternating dark and light bands called **striae,** or **striations,** in the **sarcoplasm** (cytoplasm). The entire fiber is covered by a very thin membrane, the **sarcolemma.** See Figure 18.1.

Cardiac Muscle

The muscular walls of the heart contain **cardiac muscle tissue** that contracts rhythmically without conscious control or neural stimulation. Thus, it is classified functionally as **involuntary.**

The fibers of cardiac muscle are composed of cells joined end to end and separated from each other

---

Figure 18.1 Skeletal muscle fibers.

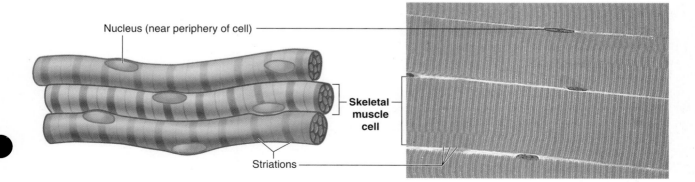

Nucleus (near periphery of cell)

Skeletal muscle cell

Striations

Figure 18.2 Cardiac muscle fibers.

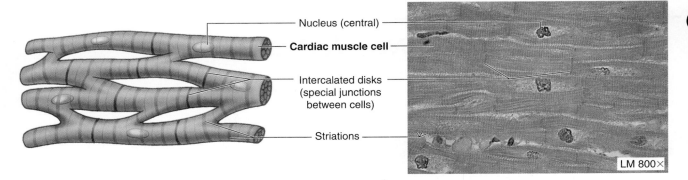

Nucleus (central)
Cardiac muscle cell
Intercalated disks (special junctions between cells)
Striations

LM 800×

by **intercalated disks.** The interconnected, branching fibers form a complex network. Each cell has a single nucleus, and striae are present in the sarcoplasm. See Figure 18.2.

Smooth Muscle

Smooth muscle tissue occurs in the walls of hollow internal organs and blood vessels. Individual cells are spindle-shaped, have a single centrally located nucleus, and lack striae in the sarcoplasm. Since this tissue is not under conscious control, it is **involuntary** in function. See Figure 18.3.

Nerve Tissue

Nerve tissue occurs in the central nervous system (brain and spinal cord) and in peripheral nerves. The primary cells of nerve tissue are **neurons,** or nerve cells. The function of neurons is to receive impulses (electrical signals) from sensory receptors or other neurons and to transmit impulses to other neurons or effectors (muscles or glands). In this way, the nervous system coordinates body functions almost instantly. **Neuroglial cells,** or glial cells, are associated with the neurons and aid neuron functions. They provide support for the neurons and prevent contact between neurons except at particular sites. The structure of neurons varies in accordance with their particular roles, but all neurons possess common features.

Neuron Structure

The basic structure of a neuron is shown in Figure 18.4. The **cell body** is the enlarged portion that contains the nucleus. The **nucleus** is centrally located with a well-defined **nucleolus.** The cytoplasm contains a cytoskeleton of microtubules and **neurofibrils** (label 5) composed of bundles of actin filaments. The meshwork of neurofibrils encompass the dark-staining **Nissl bodies** (label 2), which are composed of aggregations of rough endoplasmic reticulum (RER), and extend into the neuron processes. The cytoplasm also contains mitochondria, lysosomes, Golgi apparatus, and various inclusions, but these organelles are not shown in Figure 18.4.

Several rather short processes extend from the cell body and branch extensively to form vast numbers of **dendrites,** which are the primary sites for receiving signals (impulses) from other neurons. The number of dendrites in different types of neurons will vary from one to thousands.

Figure 18.3 Smooth muscle fibers.

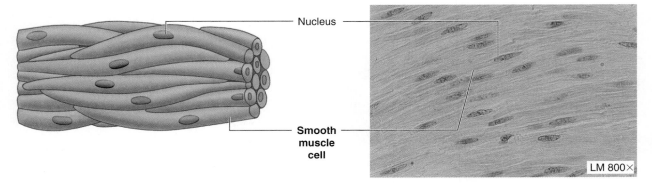

Nucleus

Smooth muscle cell

LM 800×

Figure 18.4 Neuron structure. (a) A spiral motor neuron illustrates the basic structure of a neuron. (b) A Schwann cell forms the myelin sheath and neurilemma around axions of peripheral neurons.

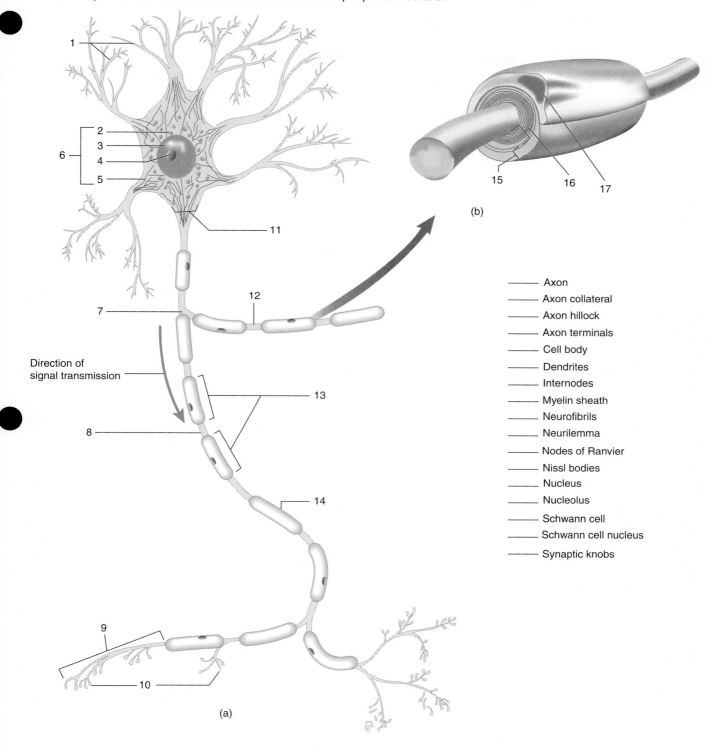

Direction of
signal transmission

(a)

(b)

—— Axon
—— Axon collateral
—— Axon hillock
—— Axon terminals
—— Cell body
—— Dendrites
—— Internodes
—— Myelin sheath
—— Neurofibrils
—— Neurilemma
—— Nodes of Ranvier
—— Nissl bodies
—— Nucleus
—— Nucleolus
—— Schwann cell
—— Schwann cell nucleus
—— Synaptic knobs

The **axon hillock,** a small mound on one side of the cell body, is where the axon originates. The **axon,** or **nerve fiber,** is a single elongated process that carries impulses from the dendrites to another neuron or an effector (muscle or gland). A neuron has only one axon. It may have side branches called **collaterals,** but they are usually few in number. The **axon terminals** (label 9) may be highly branched in order to contact multiple neurons or effectors. The tip of each terminal branch is a **synaptic knob,** which is in direct contact with a dendrite of another neuron or with an effector.

The axons of peripheral (i.e., outside the central nervous system) neurons are enclosed by a covering of **Schwann cells** (label 14) one type of neuroglial cells. In larger peripheral fibers as shown in Figure 18.4, the inner multiple wrappings of Schwann cell membranes form the **myelin sheath** (label 16). The outer layer of the Schwann cell, the part containing the nucleus and most of the cytoplasm, constitutes the **neurilemma.** Axons with a myelin sheath are said to be **myelinated;** those enclosed in Schwann cells but lacking a myelin sheath are **unmyelinated.** Schwann cells are essential for the regeneration of a damaged nerve fiber because they provide a pathway for the regrowth of the axon. Regeneration of nerve fibers is not possible in the brain and spinal cord because of the absence of Schwann cells in the central nervous system. In the brain and spinal cord, the myelin sheath of neurons is formed by *oligodendrocytes,* one of four types of neuroglial cells found only in the central nervous system. However, a neurilemma is absent; only Schwann cells form a neurilemma.

The myelin sheath and neurilemma are interrupted at regular intervals along the axon by the **nodes of Ranvier.** These nodes are tiny spaces between the Schwann cells where the axon is exposed. They play an important role in increasing the speed of impulse transmission. The axon segments between the nodes, where Schwann cells are wrapped around the axon, are called **internodes.**

Types of Neurons

Neurons vary considerably in size, shape, and function. Most neurons belong to one of three basic structural types: multipolar neurons, bipolar neurons, and unipolar neurons. See Figure 18.5.

Multipolar Neurons   These neurons have many dendrites and a single axon like the one in Figure 18.4. They are the most abundant neuron type and include most neurons in the brain and spinal cord. Spinal motor neurons are an example.

Figure 18.5   Structural types of neurons.

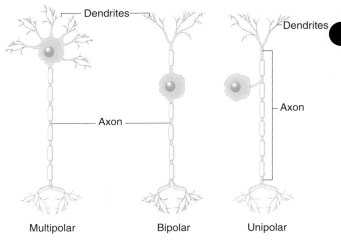

Multipolar   Bipolar   Unipolar

Bipolar Neurons   These neurons have a single dendrite and a single axon extending from opposite sides of the cell body. Olfactory neurons in the nasal cavity, some neurons in the retina of the eye, and sensory neurons in the inner ear are of this type.

Unipolar Neurons   These neurons have a single process extending from the cell body. The process quickly branches in the shape of a T with the two branches extending in opposite directions. Spinal sensory neurons are an example.

There are three functional types of neurons: sensory neurons, interneurons, and motor neurons. An example of the interaction of the three functional types of neurons is shown in Figure 18.6.

Sensory Neurons   These unipolar neurons carry impulses from **sensory receptors** (label 6) toward the central nervous system (brain and spinal cord). Thus, they are sometimes called **afferent neurons.** Note in Figure 18.6 that a **spinal nerve** is joined to the spinal cord by two roots: an **anterior,** or **ventral, root** (label 10) and a **posterior,** or **dorsal, root** (label 9). A nerve consists of bundles of axons (nerve fibres). Spinal sensory neurons enter the spinal cord via the posterior root of a spinal nerve, and their cell bodies are located within a **posterior root ganglion,** a swollen region on the posterior root. A ganglion consists of a cluster of neuron cell bodies. Only the axon tip of the sensory neuron enters the spinal cord where it forms a synapse with interneurons. A **synapse** is the junction between an axon synaptic knob and either (1) a dendrite or cell body of another neuron, or (2) an **effector,** such as a muscle or gland.

Figure 18.6 Interaction of the three functional types of neurons in the spinal cord and spinal nerves.

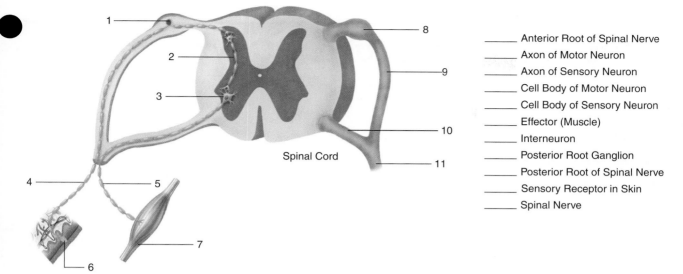

Spinal Cord

_____ Anterior Root of Spinal Nerve
_____ Axon of Motor Neuron
_____ Axon of Sensory Neuron
_____ Cell Body of Motor Neuron
_____ Cell Body of Sensory Neuron
_____ Effector (Muscle)
_____ Interneuron
_____ Posterior Root Ganglion
_____ Posterior Root of Spinal Nerve
_____ Sensory Receptor in Skin
_____ Spinal Nerve

Interneurons These multipolar neurons, sometimes called **association neurons,** occur only in the central nervous system (brain and spinal cord) and compose about 99% of the neurons in the body. They carry impulses from one part of the central nervous system to another. In Figure 18.6, the interneuron is shown linking the sensory neuron with the motor neuron within the gray matter of the spinal cord. The gray matter in the brain and spinal cord is composed mostly of interneuron cell bodies, while their myelinated axons compose the white matter. Myelin is responsible for the lighter coloration. A synapse is formed between the sensory neuron and interneuron and between the interneuron and motor neuron.

Motor Neurons These multipolar neurons carry impulses away from the central nervous system to an effector (a muscle or gland). Thus, they are also called **efferent neurons.** The cell body and dendrites of a motor neuron are located within the central nervous system, and only the myelinated axon is located outside the central nervous system. In Figure 18.6, note that the axon of a motor neuron exits the spinal cord via the anterior root of the spinal nerve and terminates at an effector.

 ASSIGNMENT

1. Label Figures 18.4 and 18.6.
2. Complete Sections A, C, D, and E of the laboratory report.
3. Examine slides of each type of muscle tissue using Figures 18.1, 18.2, and 18.3 and Figures HA-10 and HA-11 in the Histology Atlas as references.
4. Examine a slide of giant multipolar neurons using Figure HA-12 as a reference.
5. Complete the laboratory report.

Exercise 19

THE NATURE OF MUSCLE CONTRACTION

OBJECTIVES

After completing this exercise, you should be able to:

1. Describe the basic structure and function of the neuromuscular junction.
2. Describe the orientation of myosin and actin filaments in resting and contracted muscle fibers.
3. Describe the basic mechanism of muscle contraction.
4. Define all terms in bold print.

Materials

Test tube of glycerinated rabbit psoas muscle
Dropping bottles of:
glycerol
0.25% ATP in triple distilled water
0.05 M KCl
0.001 M MgCl$_2$
Microscope slides and cover glasses, 3 each
Dissecting instruments
Plastic ruler, clear and flat
Microscopes, compound and dissecting

Contraction of a muscle fiber is a complex physicochemical process that involves both chemical changes and microscopic structural changes within the muscle fiber. Furthermore, contraction is dependent upon impulses from a motor neuron to initiate the process. In this exercise you will study the interaction of a motor neuron and a muscle fiber in the process of contraction, as well as the mechanism of contraction.

Skeletal muscle fibers are innervated by motor neurons. A motor neuron and the muscle fibers it innervates constitute a **motor unit.** A large motor unit involves the innervation of many (e.g., 100–2,000) muscle fibers by a single motor neuron. Large motor units activate muscles, such as postural muscles, where precise control is unimportant. A small motor unit involves innervation of only a few (e.g., 5–30) muscle fibers activated by a single motor neuron. Small motor units activate muscles where precise control is important, such as in muscles controlling finger movements.

The junction between a neuron's axon tip and its target cell, such as another neuron, is called a

synapse. When the target cell is a muscle fiber, the synapse is called a **neuromuscular junction.** Figure 19.1 shows (1) the synaptic knobs of a motor neuron's axon attached to a muscle fiber forming a neuromuscular junction, and (2) the ultrastructure of a neuromuscular junction.

Contraction of a skeletal muscle fiber involves the interaction of (1) a motor neuron, (2) the neuromuscular junction, and (3) a muscle fiber. As an impulse (action potential) passes along an axon to the neuromuscular junction, it causes chemical changes at the neuromuscular junction that form an action potential in the muscle fiber. This action potential, in turn, initiates physicochemical changes in the muscle fiber that produce the contraction.

The Neuromuscular Junction

The specialized structures of a neuromuscular junction perform specific functions in the transmission of an impulse from a neuron's axon to a muscle fiber. Each **axon** synaptic knob (label 2) fits into a depression in the **sarcolemma** (label 3) called the **motor end plate.** The minute space between the synaptic knob and the sarcolemma is the **synaptic cleft** (label 7). Numerous folds of the sarcolemma adjacent to the synaptic cleft greatly increase its surface area.

When a neural impulse or action potential (depolarization of the neurilemma) reaches the axon synaptic knob, **synaptic vesicles** (label 8) in the axon synaptic knob release the neurotransmitter **acetylcholine** into the synaptic cleft by exocytosis (label 9). Acetylcholine quickly binds to acetylcholine receptors in the sarcolemma, producing an action potential in the muscle fiber (depolarization of the sarcolemma). The action potential (a muscle impulse) moves inward along the **transverse tubule (T-tubule) system** (label 5) to the **sarcoplasmic reticulum** (label 6), causing it to release calcium ions that enable contraction of the myofibrils. After initiating muscle contraction, acetylcholine is quickly (within 1/500 of a second) inactivated by an enzyme, **acetylcholinesterase,** from the sarcolemma, and the sarcolemma is returned to its polarized resting state. It is now capable of receiving another impulse.

Figure 19.1 The neuromuscular junction.

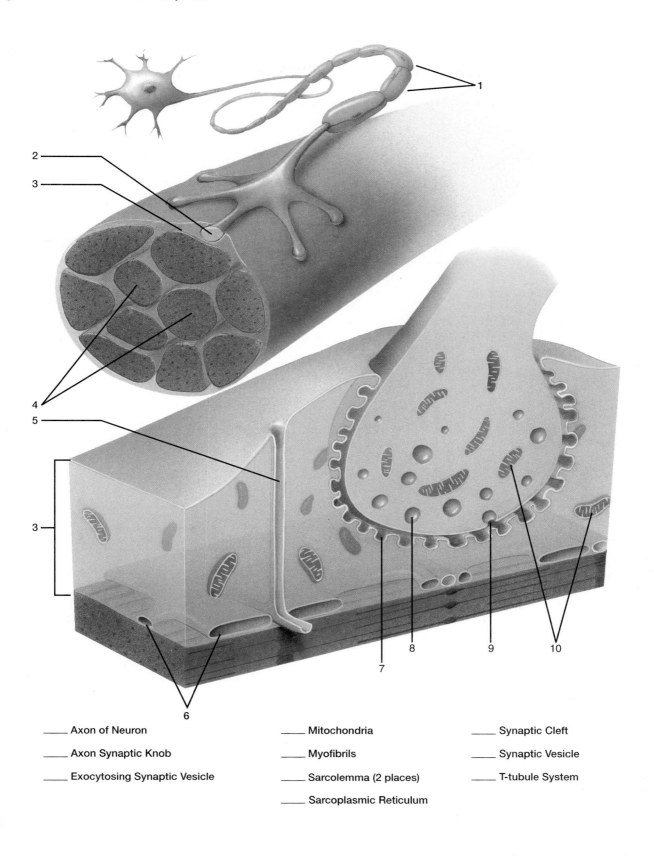

_____ Axon of Neuron

_____ Axon Synaptic Knob

_____ Exocytosing Synaptic Vesicle

_____ Mitochondria

_____ Myofibrils

_____ Sarcolemma (2 places)

_____ Sarcoplasmic Reticulum

_____ Synaptic Cleft

_____ Synaptic Vesicle

_____ T-tubule System

Figure 19.2 The ultrastructure of a muscle fiber.

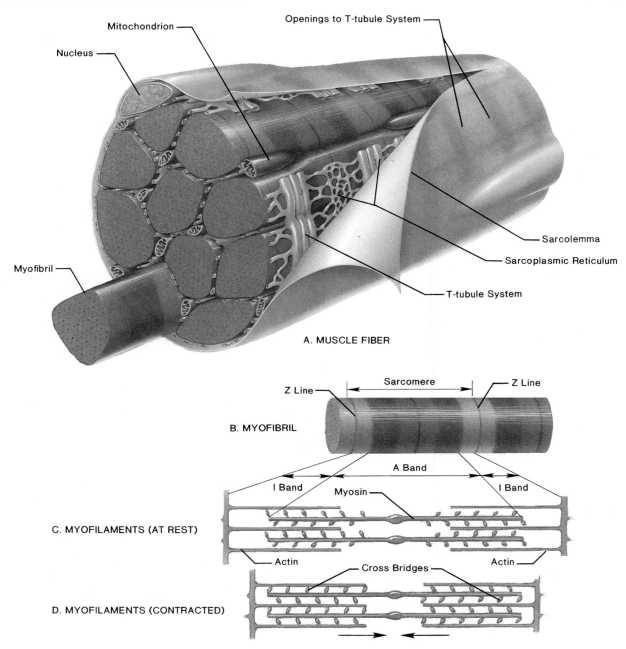

A. MUSCLE FIBER

B. MYOFIBRIL

C. MYOFILAMENTS (AT REST)

D. MYOFILAMENTS (CONTRACTED)

Energy for both neuron function and muscle contraction is supplied by ATP (adenosine triphosphate). ATP is produced by aerobic cellular respiration in numerous **mitochondria** present in neurons and muscle fibers.

Myofibril Ultrastructure

Figure 19.2 shows the ultrastructure of a muscle fiber. Illustration A shows a single muscle fiber composed of several myofibrils. Note that the sarcoplasmic reticulum and T-tubule system envelop each myofibril, that mitochondria are located between myofibrils, and that the entire muscle fiber is enveloped by the sarcolemma. Locate the peripherally located nuclei.

Illustration B shows a single myofibril and its striations (striae). The dark and light striations of a myofibril are due to the arrangement of two kinds of **myofilaments** that are composed of proteins. Illustrations C and D show the ultrastructure and arrangement of the myofilaments when the myofibril is relaxed and when it is contracted.

Myosin myofilaments are the thicker of the two, and they are the major components of the dark-colored **A band.** The thinner **actin** myofilaments extend into the A band from the **Z lines,** and they are the only myofilaments in the light-colored **I bands** on each side of an A band. The Z lines form the boundaries of a **sarcomere,** the contractile unit of a myofibril.

Compare the position of the Z lines and the myosin and actin myofilaments in the relaxed state (illustration C) and in the contracted state (illustration D).

Mechanics of Contraction

The shortening of a muscle fiber by contraction results from the interaction of actin and myosin myofilaments in the presence of calcium and magnesium ions. ATP provides the necessary energy. The most accepted explanation for myofibril contraction is known as the sliding filament theory. The sequence of events is described below.

1. As the action potential moves along the sarcolemma, transverse tubule system, and sarcoplasmic reticulum, large quantities of calcium ions are released.
2. Calcium ions cause changes that expose the chemically active sites on the actin myofilaments.
3. The cross-bridges of the myosin myofilaments attach to the active sites on the actin molecule. The cross-bridges bend, exerting a power stroke that pulls the actin myofilaments toward the center of the A band.
4. After the first power stroke, the cross-bridges separate from the first actin active sites, bond with next active sites, and produce another power stroke.
5. This ratchetlike process repeats itself until maximum contraction is attained.

ASSIGNMENT

1. Label Figure 19.1.
2. Complete Sections A through D of the laboratory report.

Experimental Muscle Contraction

Perform the following experiment in which a glycerinated rabbit psoas muscle will be induced to contract using ATP, magnesium ions, and potassium ions.

Your instructor has removed the muscle from its test tube and cut it into 2-cm segments that have been placed in a Petri dish of glycerol. These muscle segments have been teased apart to obtain thin muscle strands composed of very few fibers. The strands should be no more than 0.2 mm thick for use in the experiment.

1. Place a muscle strand in a small drop of glycerol on a clean microscope slide and add a cover glass. Examine it microscopically with the high-dry or oil immersion objective. Draw the pattern of striations of the relaxed muscle fibers in Section E of the laboratory report.
2. Place 3 or 4 of the thinnest strands on another slide in a minimal amount of glycerol. Arrange the strands straight, parallel, and close to each other.
3. Using a dissecting microscope, measure the length of the strands by placing a clear plastic ruler under the slide. Record the lengths on the laboratory report. Estimate the width of the strands.
4. While observing through the dissecting microscope, add one drop from each of the solutions: ATP, magnesium chloride, and potassium chloride. Observe the contraction.
5. After 30 seconds, remeasure and record the length of the strands. Calculate and record the percentage of contraction. Has the width of the strands changed?
6. Remove one of the contracted strands to another slide, add a cover glass, and observe the pattern of the striations using a compound microscope. Draw the pattern of striations of the contracted muscle fibers in Section E on the laboratory report. Complete the laboratory report.
7. Clean your workstation.

Exercise **20**

THE PHYSIOLOGY OF MUSCLE CONTRACTION USING THE BIOPAC© STUDENT LAB SYSTEM

OBJECTIVES

After completing this exercise, you should be able to:

1. Observe, record, and calculate motor unit recruitment, which increases the power of skeletal muscle contractions.
2. Determine maximum clench force of dominant and nondominant hands.
3. Record the maximum force produced by a clenched fist and the decline in force during fatigue.
4. Calculate the time it takes to fatigue muscles clenching the fist.

Materials

Computer system (Mac—minimum 68020;
   PC—running Windows 95/98/NT 4.0 or better;
   with 4.0 MB RAM)
BIOPAC student lab software V3.0 or higher
BIOPAC acquisition unit (MP30) with (AC100A)
   transformer
BIOPAC serial/cable (CBLSERA) or USB cable
   (USBI W) if using an USB port
BIOPAC electrode lead set (SS2L)
BIOPAC disposable vinyl electrodes (EL503),
   6 electrodes per subject
BIOPAC electrode gel (GEL1) and abrasive pad
   (ELPAD) or alcohol wipes
BIOPAC headphones (OUT1)
BIOPAC SS25 hand dynamometer (for Part 2)

In this laboratory exercise, you will investigate the electrical properties of skeletal muscle contractions using **electromyography,** a procedure that produces a recording of the skin-surface voltage generated by underlying skeletal muscle contractions. The voltage produced by muscle contraction is weak, but it can be detected by surface sensors to produce the recording, which is called an **electromyogram (EMG).**

Human skeletal muscle consists of hundreds of cylindrical skeletal muscle cells (also called fibers). Each fiber must be stimulated to contract by a somatic motor neuron. Although each muscle fiber is stimulated by only one motor neuron, each motor neuron can stimulate a number of muscle fibers. The functional unit formed by a single somatic motor neuron and all the muscle fibers that it stimulates are called a **motor unit.** As a result, each skeletal muscle consists of many motor units. The strength of muscle contraction varies depending upon how many motor units are stimulated. The more motor units that are stimulated, the stronger is the contraction by the muscle. This phenomenon is known as **motor unit recruitment.**

The primary function of muscle is to convert chemical energy into mechanical energy in order to contract. Skeletal muscles that are required to contract repetitively or continuously at their maximum strength will eventually run out of energy and will **fatigue.**

This laboratory exercise has been divided into two parts. In Part 1, you will observe a standard EMG produced by a simple fist clench while electrodes are attached to the forearm. In Part 2, you will use a **hand dynamometer** equipped with an electronic transducer. A dynamometer produces a special type of recording called a **dynagram,** which measures the force exerted during the fist clench.

This exercise is best done by students working in pairs. One student is the subject, and the other reads the directions and is the computer and equipment operator. After data are collected for the subject, students are to switch roles and repeat the procedures.

You should have no trouble using The BIOPAC Student Lab system as long as you carefully follow the directions provided. It is also a safe system, but you need to follow the directions carefully when using the electrodes and transducers. Read through all of the procedures for each activity before starting the activity. If you have questions about the procedures, consult your instructor.

Part 1–Electromyography: Standard and Integrated EMG

Setup

1. With your computer turned **on** and the BIOPAC MP30 Data Acquisition unit turned **off**, plug the electrode lead set (SS2L) into channel 3 of the BIOPAC MP30 unit. Plug the

headphones (OUT1) into the back of the unit, if desired. See Figure 20.1.

2. Turn on the MP30 data acquisition unit.

3. Attach the electrodes to the dominant forearm (the right arm if the subject is right-handed; the left if the subject is left-handed) of the subject as shown in Figure 20.2. This will be **Forearm I.** To help to ensure a good contact, make sure the area to be in contact with the electrodes is clean by wiping with the abrasive pad (ELPAD) or alcohol wipe. A small amount of BIOPAC electrode gel (GEL1) can also be used to make better contact between the sensor in the electrode and the skin. For optimal adhesion, place the electrodes on the skin at least 5 minutes before the calibration procedure.

4. Attach the electrode lead set by the pinch connectors to the electrodes as shown in Figure 20.2. Make sure the color-coded leads are attached to the correct electrodes.

5. Start the BIOPAC student lab program on your computer. Select Electromyography I, lesson L01-EMG-1 and click OK.

6. Type in a unique filename to be used to save the subject's data.

7. Click OK to end the setup.

Calibration

Calibration is necessary for best results. The following calibration steps establish the performance parameters for the hardware.

1. Click **Calibrate** on the computer screen. A dialog box will direct you through the calibration procedure. After reading the box, click **OK.**

2. Have the subject wait about 2 seconds, and then clench the fist of Forearm 1 as hard as possible, then release. Wait for the calibration to stop. It will stop automatically and will take about 8 seconds.

3. The calibration recording should look similar to Figure 20.3. If not, click **Redo Calibration** and repeat steps 1–3. If the baseline is not at zero, the subject did not wait 2 seconds before clenching his/her fist.

Figure 20.1   BIOPAC acquisition unit and cable setup.

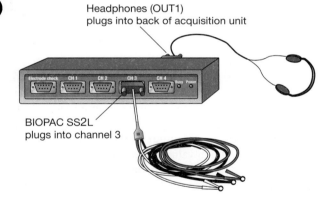

Figure 20.2   Placement of electrodes and lead attachments.

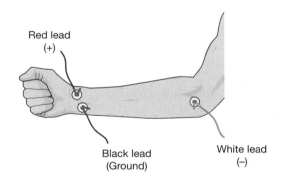

Figure 20.3   EMG calibration recording.

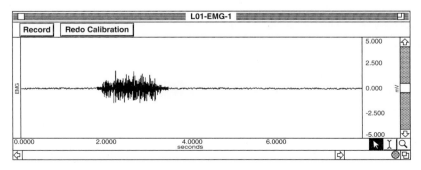

Data Recording

1. Fill out the subject profile on Laboratory Report 20. Read through this entire segment so you know what to do before starting the recording.
2. When you click **Record,** a marker labeled **Forearm 1** will be inserted. The subject is to perform a series of four Clench–Release–Wait cycles starting with a gentle clench and gradually increasing the strength of each clench so the fourth clench is at the maximum force. The pattern is: clench for 2 seconds, release, and wait for 2 seconds. Repeat this cycle four times.
3. Click on **Record** and have the subject perform a series of fist clinches as described in step 2.
4. Click **Suspend.** The recording should be similar to Figure 20.4. If not, click **Redo** and repeat, starting at step 2. If the recording is acceptable, proceed to step 5.
5. Attach three electrodes to the nondominant arm (**Forearm 2**) of the subject as shown in Figure 20.2 and attach the color-coded electrode lead set as before.
6. Click **Resume.** A marker labeled **Forearm 2** will automatically be inserted when you do this.
7. Repeat the data recording procedures, steps 2–4, for Forearm 2. Then, click **Suspend.**
8. If your recording does not look like Figure 20.4, click **Redo** and repeat the data recording for Forearm 2. If your recording looks like Figure 20.4, click **stop.**
9. If you want to listen to the EMG signal, put on the headphones and click **Listen.** Have the subject experiment by changing the clench force as you listen and watch the screen. Note that the intensity of the sound varies with the strength of contraction. When finished, click **Stop.** To listen again, click **Redo.**
10. When you are finished with the recording and listening, click **Done.**

At this point, you may proceed with the data analysis of the current file or save it for analysis later. If you save the file for later analysis, you may continue to collect data for the current subject in Part 2 of this exercise, or switch roles with your lab partner and repeat the preceding procedures for the new subject starting with calibration.

Data Analysis

1. You can either analyze the current file or save it and do the analysis later by using the **Review Saved Data mode** and selecting the unique file for the subject.
2. Study Figure 20.5 and compare it to the display window. The on-screen tools can be used to adjust the display window. Locate the channel measurement boxes located across the display window above the marker region. Each measurement box has three parts: channel number, measurement type, and result. The first two are pull-down menus that are activated when you click on them.
3. Note the channel number (CH) designations in the channel measurement boxes: CH 3 displays Raw EMG (the actual voltage recording), and CH 40 displays the Integrated EMG (that reflects the absolute intensity of the voltage).
4. Set up the display window for viewing the data for **Forearm 1.**
5. Set up the channel measurement boxes as follows:

Channel	Measurement
CH 3	min
CH 3	max
CH 3	p–p
CH 40	mean

This will give you these measurements:
> **min**—the minimum value in the selected area
> **max**—the maximum value in the selected area
> **p-p**—subtracts the minimum from the maximum value in the selected area
> **mean**—the average value in the selected area

Figure 20.4 EMGs of four fist clenches showing increasing contractions at each clench.

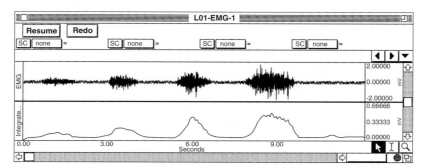

Figure 20.5 A sample display of a subject's data file in the Review Saved Data Mode. On-screen tools may be used for adjusting the display window.

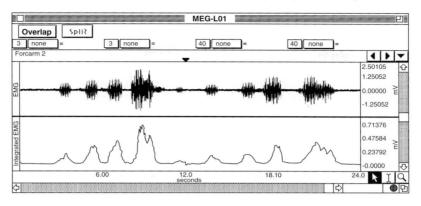

Figure 20.6 Selecting the first clench for analysis.

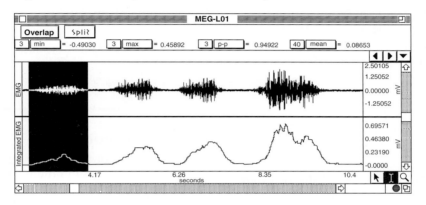

6. After clicking the I beam curser box to activate the select function, select the area of the first EMG cluster as shown in Figure 20.6.
7. Record the values (results) in the measurement boxes in the table in Part I A of Laboratory Report 20.
8. Repeat this procedure for the other three EMG clusters and record the data in Part I A Laboratory Report 20.
9. Scroll to the second recording segment, which begins with a marker labeled Forearm 2. Analyze the EMG clusters for Forearm 2 by repeating steps 6–8.
10. Complete Part I of Laboratory Report 20.

Part 2–Electromyography: Motor Unit Recruitment and Fatigue

Set Up

1. With your computer turned **on** and the BIOPAC MP30 unit turned **off,** plug the electrode lead set (SS2L) into Channel 3 and BIOPAC SS25 hand dynamometer into Channel 1. Plug the headphones (OUT 1) into the back of the unit if desired. See Figure 20.7.

Figure 20.7 BIOPAC acquisition unit and cable setup for using the hand dynamometer.

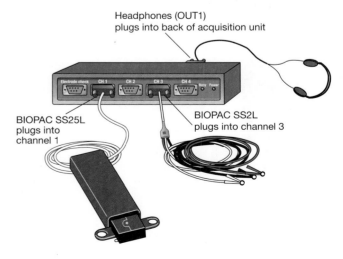

Headphones (OUT1) plugs into back of acquisition unit

BIOPAC SS25L plugs into channel 1

BIOPAC SS2L plugs into channel 3

2. Turn on the BIOPAC MP30 data acquisition unit.
3. Attach the electrodes to the dominant forearm (the right arm if the subject is right-handed; the left if the subject is left-handed) of the subject as shown in Figure 20.2. This will be **Forearm I.** To help to ensure a good contact, make sure the area to be in contact with the electrodes is clean by wiping with the abrasive pad (ELPAD) or alcohol wipe. A small amount of BIOPAC electrode gel (GEL1) may also be used to make better contact between the sensor in the electrode and the skin. For optimal adhesion, the electrodes should be placed on the skin at least 5 minutes before the calibration procedure.
4. Attach the electrode lead set by the pinch connectors as shown in Figure 20.2. Make sure the proper color lead is attached to the appropriate electrode.
5. Start the BIOPAC student lab program for Electromyography I, lesson L02-EMG-2.
6. Type in a unique filename to be used to save the subject's data, then click **OK.**

Calibration

1. Click **Calibrate.** Place the hand dynamometer on the table and click **OK.** This establishes a zero-force calibration before you continue the calibration sequence.
2. Grasp the hand dynamometer with the dominant hand (Forearm 1) as close to the dynagrip crossbar as possible *without touching the crossbar.* See Figure 20.8. Note the position of your hand and try to replicate this hand position during the recording phase.

Figure 20.8 Holding the hand dynamometer.

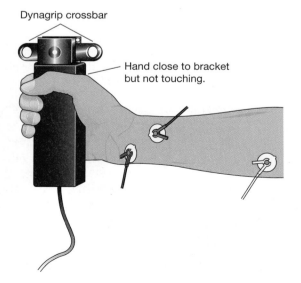

Dynagrip crossbar

Hand close to bracket but not touching.

3. Follow the instructions on the two successive pop-up windows and click **OK** when ready for each.
4. Wait two seconds then clench (squeeze) the hand dynamometer as hard as possible for 2–4 seconds, then release.
5. Wait for the calibration to stop. This will take about 8 seconds.
6. If the calibration recording does not resemble Figure 20.9, click **Redo Calibration** and repeat the calibration steps. If the calibration recording did not start with a zero baseline, the subject didn't wait 2 seconds before clenching his/her fist.
7. Before proceeding to the recording procedure, you need to determine the force increments that you will use. Using the maximum force from the calibration, the software sets the increments as follows:

Maximum Force Calibration	Assigned Increment
0 – 25 kg	5 kg
26 – 50 kg	10 kg
> 51 kg	20 kg

Data Recording

1. Prepare for the recording by reading the entire section so you know what you are to do before starting. When recording, stop each recording segment as soon as possible to prevent a waste of recording memory.
2. When you click **Record,** a marker labeled **Forearm 1, Increasing Clench Force,** will be inserted. The screen will display only the hand dynamometer channel and a grid showing your assigned increment as a division scale. This enables you to visually observe and control the contraction force level. The subject is to perform a series of four to five clench–release–wait cycles starting with the assigned increment level and increasing the strength of each clench by the assigned increment until the maximum force is attained. The pattern is:

a. Clench for 2 seconds, release, and wait for 2 seconds.
b. Repeat three or four cycles, increasing the clench force at each cycle by the assigned increment level. For example, if your assigned increment level is 10 kg, start at a force of 10 kg and repeat cycles at 20 kg, 30 kg, and 40 kg. Try to maintain the contraction force constant during each fist clench in order to produce plateaus for each clench.

Figure 20.9 Calibration recording using the hand dynamometer.

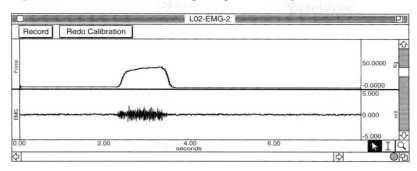

Figure 20.10 Dynagram of five fist clenches showing increasing contraction at each clench.

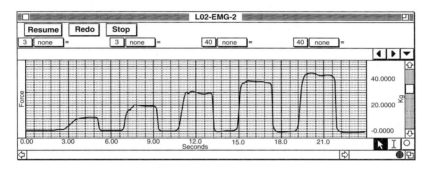

Figure 20.11 Dynagram showing maximal contraction force and gradual fatigue.

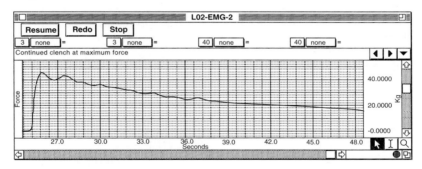

3. Click **Record** and perform four to five clench–release–wait cycles as described previously.

4. Click **Suspend.** The recording should resemble Figure 20.10, showing multiple plateaus with the clench strength increasing at each cycle indicating **Motor Unit Recruitment.** If not, click **Redo** and repeat, starting at step 2.

5. Click **Resume.** A marker labeled Forearm 1, Continued Clench at Maximum Force, will be inserted.

6. Clench the hand dynamometer with your maximum force. Record the maximum force and try to maintain it. As the forearm fatigues, the force will gradually decline.

7. When the maximum clench force has decreased by more than 50%, click **Suspend.** The recording should resemble Figure 20.11 showing gradual **Fatigue.** If not, click **Redo** and repeat steps 5–7.

8. If correct, click **Stop.**

9. If you want to listen to the EMG signal, put on the headphones and click **Listen.** Have the subject experiment by changing the clench force as you watch the screen. When finished, click **Stop.**

10. Attach the electrode lead set to the electrodes on the nondominant forearm of the same subject. Be sure the color-coded leads are attached to the proper electrodes as in Figure 20.2. Click **Forearm 2.** This will return you to the **Calibration**

sequence. Proceed to calibrate and record data for Forearm 2 as you did for Forearm 1.

11. When finished, click **Done.**

You can either analyze the current file or save it and do the analysis later using the **Review Saved Data mode** and select the correct file. The file with the extension "1-L02" is for Forearm 1 and "2-L02" is for Forearm 2. If you save the file for later analysis, switch roles with your lab partner and repeat the procedures for Part 1 and Part 2 with the new subject starting with calibration.

When finished recording, remove the lead set from the electrodes and the electrodes from the forearms. Discard the electrodes and wash the electrode gel from the skin with soap and water. The electrodes may leave a slight ring on the skin for a few hours.

Data Analysis

1. Enter the **Review Saved Data mode** and select the correct file. The file with the extension "1-L02" is for Forearm 1 and "2-L02" is for Forearm 2.
2. Note the channel number designations in the channel measurement boxes.
   CH 1 displays Force (kg)
   CH 3 displays Raw EMG (mV)
   CH 40 displays the Integrated EMG (mV).
3. Use the on-screen tools (Figure 20.5) to set up your display for viewing the increasing clench force data showing motor unit recruitment. The grid increments should use the same force increments that were used in the recording.
4. Set up the channel/measurement boxes as follows:

Channel	Measurement
CH 1	mean,
CH 3	p–p
CH 40	mean

p-p—subtracts the minimum from the maximum value in the selected area.

mean—the average value in the selected area.

5. Use the I-beam cursor to select an area on the plateau phase of the first clench. Record the measurements (results) from the channel measurement boxes on Laboratory Report 20 in the table for Motor Unit recruitment (II Part A 1).
6. Repeat step 5 on the plateau for each successive clench.
7. Scroll to the second recording segment showing maximum force followed by fatigue.
8. Set up the channel/measurement boxes as follows:

Channel	Measurement
CH 1	value (the highest value in the selected area)
CH 40	delta T (the amount of time in the selected area)

9. Use the I-beam cursor to select the point of maximal clench force at the start of segment 2. Record the measurements on Laboratory Report 20 in the table for Fatigue (II Part A 2).
10. Calculate 50% of the maximum clench force from step 9. Record the calculation on Laboratory Report 20 in the table for Fatigue (II Part A 2).
11. Visually, estimate the point on the record that is 50% of the maximal force. Use the I-beam cursor to click on several points in this area, while reading the values in the measurement box, until you locate the 50% point. Leave the cursor at this point.
12. Select the area from the point of 50% clench force back to the point of maximum clench force by dragging the cursor. Note the time to fatigue (CH 40/delta T) and record the time on Laboratory Report 20 in the table for Fatigue (II Part A 2).
13. Repeat the data analysis starting with step 1 for Forearm 2.
14. Exit the program.

Exercise 21

THE PHYSIOLOGY OF MUSCLE CONTRACTION USING THE INTELITOOL PHYSIOGRIP™

OBJECTIVES

After completing this exercise, you should be able to:

1. Describe the parts of a simple twitch myogram.
2. Interpret your myograms showing threshold and maximal stimuli, spatial summation, temporal summation and tetanus, a simple twitch, and fatigue.
3. Explain the mechanisms of spatial summation, temporal summation, tetanus, simple muscle twitch, and fatigue.
4. Define all terms in bold print.

Materials

    Computer system (PC running Windows 3.1 or better;
       MacIntosh running system 6.0.8 or higher)
    McADDAM II (Multiple Channel Analog-to-Digital
       Data Acquisition Module) interface unit
    Physiogrip™ transducer on wooden base
    Physiogrip™ program disk
    Physiogrip™ *Lab Manual*
    Physiogrip™ *User Manual*
    Electronic stimulator
    Flat-plate electrode, cable, and rubber strap
    Stimulator probe and cable
    Safety isolated voltage reducer
    Transducer cables
    USB cable
    Electrode gel
    Alcohol wipes
    Sphygmomanometer (for investigation of fatigue
       and muscle contraction)

You will study muscle contraction phenomena here utilizing an Intelitool Physiogrip™ controlled by computer software. Electrical stimuli will be applied to the median nerve in the right forearm that controls flexion of the third finger. With the third finger on the trigger of the Physiogrip™, flexor response is fed into the computer for data accumulation and evalutation. You will study the following phenomena of muscle contraction: 1) threshold stimulus, spatial summation, and maximal stimulus; 2) temporal summation and tetany; 3) single muscle twitch; and 4) fatigue and muscle contraction.

The instructions provided here assume that you have a basic knowledge of the computer platform you are using. Note in the materials list that among the items used in this experiment are copies of two Physiogrip™ manuals that are supplied with the equipment. Although you may not need them, keep the manuals available for reference.

You will be working with a lab partner. One of you will be the subject, and the other will read the instructions and operate the computer and equipment. After completing all experiments, switch roles and repeat and experiments.

Equipment Setup

The equipment you will use has been set up for you, and the software has been installed on the computer. Take a few moments to become familiar with the setup.

1. The McADDAM II Data Acquisition Module is connected to the computer with an USB cable and to the Physiogrip™ transducer via a transducer cable.
2. A safety isolated voltage reducer is plugged into the stimulator output, and a transducer cable connects the voltage reducer to the McADDAM II module.
3. The cable of the stimulator probe is plugged into the positive (red) output jack of the voltage reducer (Figure 21.1).
4. The cable of the flat plate electrode is plugged into the negative (black) output jack of the voltage reducer (Figure 21.1).
5. The printer is connected to the computer via a serial port cable.

This setup enables electrical input to the McADDAM II module from the stimulator (via the voltage reducer) and Physiogrip™ transducer. The McADDAM II module converts the data from analog to digital and inputs it to the computer where it is displayed on the screen. Everything is controlled by the Physiogrip™ software. Arrange the various components in a manner that will make it convenient for you to use all pieces of equipment.

Figure 21.1 Connections of the safety isolated voltage reducer to the stimulator, stimulus probe, flat-plate electrode, and the McADDAM II™ interface module. Courtesy of Phipps & Bird/Intelitool.

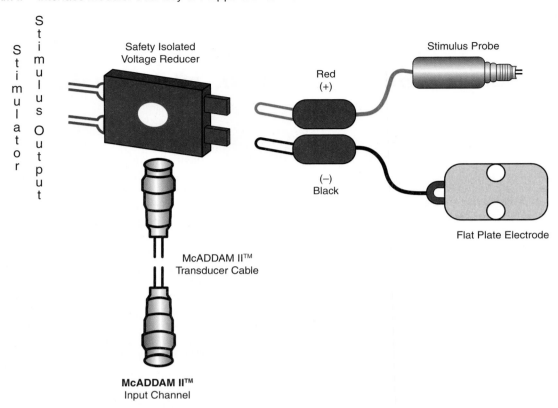

Calibration

1. Power up the McADDAM II and the computer. Open **Intelitool Apps** and click on the **Physiogrip** lab.
2. Select **File** and then **New.** The **Setup** window should appear. Make sure all the connections are correct.
3. Click **Calibrate Transducer** and follow the instructions on the screen. Click **OK.** Make sure Physiogrip transducer trigger is all the way out and don't touch it. Click **OK,** then squeeze the trigger all the way in. Click OK and proceed. If the screen indicates that the transducer needs adjustment, inform your instructor.
4. Click **Calibrate Stimulator,** follow the instructions on the screen, and click **OK.** Set up stimulator as follows:
   Mode: Continuous
   Events/sec: 2
   Duration: 50 ms
   Amplitude: 100 V
   Power: On.
5. Click **OK.** Calibration will take about 4 seconds. Click **OK** to proceed.

Locating the Motor Point

Before beginning any of the experiments, you will need to find the motor point for the *flexor digitorum superficialis* muscle, which flexes the third finger.

> *Caution: If you have a history of heart problems, do NOT participate as the subject in these experiments.*

To locate the motor point, you (the subject) will have to use your left hand to probe around on the anterior surface of your right forearm with the stimulator probe with the stimulator set at a certain voltage, frequency, and duration. After you locate the motor point, almost all discomfort from the stimulator will disappear, and a good flexion of the third finger will occur. Proceed as follows to locate the motor point on the median nerve (point A, Figure 21.2).

1. Double-check to be sure that all connections are correct.
2. Vigorously wipe the back of the subject's right hand with an alcohol pad to remove loose cells and oils. Put a generous amount of

Figure 21.2 Placement of the negative (-) flat-plate electrode on the back of the right hand and of the positive (+) stimulus probe on the median nerve at the motor point (area A) for the *flexor digitorum superficialis* muscle.

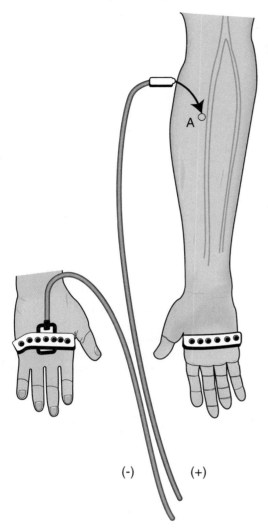

(-)   (+)

To safety isolated voltage reducer

electrode gel under the flat plate electrode and use the rubber strap to attach it as indicated in Figure 21.2.

3. With the stimulator OFF, have your partner set the stimulator as follows:

    Mode: Continuous
    Events/sec: 1
    Amplitude: 60 v
    Duration: 2 ms

4. Place a dab of electrode gel on the end of the stimulator probe. Do *not* get the gel on the sides of the stimulator probe. You will need to refresh the gel under the electrode plate and on the end of the stimulator probe throughout these experiments.

5. Turn ON the stimulator.

> *Caution: Avoid touching the metal tip of the stimulator probe with your left hand—doing so will cause current to flow between your left arm and your right arm (i.e., through your chest). For the same reason, avoid getting electrode gel on the sides of the probe where it is likely to contact your left hand.*

6. Holding the stimulator probe with your left hand, move it around on area A (Figure 21.2) of your right forearm to locate the motor point. Note that stimulation of area A causes your third finger to flex.

    If moving the probe around area A does not elicit flexion of your third finger, have your partner gradually increase the voltage, but not more than 90 v. You may have to press rather hard with the probe to get maximum finger flexion. If you still do not get a response, check the connections and stimulator to be sure that they are set up correctly.

7. Once you have located the motor point, try to relax your hand and arm to avoid any voluntary contractions. When you are doing an experiment, voluntary movements will invalidate the results.

Threshold Stimulus, Spatial Summation, and Maximal Stimulus

In this section, you will first determine the minimal electrical stimulation that produces a response. This is the *threshold (liminal) stimulus.* Next, you will demonstrate *spatial summation,* which is also called *motor unit recruitment.* Spatial summation is the increasing strength of contraction due to an increasing number of motor units that are stimulated to contract. Finally, you will determine the magnitude of stimulus above which there is no further increase in the strength of contraction. This is the *maximal stimulus.* Proceed as follows to demonstrate these phenomena.

1. Start a new data file by clicking on **File** and then **New.** This will cause the Data Acquisition Window to appear on the screen. Before starting to collect data, you need to become familiar with the Data Acquisition Window. Most of the window is occupied by the **Graphic Display Area** where stimulus and muscle contraction data are plotted.

    To the left of the Display Area is the **File Information and Setup Area.** At the top are three boxes that allow you to turn on and off certain features. Checking **Stimulator Channel** collects data from the stimulator as well as from the transducer. Checking **Auto Stop** prevents the over-writing of data. Checking **Metronome** provides a visual and audible pace setter used in fatigue experiments.

The **File Setup Area** provides two pull-down menus that allow you to select the sampling rate in Hz and the number of samples in a file. Below these pull-down menus is information that changes based upon what you have selected. At the bottom, the **File Space** indicator informs you how much space has been used in the file.

The **Control Button Area** is located above the Display Area. The **Start Button** begins data acquisition and becomes a **Stop Button** that stops data acquisition. The **Pause Button** allows you to temporarily stop data acquisition. The **Continue Button** allows you to restart data acquisition. After data has been collected, the **Analyze Button** moves the data to the **Data Analysis Window.**

2. Set the Controls in the File Information and Setup Area as follows:

    Stimulator Channel: ON

    Auto Stop: ON

    Metronome: OFF

    Samples per sec: 200

    File Size: 16 k

3. Without applying a stimulus, grasp the pistol grip so that when you are relaxed (no stimulus), the natural tonus in your third finger is putting a very slight pressure on the trigger. The trigger should be displaced 3–5 mm (3–5 increments on the screen). In other words, before a stimulus is applied, you should have enough pressure on the trigger to lift the plot up off the bottom of the screen. Rest your second finger on the shelf on the right side of the pistol grip. Try to relax the muscles in your arm so that you don't dampen the response. With the stimulator frequency set to 1 stimulus per second and the duration at 15.0 ms, have your lab partner turn the stimulator voltage down to well below the point where you get a response. This will be around 30 v. While stimulating the motor point location on your forearm, have your lab partner gradually increase the voltage until you get a slight response. This is the *threshold (liminal) stimulus.*

4. Now have your partner increase the stimulus voltage gradually. On the screen, you should see a stair-stepping in the strength of contractions corresponding to the increase in stimulus strength as seen in Figure 21.3. This is *spatial summation.* When the stimulus is increased to the point where all motor units are responding (maximal contraction), turning up the voltage further will increase your discomfort but will

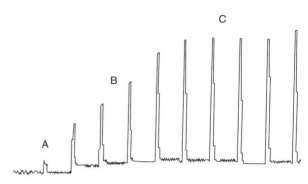

Figure 21.3 A Physiogrip™ myogram showing (A) minimal contraction by the threshold stimulus, (B) spatial summation, and (C) maximum contraction produced by the maximal stimulus.

not increase the strength of response. Stop increasing the voltage. Do *not* continue to turn the voltage up to the point of real pain. If the responses are so strong that the trigger is reaching the end of its travel, increase the spring tension or use a stronger spring and start over.

5. Click **Stop** Data Acquisition and click **Analyze.**

6. Measure the height of the contraction produced by the threshold stimulus by clicking on **Marker Δ1** and then click on the myogram where the contraction first occurred. Note the voltage of the threshold stimulus and height of the minimal contraction in the boxes above Marker Δ1. Measure the height of the maximal contraction by clicking on **Marker Δ2** and then on the first maximal contraction. Note the height of the maximal contraction and the voltage of the maximal stimulus in the boxes above Marker Δ2. (The voltages on the screen are the actual voltages applied as stimuli even though the stimulator is set at a higher voltage. This reduction in voltage results because the subject puts an electrical "load" on the stimulator.)

7. Record these threshold and maximal stimulus voltages on Laboratory Report 21 section A.

8. Print copies of the myogram for you and your partner. Attach the myogram under section A on Laboratory Report 21.

Temporal Summation and Tetanus

Temporal summation in skeletal muscle occurs when the muscle receives a series of stimuli in very rapid succession. If the frequency of stimulation is increased enough, *tetanization* will occur. See Figure 21.4. Proceed as follows to demonstrate these two phenomena:

1. Set up the Physiogrip™ and stimulator and leads as in the previous section.

2. Find the motor point for the *flexor digitorum* muscle again. (area A in Figure 21.2).
3. Start a new data file by selecting **New** from the **File** menu.
4. Set the Controls in the File Information and Setup Area as follows:

   Stimulator Channel: ON
   Auto Stop: ON
   Metronome: OFF
   Samples per second: 200
   File Size: 16 k

   Grasp the pistol grip so that when you are relaxed (no stimulus), the natural tonus in your finger is putting a very slight pressure on the trigger. The trigger should be displaced 3–5 mm (3–5 increments on the screen). In other words, before a stimulus is applied, you should have enough pressure on the trigger to lift the plot up off the bottom of the screen. Rest your second finger on the shelf on the right side of the pistol grip.
5. With the stimulus duration set at 15 ms, and the frequency set at 1 per sec, set the stimulus voltage to a level where the contraction from each stimulus takes up about 25% of the display screen vertically. The stimulator should be on continuous or multiple mode.
6. Be very careful not to lose the motor point and have your partner click **Start.** Rapidly proceed to the following steps.
7. Have your partner gradually increase the frequency of stimulation up to a tetanic contraction. This should be accomplished within 10 seconds or so. Do this several times, sustaining tetanic contraction for 1–2 seconds. Your chart should resemble Figure 21.4. It may take several tries to get it right. If the tetanic contraction causes a contraction that is so strong that the pistol grip trigger is reaching the

end of its travel, increase the spring tension or use a stronger spring and repeat the procedure.
8. Click **Stop** Data Acquisition and click **Analyze.**
9. Determine and record the frequency of stimuli and the magnitude of the response just before the onset of summation by using **Marker Δ1** and **Marker Δ2.** The time interval information is in the boxes below the Display Area. The best approximation for determining stimulus frequency is to measure the interval between the beginning of one response and the beginning of the next response. Determine and record the frequency of stimulation and magnitude of the response at the point where tetanic contraction was reached by using **Marker Δ1** and **Marker Δ2.** Here, the frequency of stimulation is best determined by measuring the interval between the peak of one response and the peak of the next response.
10. Record these frequencies on Laboratory Report 21 section B.
11. Print copies of the myogram for you and your partner. Attach the myogram under section B on Laboratory Report 21.

Single Muscle Twitch

We are now ready to study the myogram (tracing) of a single muscle twitch and determine its characteristics. All muscle twitches display three phases: a latent period, a contraction phase, and a relaxation phase. Each of these phases has a characteristic time duration, which may vary from muscle to muscle (Figure 21.5).

The **latent period** is a very short time lapse between the time of stimulation and the start of

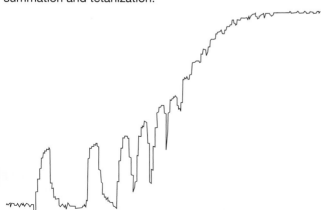

Figure 21.4 A Physiogrip™ myogram showing temporal summation and tetanization.

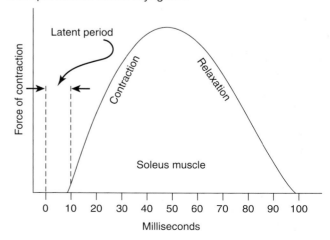

Figure 21.5 A myogram of a single muscle twitch of the human *soleus* muscle illustrates the components of a simple muscle twitch myogram.

contraction. In most muscles of the body it lasts about 10 milliseconds.

In the **contraction phase,** the muscle shortens because of chemical changes that occur within the muscle fibers. The time duration here varies from muscle to muscle.

After contraction, the muscle goes into a **relaxation phase** during which the muscle returns to its former length. The duration of this phase will also vary to some extent from muscle to muscle.

Figure 21.6 illustrates what the trace will look like with a Physiogrip™. Proceed as follows to demonstrate a single muscle twitch.

1. Set up the Physiogrip™, stimulator, and leads as in the previous section.
2. Find the motor point for the *flexor digitorum* muscle again.
3. Start a new data file by selecting **New** from the **File** menu.
4. Set the Controls in the File Information and Setup Area as follows:

       Stimulator Channel: ON
       Auto Stop: ON
       Metronome: OFF
       Samples per sec.: 200
       File Size: 16 k

5. Set the stimulator voltage to the maximal voltage as recorded in section A of Laboratory Report 21, the frequency to 1 per second and duration to 15 ms. The stimulator should be on continuous or multiple mode while you find the motor point.
6. Grasp the pistol grip so that when you are relaxed (no stimulus), the natural tonus in your finger is putting a very slight pressure on the trigger. The trigger should be displaced 3–5 mm (3–5 increments on the screen). In other words, before a stimulus is applied, you should have enough pressure on the trigger to lift the plot up off the bottom of the screen. If

you are using your third finger, rest your second finger on the shelf on the right side of the pistol grip. Turn the stimulator voltage up slowly until you are getting a good response so that each twitch uses about 75% of the trigger's travel.
7. Be very careful not to lose the motor point and have your partner click **Start.** Let the software run to monitor the next two steps.
8. After collecting a good number of high-quality twitches, click **Stop** Data Acquisition and click **Analyze.**
9. Study the double plot. The stimulator graph (at the top of the Display Area) will indicate the exact moment when the stimulator fired. With the **Velocity** plot on, it should be apparent that during the contraction period of the muscle twitch, the muscle was moving (contracting) faster than it was moving (relaxing) during the relaxation period. You can observe this phenomenon by noting that, corresponding to each muscle twitch, the peak of the positive velocity (contraction) is greater (farther from the zero line) than is the peak of the negative velocity (relaxation).
10. Notice that at the beginning of each twitch there is a period in which the muscle has not yet shown an isotonic response to the applied stimulus. Use Markers Δ1 and Δ2 to measure the length of the latent period from several twitches and record the average on Laboratory Report 21 section C.
11. Measure the length of the contraction period from several twitches and record the average on Laboratory Report 21 section C.
12. Measure the length of a relaxation period from several twitches and record the average on Laboratory Report 21 section C.
13. Measure the length of an entire twitch from stimulation to relaxation from several twitches and record the average on Laboratory Report 21 section C.
14. Print copies of the myogram of a single twitch for you and your partner. Attach the myogram under section C on Laboratory Report 21.

Fatigue and Muscle Contraction

Materials
Physiogrip™ setup from previous experiment
Sphygmomanometer

In this experiment we will study the effect of blood supply on fatigue in a muscle. Muscle fatigue is caused by the depletion of oxygen, glucose, and other raw materials, or by the accumulation of waste products such as lactic acid, or by both. All of these factors have a depressing effect on the neuromuscular junction.

Figure 21.6 A Physiogrip™ myogram of a single muscle twitch.

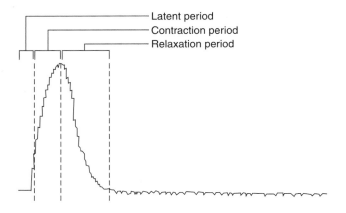

- Latent period
- Contraction period
- Relaxation period

Because all of these materials are brought to or removed from the muscle by blood, the rate of fatigue for a given rate and force of contraction will vary with the quantity and quality of the blood supply to the muscle. To restrict the blood flow to *flexor digitorum* muscle during part of the experiment, your partner will apply a sphygmomanometer (blood pressure cuff) to your upper arm.

1. Set up the Physiogrip™ and stimulator and leads as in the previous section.
2. Install the strongest spring supplied onto the pistol grip.
3. Start a new data file (select **New** from the **File** menu).
4. Set the Controls in the File Information and Setup Area as follows:
   Stimulator Channel: ON
   Auto Stop: ON
   Metronome: ON
   Rate: 4.0 Hz
   Samples per sec.: 20
   File Size: 32 k
5. Grasp the pistol grip so that your pointer finger (second finger) is on the trigger.
6. Have your partner click **Start** and immediately begin squeezing the trigger all the way in and letting it all the way out two times per second. (If you always have the trigger in when the metronome is to the right, and out when it is to the left, you will be squeezing in at the rate of two times per second.)
7. Continue to rhythmically squeeze until you can no longer squeeze the trigger *all the way in,* then immediately press the **Pause** button. Your myogram should look similar to Figure 21.7.

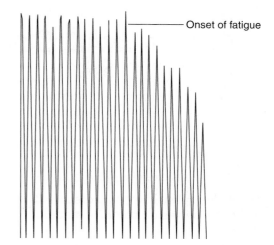

Figure 21.7 A Physiogrip™ myogram showing the onset of fatigue.

Onset of fatigue

8. Rest your arm for at least 5 minutes. Place a sphygmomanometer cuff on your arm just above the elbow. Inflate the cuff to 140 mm Hg and repeat as in step 7. To re-start data acquisition, press the **Continue** button.
9. Squeeze the trigger of the Physiogrip™ all the way in, at a rate of two times per second as before. When you can no longer squeeze the trigger *all the way in,* this time immediately press the **Stop** button and then the **Analyze** button. *Deflate the sphygmomanometer cuff.*
10. Using Markers Δ1 and Δ2, measure the time it took for fatigue to set in for each data file and record this on Laboratory Report 21 section D.
11. Print copies of the myogram for you and your partner. Attach the myogram under section D on Laboratory Report 21.

Exercise 22

THE SPINAL CORD AND REFLEX ARCS

Materials
  Model of spinal cord in cross section
  Prepared slides of spinal cord, x.s.
  Reflex hammers

Automatic stereotyped responses to stimuli enable rapid adjustment to environmental conditions. These responses are called **reflexes.** Reflexes play an important role in maintaining *homeostasis*. There are two types of reflexes. **Somatic reflexes** involve skeletal muscle responses. **Visceral reflexes** involve responses by smooth muscle, cardiac muscle, or glands. For example, they control heart rate, blood pressure, body temperature, water balance, and other vital physiological conditions. The neural pathway involved in a reflex is known as a **reflex arc,** and it always involves the central nervous system—either the brain or the spinal cord. For example, a spinal reflex involves receptors, spinal nerves, spinal cord, and effectors.

The Spinal Cord

The spinal cord extends downward from the brain, within the vertebral canal. In infants, the spinal cord extends the full length of the vertebral canal, but in adults it ends near the second lumbar vertebra. This difference results from the differences in growth rates of the spinal cord and the axial skeleton. Posterior portions of the axial skeleton have been removed in Figure 22.1 to reveal the spinal cord and spinal nerves.

Extending downward from the end of the spinal cord are the proximal ends of spinal nerves serving body regions below the cord. This aggregation of nerves is called the **cauda equina,** or horse's tail. The **dural sac** (dura mater) ends within the sacrum; only a portion of it is shown (label 14) for better observation of the nerves.

The Spinal Nerves

The spinal nerves emerge from the spinal cord through the intervertebral foramina on each side of the vertebral column. There are eight cervical, twelve thoracic, five lumbar, and five sacral pairs. These thirty pairs of nerves, plus one pair of coccygeal nerves, make a total of thirty-one pairs of spinal nerves.

As shown in the sectional view in Figure 22.1, each spinal nerve is joined to the spinal cord by **anterior** (or ventral) and **posterior** (or dorsal) **roots.** The enlargement on the posterior root is a **posterior root ganglion** that contains cell bodies of sensory neurons. Note in the sectional view that the **gray matter** (label 19), which is composed of unmyelinated nerve fibers and neuron cell bodies, is encompassed by **white matter,** which is composed of myelinated nerve fibers.

Some spinal nerves combine to form complex networks called **plexuses** on each side of the spinal cord. In a plexus, the nerve fibers are rearranged and combined so that fibers innervating a particular body part occur in the same nerve, although the fibers may originate in different spinal nerves.

Cervical Nerves

There are eight pairs of cervical nerves. The first pair (C_1) exit the spinal cord between the skull and the atlas. The eighth pair (C_8) exit between the seventh cervical vertebra and the first thoracic vertebra. The first four cervical nerves unite to form the **cervical plexus** (label 16), which serves the muscles and skin of the neck. Cervical nerves 5 through 8 unite with the first thoracic nerve (T_1) to form the **brachial plexus,** which is located in the shoulder region. Nerves serving the arm and hand arise from this plexus.

Figure 22.1 The spinal cord and spinal nerves.

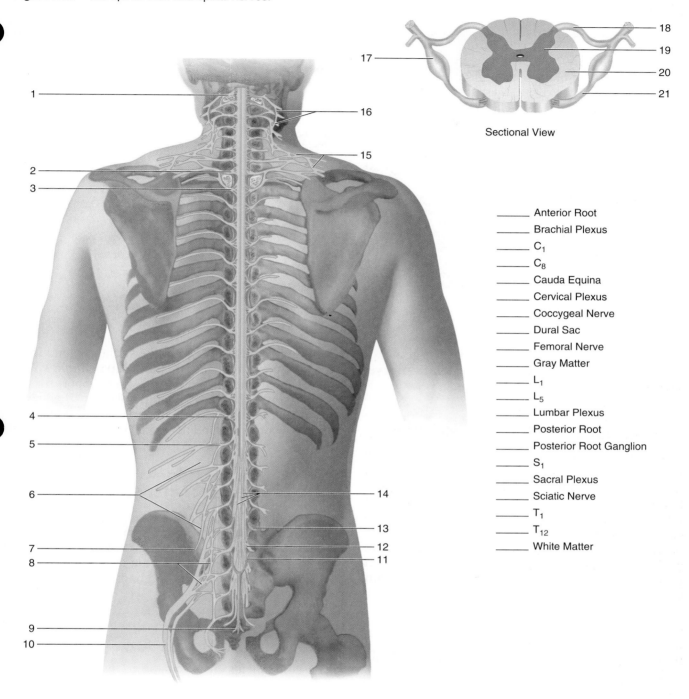

Sectional View

_____	Anterior Root
_____	Brachial Plexus
_____	C_1
_____	C_8
_____	Cauda Equina
_____	Cervical Plexus
_____	Coccygeal Nerve
_____	Dural Sac
_____	Femoral Nerve
_____	Gray Matter
_____	L_1
_____	L_5
_____	Lumbar Plexus
_____	Posterior Root
_____	Posterior Root Ganglion
_____	S_1
_____	Sacral Plexus
_____	Sciatic Nerve
_____	T_1
_____	T_{12}
_____	White Matter

Thoracic Nerves

There are twelve pairs of thoracic nerves, T_1 through T_{12}. Each nerve emerges from the spinal cord just below its corresponding vertebra.

Lumbar Nerves

There are five pairs of lumbar nerves. Lumbar nerves L_1 (label 5) through L_4 unite to form the **lumbar plexus** (label 6) from which the large **femoral nerve** (label 7) emanates to supply the thigh and leg.

Sacral and Coccygeal Nerves

There are five pairs of sacral nerves and one pair of coccygeal nerves (label 9). The **sacral plexus** (label 8) is formed by lumbar nerves L_4 and L_5

Figure 22.2 The spinal cord and meninges.

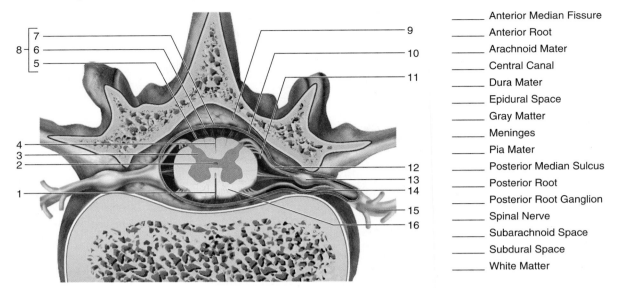

_____ Anterior Median Fissure
_____ Anterior Root
_____ Arachnoid Mater
_____ Central Canal
_____ Dura Mater
_____ Epidural Space
_____ Gray Matter
_____ Meninges
_____ Pia Mater
_____ Posterior Median Sulcus
_____ Posterior Root
_____ Posterior Root Ganglion
_____ Spinal Nerve
_____ Subarachnoid Space
_____ Subdural Space
_____ White Matter

and sacral nerves S_1 through S_3. The large nerve emanating from this plexus is the **sciatic nerve,** which serves the thigh, leg, and foot.

The Spinal Cord in Cross Section

The cross-sectional structure of the spinal cord within the vertebral canal is shown in Figure 22.2. The H-shaped pattern of the gray matter is surrounded by white matter that is composed mostly of myelinated neuron fibers. Two fissures occur on the midline. The **anterior median fissure** is wider than the **posterior median sulcus.** The small **central canal** lies on the median line in the gray matter. It extends the length of the spinal cord and is continuous with the ventricles of the brain.

The Meninges

Surrounding the spinal cord (and also the brain) are three membranes, the **meninges** (meninx, singular). The outermost membrane, the **dura mater,** also covers the spinal nerve roots. It consists of fibrous connective tissue and is the toughest of the meninges. Between the dura mater and the vertebrae is an **epidural space** that contains areolar connective tissue, adipose tissue, and blood vessels.

The innermost membrane is the **pia mater** (label 5), a thin membrane tightly adhered to the spinal cord. The third membrane, the **arachnoid mater** (label 6), lies between the pia mater and the dura mater. It is separated from the dura mater by the very thin **subdural space** (label 10), which contains

a film of serous fluid. Web-like strands extend from the arachnoid mater across the **subarachnoid space** (label 11) to the pia mater providing support for the spinal cord. The subarachnoid space is filled with **cerebrospinal fluid,** which serves as a protective shock absorber around the spinal cord.

ASSIGNMENT

1. Label Figures 22.1 and 22.2.
2. Examine a prepared slide of spinal cord, x.s., and locate the parts shown in Figure 22.2. Compare your observations with Figure HA-12. Use high power to locate neuron cell bodies in the gray matter.

The Somatic Reflex Arc

The components of a somatic reflex arc involving the spinal cord are shown in Figure 22.3. Arrows indicate the path of impulse transmission. These components are (1) a **receptor** that receives stimuli and forms impulses, such as a touch receptor in the skin; (2) a **sensory neuron** that carries impulses into the spinal cord via a spinal nerve and its posterior root; (3) an **interneuron** that transmits impulses to a motor neuron within the gray matter of the spinal cord; (4) a **motor neuron** that transmits impulses to an effector via the anterior root of a spinal nerve; and (5) an **effector,** a muscle that contracts when activated by the impulses.

Figure 22.3 The spinal somatic reflex arc.

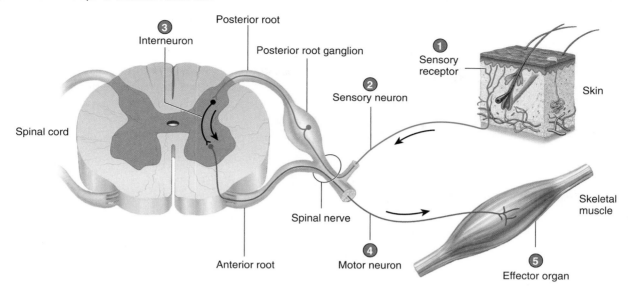

Reflex arcs may lack an interneuron, as in the patellar (knee jerk) reflex, or they may have more than one interneuron. In general, the fewer synapses involved, the more rapid is the reflex action.

The Visceral Reflex Arc

Reflexes of the viscera are controlled by the **autonomic nervous system** and follow a different pathway than somatic reflexes. Visceral reflex arcs differ from somatic reflex arcs by the presence of two efferent neurons instead of one. The two efferent neurons synapse outside the central nervous system in an autonomic ganglion. Figure 22.4 shows the major components of a visceral reflex arc.

There are two types of autonomic ganglia. **Vertebral autonomic ganglia** (label 5) are united to form two chains; one chain lies along each side of the vertebral column. **Collateral ganglia** (label 2) are located closer to the visceral organs.

Stimulation of a receptor in a visceral organ initiates impulses that are carried by a **visceral afferent (sensory) neuron** (label 3). This afferent neuron passes through a vertebral autonomic ganglion and enters the spinal cord via the posterior root of a spinal nerve. It synapses with a **preganglionic efferent neuron** (label 4) within the gray matter of the spinal cord. This first efferent neuron exits the spinal cord via the anterior root of a spinal nerve and synapses with a **postganglionic efferent neuron** (label 1) at either a vertebral autonomic ganglion or a collateral ganglion. This second efferent neuron carries impulses to the visceral organ where the response occurs.

Recall that the autonomic nervous system consists of two divisions. The **sympathetic division** is involved with spinal nerves of the thoracic and lumbar regions. The **parasympathetic division** involves the cranial and sacral nerves. Most viscera are innervated by nerves from both divisions, which enables a dynamic functional balance to be maintained through stimulation by one division and inhibition by the other.

ASSIGNMENT

1. Label Figure 22.4.
2. Complete Sections A through D of the laboratory report.

Diagnostic Somatic Reflexes

Reflex tests are standard clinical procedures used by physicians in evaluating neural functions. Abnormal reflex responses may indicate pathology of the involved portion of the nervous system. Interpretation of the results requires considerable experience. Our purpose in performing three common tests is not to diagnose but to observe the reflexes and to understand why they are used.

In a clinical setting, the first thing clinicians look for is equal responses on the right and left sides of the body. Secondly, grossly decreased or increased responses are noted even if they are equal bilaterally.

Figure 22.4 The visceral reflex arc.

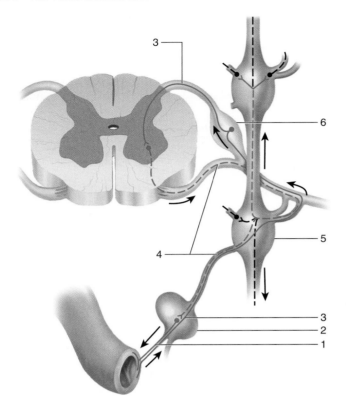

_____ Collateral Ganglion
_____ Postganglionic Efferent Neuron
_____ Preganglionic Efferent Neuron
_____ Posterior Root Ganglion
_____ Vertebral Autonomic Ganglion
_____ Visceral Afferent Neuron (2 places)

Reflex Responses

Abnormal responses may be either **hypoflexia** (hypotonic or hypoactive), a diminished response, or **hyperflexia** (hypertonic or hyperactive), an exaggerated response. Hypoflexia may result from several causes such as malnutrition, neuronal lesions, aging, or deliberate control. Hyperflexia may be due to a loss of inhibition in motor control areas. Some reflexes occur only if pathological conditions exist.

Figure 22.5 shows the procedures to be used in performing the reflex tests. Each involves striking a tendon to stimulate a muscle's stretch receptors, which normally causes a quick contraction in response. Use the following scale in interpreting the responses of your partner for each reflex. Record your results on the laboratory report.

++++ very strong response; often indicative of pathology
+++ stronger than average response
++ normal or average response
+ weaker than normal response
0 no response

Biceps Reflex   This reflex causes flexion of the forearm. It is elicited by holding the subject's elbow, with the thumb placed on the tendon of the biceps brachii, and striking a sharp blow to the first joint of your thumb with the reflex hammer. Reinforce the response by asking the subject to squeeze his or her thigh with the other hand during the test. Test both arms. This reflex involves the C_5 and C_6 spinal nerves.

Figure 22.5 Methods used for testing four types of somatic reflexes.

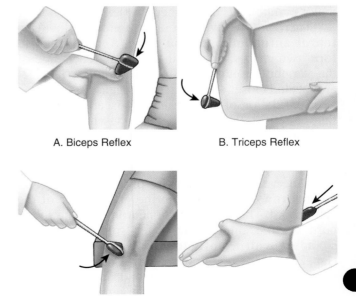

A. Biceps Reflex

B. Triceps Reflex

C. Patellar Reflex

D. Achilles Reflex

Triceps Reflex This reflex produces a slight extension of the arm in normal individuals. To demonstrate this reflex, flex the subject's arm at the elbow, holding the wrist as shown in Figure 22.5 with the palm toward the body. Strike the triceps brachii tendon above the elbow with the pointed end of the reflex hammer. Use the same reinforcement technique as in the biceps reflex. Test both arms. This reflex involves the C_7 and C_8 spinal nerves.

Patellar Reflex The subject must be seated on the edge of a table with the leg suspended over the edge. To elicit a response, strike the patellar tendon just below the kneecap. Reinforce the response by having the subject interlock the fingers of both hands and pull each hand against the other isometrically. Test both legs. This reflex involves spinal nerves L_2, L_3, and L_4.

Sometimes a series of jerky contractions occurs after a normal response. This usually indicates damage within the central nervous system.

Achilles Reflex This reflex produces plantar flexion. To elicit the response, hold the foot with one hand in a slightly dorsiflexed position and strike the Achilles tendon with the reflex hammer. Reinforce the response as in the patellar reflex. Test both feet. This reflex involves spinal nerves S_1 and S_2. A diminished response may indicate hypothyroidism.

ASSIGNMENT

Complete the laboratory report.

BRAIN ANATOMY: EXTERNAL

OBJECTIVES

After completing this exercise, you should be able to:

1. Identify the external structures of sheep and human brains on preserved specimens, charts, or models.
2. Describe the function of the cranial nerves.
3. Identify the major functional areas of the human brain.
4. Define all terms in bold print.

Materials
Colored pencils
Human brain models
Sheep brains, preserved
Protective, disposable gloves
Dissecting instruments and trays

Before You Proceed

Consult with your instructor about using protective disposable gloves when performing portions of this exercise.

The Meninges

Both the brain and spinal cord are covered by three protective membranes, the **meninges.** Figure 23.1 shows the relationships of the meninges in a frontal section through the cranium and brain.

The **pia mater** (label 6) is the thin, innermost meninx adhered to the brain surface. The **arachnoid mater** is the intermediate meninx with delicate fibers extending from its inner surface through the **subarachnoid space** (label 8) to the pia mater. This space contains **cerebrospinal fluid,** which serves as a protective shock absorber for the brain. The outermost meninx is the tough, fibrous **dura mater** (label 4),

Figure 23.1 The meninges. (a) The meninges are protective membranes that envelope the brain and spinal cord; (b) detail of the meninges enveloping the brain.

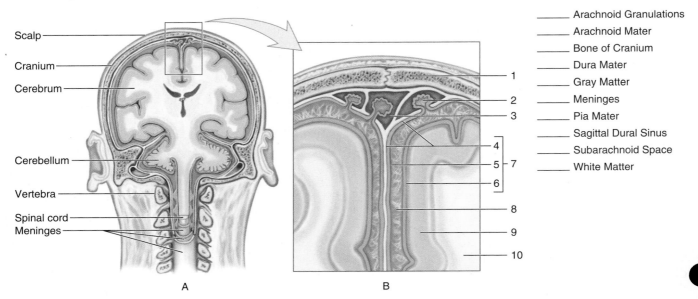

Scalp
Cranium
Cerebrum
Cerebellum
Vertebra
Spinal cord
Meninges

A

B

1
2
3
4
5 7
6
8
9
10

_____ Arachnoid Granulations
_____ Arachnoid Mater
_____ Bone of Cranium
_____ Dura Mater
_____ Gray Matter
_____ Meninges
_____ Pia Mater
_____ Sagittal Dural Sinus
_____ Subarachnoid Space
_____ White Matter

Figure 23.2 Lateral aspects of sheep and human brains.

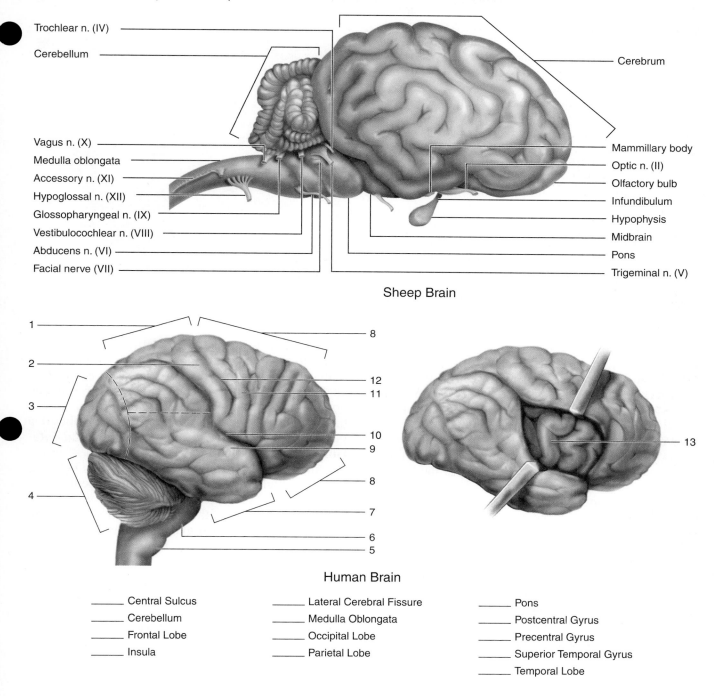

Trochlear n. (IV)

Cerebellum

Cerebrum

Vagus n. (X)

Medulla oblongata

Accessory n. (XI)

Hypoglossal n. (XII)

Glossopharyngeal n. (IX)

Vestibulocochlear n. (VIII)

Abducens n. (VI)

Facial nerve (VII)

Mammillary body

Optic n. (II)

Olfactory bulb

Infundibulum

Hypophysis

Midbrain

Pons

Trigeminal n. (V)

Sheep Brain

Human Brain

_____ Central Sulcus	_____ Lateral Cerebral Fissure	_____ Pons
_____ Cerebellum	_____ Medulla Oblongata	_____ Postcentral Gyrus
_____ Frontal Lobe	_____ Occipital Lobe	_____ Precentral Gyrus
_____ Insula	_____ Parietal Lobe	_____ Superior Temporal Gyrus
		_____ Temporal Lobe

which is attached to the inner surface of the cranial bones. Within the dura mater at the superior midline is a **sagittal dural sinus** that receives venous blood from the brain as it returns to the heart. The cerebrospinal fluid diffuses into the venous blood from the **arachnoid granulations (villi)** that project into the dural sinus. Note that the meninges extend into the longitudinal fissure that separates the cerebral hemispheres. Also notice that the section of cerebrum reveals that **gray matter** (label 9) is located exterior to the **white matter.**

Major Parts of the Brain

Locate the parts of the sheep and human brains shown in Figure 23.2 as you study this section.

The Cerebrum

In both sheep and human, the cerebrum is the largest portion of the brain. The surface of the cerebrum has numerous ridges or convolutions called **gyri** that increase the surface area of the gray matter. Gyri are bordered by shallow furrows called **sulci.** Deeper furrows are called **fissures.**

The cerebrum is divided into left and right **cerebral hemispheres** by the **longitudinal cerebral fissure** (label 20, Figure 23.4), which extends along the median line. Each hemisphere consists of four lobes (see Figure 23.2) that are named for the cranial bones under which they lie.

The **frontal lobe** is the most anterior lobe. It is separated from the **parietal lobe** by the **central sulcus** (label 12, Figure 23.2), and its inferior margin is the **lateral cerebral fissure** (label 10). The central sulcus divides the **precentral gyrus** of the frontal lobe from the **postcentral gyrus** of the parietal lobe. The **occipital lobe** is the most posterior lobe. Its anterior margin lies at the **parieto-occipital fissure,** which is distinct on the medial, but not on the lateral, cerebral surface. Therefore, its position is indicated by a dotted line that extends inferiorly. The **temporal lobe** lies inferior to the lateral cerebral fissure and anterior to the parieto-occipital fissure. Just inferior to the lateral cerebral fissure is the **superior temporal gyrus** of the temporal lobe. The separation of the temporal lobe from the occipital and parietal lobes is shown by dotted lines in Figure 23.2 because the margins of these lobes are not clear in a lateral view. If the temporal and frontal lobes are retracted at the lateral cerebral fissure (lower illustration in Figure 23.2), a small mass of cortex, the **insula,** is exposed. Its function is largely unknown, but it seems to be involved in integrating sensory input from visceral receptors.

The cerebrum integrates and interprets incoming sensory impulses as sensations and initiates conscious motor responses. It is also responsible for intelligence, will, and memory. The functional areas of the cerebrum will be considered later in this exercise.

Each cerebral hemisphere controls the opposite side of the body. The right cerebral hemisphere controls the left side of the body, and the left cerebral hemisphere controls the right side of the body. This pattern of control results because nerve fibers passing between the cerebral hemispheres and the spinal cord cross to the opposite side in the brain stem.

The Cerebellum

The cerebellum is located below the occipital and temporal lobes of the cerebrum posterior to the brain stem. It is divided into two hemispheres by a medial constriction. It provides subconscious control of muscular contractions enabling muscular coordination, balance, and posture.

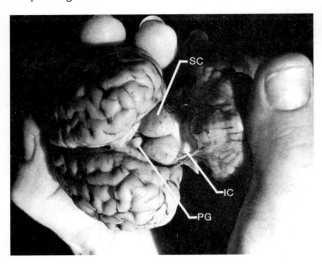

Figure 23.3 Exposing structures of the diencephalon and midbrain: SC—superior colliculus; IC—inferior colliculus; PG—pineal gland.

The Midbrain

The midbrain is a short section of the brain stem superior to the pons and is not visible externally in the human brain. It is necessary to depress the cerebellum as shown in Figure 23.3 to expose four rounded bodies known as the *corpora quadrigemina* of the midbrain. The larger pair are the **superior colliculi,** a visual reflex center, and the smaller pair are the **inferior colliculi,** an auditory reflex center. In addition, the midbrain is a pathway for nerve fibers between lower and higher brain centers. The small median projection above the superior colliculi is the **pineal gland,** which is part of the diencephalon (described in Exercise 24). The pineal gland secretes the hormone melatonin and is involved in sleep-wake patterns.

The Pons

The pons is a rounded bulge on the anterior portion of the brain stem. It contains fibers that connect the cerebellum and medulla with higher parts of the brain.

The Medulla Oblongata

The medulla oblongata is the lowest portion of the brain stem and is continuous with the spinal cord. The medulla contains centers that control the heart rate, breathing rate and depth, and diameter of blood vessels.

The Hypothalamus

The ventral portion of the brain that lies between the cerebral hemispheres and the midbrain is the **hypothalamus.** It provides autonomic regulation of body functions, such as temperature, water and electrolyte balance, sleep, and hunger.

Figure 23.4 Inferior aspects of the sheep and human brains.

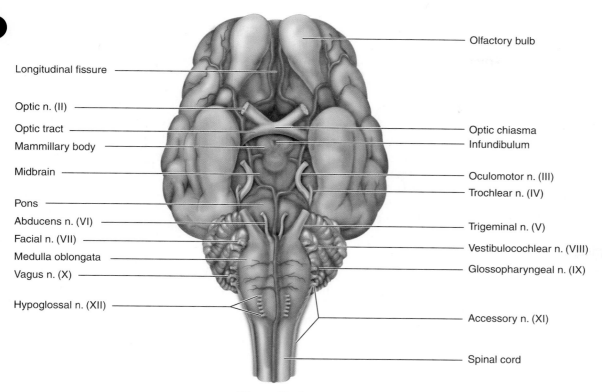

Olfactory bulb

Longitudinal fissure

Optic n. (II)

Optic tract

Mammillary body

Midbrain

Pons

Abducens n. (VI)

Facial n. (VII)

Medulla oblongata

Vagus n. (X)

Hypoglossal n. (XII)

Optic chiasma

Infundibulum

Oculomotor n. (III)

Trochlear n. (IV)

Trigeminal n. (V)

Vestibulocochlear n. (VIII)

Glossopharyngeal n. (IX)

Accessory n. (XI)

Spinal cord

Sheep Brain

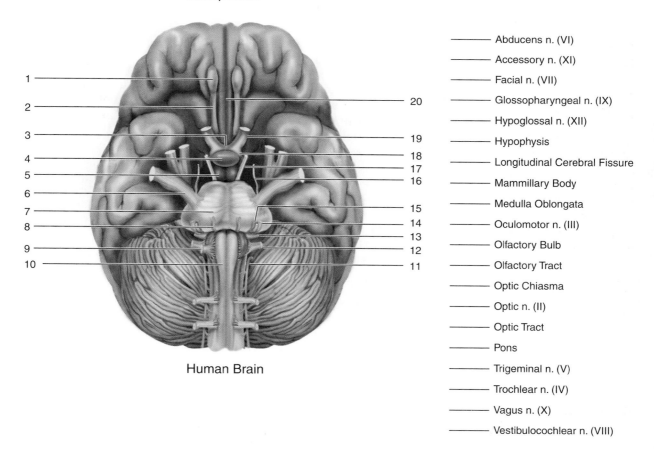

1

2

3

4

5

6

7

8

9

10

20

19

18

17

16

15

14

13

12

11

Abducens n. (VI)

Accessory n. (XI)

Facial n. (VII)

Glossopharyngeal n. (IX)

Hypoglossal n. (XII)

Hypophysis

Longitudinal Cerebral Fissure

Mammillary Body

Medulla Oblongata

Oculomotor n. (III)

Olfactory Bulb

Olfactory Tract

Optic Chiasma

Optic n. (II)

Optic Tract

Pons

Trigeminal n. (V)

Trochlear n. (IV)

Vagus n. (X)

Vestibulocochlear n. (VIII)

Human Brain

The **hypophysis,** or **pituitary gland,** hangs from the ventral surface of the hypothalamus by a short stalk, the **infundibulum.** See Figure 23.4.

Another external feature is the pair of rounded eminences, the **mammillary bodies** (label 5 in Figure 23.4), that lie just posterior to the base of the infundibulum.

ASSIGNMENT

1. Label Figures 23.1 and 23.2.
2. In Figure 23.2, color the four lobes of the cerebral hemisphere in the human brain for easy recognition: frontal—red; parietal—yellow; temporal—blue; occipital—green.
3. Complete Sections A through D of the laboratory report.

The Cranial Nerves

There are twelve pairs of cranial nerves that arise from various parts of the brain and pass through foramina to innervate parts of the head and trunk. Each is designated by name and number. Most cranial nerves are **mixed nerves** in that they carry both sensory and motor neuron processes. A few carry only sensory fibers.

As you study this section, locate the cranial nerves first on the sheep brain and then on the human brain in Figure 23.4.

I Olfactory This cranial nerve contains only sensory fibers for the sense of smell. It is not shown on the illustrations since it passes from the olfactory epithelium to the **olfactory bulb** (label 1) of the brain. It synapses with neurons in the olfactory bulb that carry the impulses on to the olfactory areas of the cerebrum. In humans, an **olfactory tract** connects the bulb with the brain.

II Optic This sensory nerve functions in vision and contains axons of sensory nerve cells in the retina. Half of the axons cross over to the other side of the brain as they pass through the **optic chiasma** (label 3). They continue through the **optic tracts** (label 18) to the thalamus and on to the visual centers of the cerebrum.

III Oculomotor This motor nerve emerges from the midbrain and innervates the levator palpebrae muscle, which raises the eyelid, and four extrinsic muscles of the eyeball: the superior, medial, and inferior rectus muscles and the inferior oblique. Its autonomic fibers control the iris of the eye and the focusing of the lens.

IV Trochlear This motor nerve arises from the midbrain and innervates the superior oblique, an extrinsic muscle of the eyeball.

V Trigeminal This is the largest cranial nerve, and it innervates much of the mouth and face. It emerges from the anterior part of the pons and subsequently divides into ophthalmic, maxillary, and mandibular branches.

The **ophthalmic division** is a sensory nerve that innervates the skin of the scalp, forehead, eyelid, and nose, the lacrimal (tear) gland, and parts of the eye. The **maxillary division** carries sensory impulses from the upper teeth, gums, and lip. The **mandibular division** is a mixed nerve innervating the lower teeth, gums, and lip, chewing muscles, lower part of the tongue, and lower part of the face.

VI Abducens This small motor nerve (label 8) originates from the posterior part of the pons and innervates the lateral rectus, an extrinsic muscle of the eyeball.

VII Facial This mixed nerve (label 15) arises from the lower part of the pons. It innervates facial muscles, salivary glands, and the anterior two-thirds of the tongue, including taste buds.

VIII Vestibulocochlear This sensory nerve arises from the medulla and consists of two branches innervating the inner ear. The **vestibular branch** carries impulses from the balance receptors. The **cochlear branch** carries impulses from the sound receptors.

IX Glossopharyngeal This mixed nerve (label 13) emerges from the medulla posterior to the vestibulocochlear. It innervates much of the mouth lining, posterior part of the tongue, and pharynx. It carries motor impulses for swallowing reflexes, salivary secretions, and respiration. It carries sensory impulses to the brain from chemo- and pressure receptors in carotid arteries, touch, pressure, and pain receptors of the tongue and pharynx, plus taste buds from the posterior third of the tongue.

X Vagus This mixed nerve (label 12) arises from the medulla and descends into the neck, thorax, and abdomen to innervate the pharynx, larynx, and the thoracic and abdominal viscera. It is a major nerve of the autonomic nervous system. It carries motor impulses for swallowing, speech, and regulation of abdominal and thoracic visceral functions. It carries sensory impulses from receptors in abdominal and thoracic viscera, including chemo- and pressure receptors in the aortic arch.

XI Accessory This motor nerve arises from both the medulla and spinal cord. The **cranial branch** innervates muscles of the pharynx and larynx. The **spinal branch** innervates the trapezius and sternocleidomastoid muscles.

XII Hypoglossal This motor nerve arises from three origins on the medulla and innervates muscles of the tongue, which are involved in food manipulation, speech, and swallowing. In Figure 23.4 each origin appears as a tree stump with the roots entering the medulla.

ASSIGNMENT

1. Label Figure 23.4.
2. Obtain a preserved sheep brain and place it on a dissecting tray for study. Locate the external features as shown in Figures 23.2, 23.3, and 23.4.
3. Locate the external features studied in this exercise on the model of a human brain.
4. Complete Section E of the laboratory report.

Functional Areas of the Cerebrum

The functional areas of the cerebrum are known largely through studies of patients with brain lesions—local injuries of various types—plus some electrochemical experiments on living subjects. The major functional areas are shown in Figure 23.5 and discussed below. Keep in mind that the brain functions as a whole unit with great interaction among its parts. The major functional areas noted below do not

function in isolation but are always integrated with other brain areas to produce the appropriate sensation or action. **Association areas** are parts of the cerebrum that integrate inputs from other parts of the cerebrum to determine appropriate interpretations of sensory input and to determine appropriate action for motor output. Association areas occupy about 75% of the cerebrum. Refer to Figure 23.5 as you study this section.

Frontal Lobe

The **somatomotor (primary motor) area** (label 1) is on the precentral gyrus, which is located just anterior to the central sulcus. It enables conscious control of skeletal muscles in the movement of body parts. The **premotor area** is a large association area in the frontal lobe anterior to the somatomotor area. It organizes and coordinates learned and skilled motor functions, such as typing, and transmits this information to the somatomotor area, which stimulates the muscle contractions. Anterior to the premotor area is the rather large **prefrontal association area.** It integrates information from sensory and motor regions of the cortex and other association areas. It allows an assessment of the world around. In many ways, integration in this important association area enables characteristics associated with being human: personality, planning, emotional control, moral judgment, critical thinking, and appropriate

Figure 23.5 Major functional areas of the cerebrum.

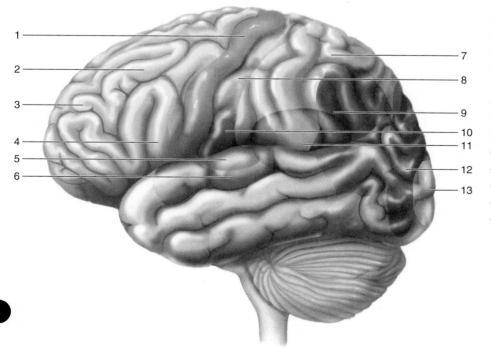

_____ Auditory Association Area
_____ General Interpretation Area
_____ Motor Speech Area
_____ Prefrontal Area
_____ Premotor Area
_____ Primary Auditory Area
_____ Primary Somatomotor Area
_____ Primary Somatosensory Area
_____ Primary Taste Area
_____ Primary Visual Area
_____ Sensory Speech Area
_____ Somatosensory Association Area
_____ Visual Association Area

behaviors. A **motor speech (Broca's) area** (label 4) is located in the left frontal lobe just superior to the lateral cerebral fissure and anterior to the somatomotor area. It receives input from the sensory speech area in the parietal lobe and generates programs for the muscles of the larynx, tongue, and lips to produce speech. These programs are transmitted to the somatomotor area, which executes them as speech.

Parietal Lobe

The **somatosensory (primary sensory) area** lies on the postcentral gyrus, which is just posterior to the central sulcus. It receives sensory impulses from receptors in skin and skeletal muscles and determines the location of the stimulated receptors. A **primary taste area** is located at the inferior end of the somatosensory area, and it is responsible for taste sensation. Sensory input to the somatosensory area is processed and analyzed by the **somatosensory association area,** which is located just posterior to the somatosensory area. It enables comprehension of the sensory input. The **sensory speech (Wernicke's) area** (label 11) is located at the junction of the parietal, temporal, and occipital lobes in one cerebral hemisphere, usually the left. It is responsible for understanding written and spoken language and transmits this information to the **general interpretation area** (label 9), a poorly defined region just superior and posterior to the sensory speech area. The general interpretation area involves parietal, temporal, and occipital lobes and usually occurs only in the left hemisphere. It integrates input from all sensory areas, interprets the information, and transmits the interpretation to the frontal lobe, where appropriate action is determined and taken. Lesions in the sensory speech or general interpretation area may prevent understanding of written or spoken words and appropriate speech.

Temporal Lobe

The **primary auditory area** (label 5) lies on the superior portion of the superior temporal gyrus, just inferior to the lateral cerebral sulcus. It receives impulses from sound receptors of the cochlea in the inner ear and interprets them as to pitch, loudness, and rhythm. The surrounding **auditory association area** allows further perception of specific sounds, such as speech and music, and it enables us to remember specific music and the sound of a person's voice. It is an essential area for understanding speech. The **primary olfactory area** is located on the medial surface of the temporal lobe (not shown in Figure 23.5). It receives impulses from olfactory receptors in the nasal epithelium and interprets them as specific odors.

Occipital Lobe

The **primary visual area** is located in the posterior portion of the occipital lobe. It receives impulses from light receptors in the retinas of the eyes and interprets them as visual images. The surrounding **visual association area** associates the sensory input with past experience, enabling recognition of specific objects such as a cat or a rose, and provides meaning for what we see.

ASSIGNMENT

1. Label Figure 23.5.
2. Complete the laboratory report.

Exercise 24

BRAIN ANATOMY: INTERNAL

OBJECTIVES

After completing this exercise, you should be able to:

1. Identify the internal structures of human and sheep brains on preserved specimens, charts, or models.
2. Describe the circulation of cerebrospinal fluid.
3. Define all terms in bold print.

Materials

Colored pencils
Model of human brain, sagittal section
Sheep brains, preserved
Dissecting instruments and tray
Long, sharp knife
Protective disposable gloves

Before You Proceed

Consult with your instructor about using protective disposable gloves when performing portions of this exercise.

The internal structure of the brain can best be observed by examining midsagittal and coronal sections cut through it. First use the labeled illustration of the sheep brain in Figure 24.1 and the written descriptions to locate and label the structures in the human brain in Figures 24.2 and 24.3. Then, use additional descriptions to label Figure 24.4. After you have gained an understanding of the internal structure of the brain by labeling the figures, you will perform a dissection of a sheep brain.

The Cerebrum

Note in the figures that the convolutions on the medial surface of the cerebral hemisphere have not been disturbed by the midsagittal section. The only part of the cerebrum that has been cut is the **corpus callosum** (label 20 in Figure 24.2). This large, curved band of white matter consists of neuron fibers connecting the two cerebral hemispheres. Locate the **sagittal sinus** in the dura mater along the midline and the **parieto-occipital fissure** (label 1).

The Cerebellum

Observe the pattern of gray and white matter in the cut surface of the cerebellum. The functions of the cerebellum are at the subconscious level. It receives impulses from motor and visual centers in the brain, balance receptors in the inner ear, and muscles of the body. This information is integrated and impulses are sent to muscles to maintain posture, balance, and muscle coordination. Unlike the situation in the cerebrum, each hemisphere of the cerebellum coordinates muscle function on its own side of the body.

The Diencephalon

Structures between the cerebrum and the brain stem constitute the **diencephalon,** or interbrain. It contains the fornix, thalamus, hypothalamus, and third ventricle.

The Fornix

The band of white matter inferior to the corpus callosum is the **fornix** (label 19 in Figure 24.2). It contains fibers associated with the sense of smell and is much better developed in sheep than in humans. The small **pineal gland** (label 3) lies on the midline. It secretes a hormone, melatonin, whose function in humans is obscure, but it is involved in sleep/wakefulness cycles.

The Thalamus

The **thalamus** consists of two ovoid masses of gray matter, one in each cerebral hemisphere, that are joined by an isthmus of neural tissue called the **intermediate mass of the thalamus** (label 18). The intermediate mass is the only part of the thalamus

125

Figure 24.1 Midsagittal section of the sheep brain.

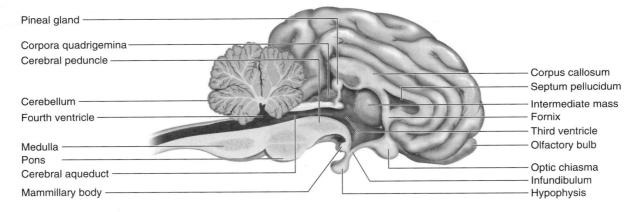

Pineal gland
Corpora quadrigemina
Cerebral peduncle
Cerebellum
Fourth ventricle
Medulla
Pons
Cerebral aqueduct
Mammillary body

Corpus callosum
Septum pellucidum
Intermediate mass
Fornix
Third ventricle
Olfactory bulb
Optic chiasma
Infundibulum
Hypophysis

Figure 24.2 Midsagittal section of the human brain.

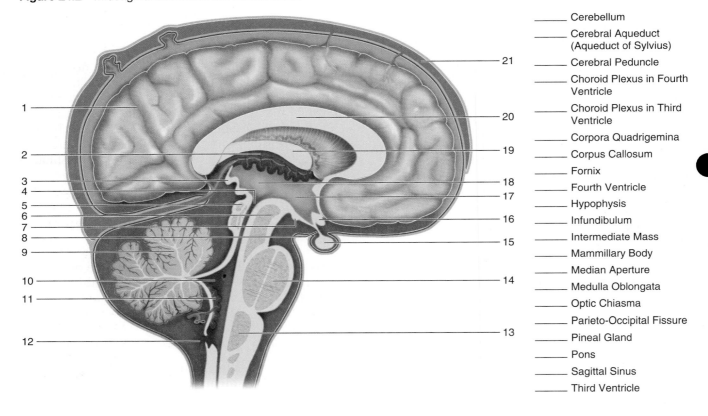

_____ Cerebellum
_____ Cerebral Aqueduct (Aqueduct of Sylvius)
_____ Cerebral Peduncle
_____ Choroid Plexus in Fourth Ventricle
_____ Choroid Plexus in Third Ventricle
_____ Corpora Quadrigemina
_____ Corpus Callosum
_____ Fornix
_____ Fourth Ventricle
_____ Hypophysis
_____ Infundibulum
_____ Intermediate Mass
_____ Mammillary Body
_____ Median Aperture
_____ Medulla Oblongata
_____ Optic Chiasma
_____ Parieto-Occipital Fissure
_____ Pineal Gland
_____ Pons
_____ Sagittal Sinus
_____ Third Ventricle

visible in a sagittal section. The thalamic masses form the lateral walls of the third ventricle.

The thalamus receives sensory impulses coming to the brain and relays them to appropriate sensory centers of the cerebrum. It also provides an uncritical sensory awareness.

The Hypothalamus

The **hypothalamus** (not labeled in Figure 24.2) is a small region that extends about 2 cm into the brain from its ventral surface anterior to the brain stem. A short stalk, the **infundibulum** (label 8), hangs from its ventral surface and supports the **hypophysis (pituitary gland),** which fits into a depression in the sella turcica of the sphenoid bone. Posterior to the infundibulum is the left **mammillary body** (label 7). The **optic chiasma,** where half of the fibers of the optic tracts cross, lies just anterior to the infundibulum.

The hypothalamus is a control center for the autonomic nervous system and regulates a variety of functions, such as body temperature, hunger, and water balance.

The Brain Stem

The midbrain, pons, and medulla form the brain stem.

The Midbrain

The short **cerebral aqueduct,** or aqueduct of Sylvius (label 4, Figure 24.2), runs longitudinally through the midbrain and connects the third and fourth ventricles. Anterior to this duct is a **cerebral peduncle.** The cerebral peduncles contain the main motor tracts between the cerebrum and lower parts of the brain stem. Posterior to the aqueduct, locate the left portions of the **corpora quadrigemina,** reflex centers for head, eye, and body movements in response to visual and auditory stimuli.

The Pons and Medulla

The **pons** lies between the midbrain and medulla and is easily recognized by its bulblike anterior protrusion on the brain stem. It consists primarily of fibrous tracts connecting lower and higher brain centers. The **medulla** is the lowest part of the brain stem and is a connecting link between the brain and the spinal cord. It also contains centers that control heart and breathing rates.

Within the brain stem is a network of fibers, the **reticular formation** (not shown in the figures), that generates impulses that keep the cerebrum awake and alert. Sleep results from its decreased activity, and unconsciousness results if it ceases to function.

The Ventricles and Cerebrospinal Fluid

There are four cavities, or ventricles, in the brain which are filled with cerebrospinal fluid (CSF). The two **lateral ventricles** are located within the cerebral hemispheres and are separated by a thin membrane, the septum pellucidum (see Figure 24.1). The third and fourth ventricles are on the midline. The **third ventricle** (label 17 in Figure 24.2) lies between the lateral masses of the thalamus; the

Figure 24.3   Origin and circulation of cerebrospinal fluid.

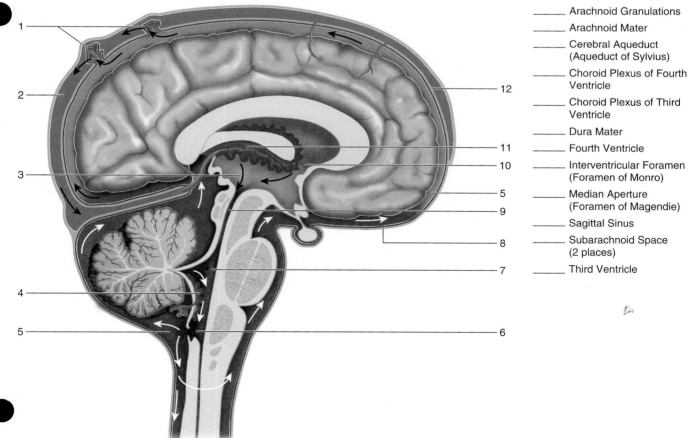

_____ Arachnoid Granulations

_____ Arachnoid Mater

_____ Cerebral Aqueduct (Aqueduct of Sylvius)

_____ Choroid Plexus of Fourth Ventricle

_____ Choroid Plexus of Third Ventricle

_____ Dura Mater

_____ Fourth Ventricle

_____ Interventricular Foramen (Foramen of Monro)

_____ Median Aperture (Foramen of Magendie)

_____ Sagittal Sinus

_____ Subarachnoid Space (2 places)

_____ Third Ventricle

fourth ventricle (label 10) is between the brain stem and the cerebellum.

Each ventricle contains a **choroid plexus,** a mass of specialized capillaries, that secretes the cerebrospinal fluid. The choroid plexuses of the third (label 2) and fourth (label 11) ventricles are shown in Figure 24.2. Locate these plexuses in Figure 24.3, which depicts the circulation of the cerebrospinal fluid.

The CSF passes from each lateral ventricle into the third ventricle through the **interventricular foramen,** or foramen of Monro (label 10 in Figure 24.3). This foramen is located anterior to the fornix and the choroid plexus of the third ventricle. From the third ventricle, the cerebrospinal fluid flows through the narrow **cerebral aqueduct,** aqueduct of Sylvius

(label 9 in Figure 24.3) that passes through the midbrain into the fourth ventricle. The fourth ventricle is continuous with the central canal of the spinal cord.

The cerebrospinal fluid then passes from the fourth ventricle through the **median aperture** (foramen of Magendie) and two other small openings into the **subarachnoid space** below the cerebellum. Some of the fluid moves upward within the subarachnoid space around the cerebellum and cerebrum and diffuses into the sagittal sinus through **arachnoid granulations** (label 1 in Figure 24.3). The remaining fluid flows downward in the subarachnoid space on the posterior surface of the spinal cord, upward on its anterior surface, continues upward around the cerebrum, and finally diffuses into the sagittal sinus.

Figure 24.4   Frontal sections of sheep and human brains.

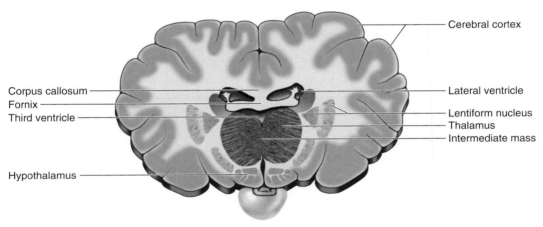

A. Sheep Infundibular Section

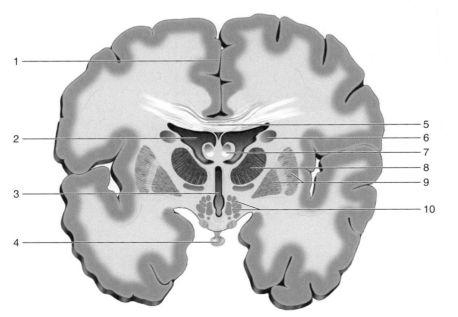

B. Human Infundibular Section

_____	Caudate Nucleus
_____	Corpus Callosum
_____	Fornix
_____	Hypophysis
_____	Hypothalamus
_____	Lateral Ventricle
_____	Lentiform Nucleus
_____	Longitudinal Cerebral Fissure
_____	Thalamus
_____	Third Ventricle

ASSIGNMENT

1. Label Figures 24.2 and 24.3.
2. Use colored pencils to color code: the parietal lobe of the cerebrum; the cerebellum; and the diencephalon.
3. Complete Sections A through D of the laboratory report.

Coronal Sections

Using Figure 24.4A as a guide, label Figure 24.4B, a frontal section of the human brain. Locate the triangular **lateral ventricles.** The white matter forming the roof of the lateral ventricles is the corpus callosum. The two ovoid masses of gray matter between the ventricles form the **fornix.** Inferior to the fornix is the **third ventricle.** On each side of the third ventricle are the lateral masses of gray matter forming the **thalamus,** and they are joined by the **intermediate mass** (not shown in Figure 24.4B), which passes through the third ventricle.

Masses of gray matter within the brain are called nuclei. The **caudate nuclei,** located superior to the thalamus in the walls of the lateral ventricles, and the **lentiform nuclei** (label 9) constitute the **basal ganglia.** They are involved in muscle coordination. Note the numerous nuclei of the **hypothalamus** on each side of the inferior portion of the third ventricle. What functions do they regulate?

ASSIGNMENT

Label Figure 24.4B.

Sheep Brain Dissection

Obtain a preserved sheep brain and review its external features shown in Figures 23.2 and 23.4. To view structures shown in Figure 24.1, you will make a midsagittal section of the brain. Place the brain dorsal surface up on your dissecting tray. Place a long, sharp knife (not a scalpel) in the longitudinal fissure and slice through the brain to form equal left and right halves. Examine the cut surface and locate all of the structures labeled in Figure 24.1.

To better understand the relationships of the thalamus and ventricles, you will make and examine a frontal section through the infundibulum of a whole sheep brain. Obtain a sheep brain and place it ventral surface up on your dissecting tray. Use the knife to cut a frontal section by starting at the infundibulum and cutting perpendicularly through the longitudinal axis of the brain.

Compare your frontal section with Figure 24.4A and locate the labeled structures. Note the distribution of gray and white matter in the cerebrum. How do their functions differ?

When finished with your study, dispose of the brain as directed by your instructor. Clean your instruments, tray, and workstation.

ASSIGNMENT

1. Locate the structures shown in Figure 24.2, 24.3, and 24.4B on models of the human brain.
2. Complete the laboratory report.

Exercise 25

THE EYE

OBJECTIVES

After completing this exercise, you should be able to:

1. Identify the parts of the eye and its accessory structures on a preserved eye, chart, or model.
2. Perform and interpret selected visual tests.
3. Define all terms in bold print.

Materials

Beef eye, preferably fresh
Dissecting instruments and tray
Meterstick or tape measure
Ishihara color blindness test plates
Laboratory lamp
Snellen eye chart
Prepared slide of retina sectioned through the fovea centralis
Protective disposable gloves

Before You Proceed

Consult with your instructor about using protective disposable gloves when performing portions of this exercise.

The Lacrimal Apparatus

The eye, lacrimal gland, and extrinsic eye muscles lie protected in the orbital cavity. The eye is protected anteriorly by the **eyelids.** A thin membrane, the **conjunctiva,** lines the inner surface of the eyelids and the anterior surface of the eye. It contains many blood vessels and pain receptors except where it covers the cornea.

The lacrimal apparatus is shown in Figure 25.1. The **lacrimal gland** is located in the upper lateral portion of the eye orbit. It secretes a lubricating fluid that keeps the anterior surface of the eye and the conjunctiva moist. The fluid reaches the eye surface via 6 to 12 tiny **lacrimal ducts** and flows downward and medially over the anterior surface of the eye. At the medial corner of the eye, the fluid enters two tiny openings, the **lacrimal puncta,** and passes into the **superior** and **inferior lacrimal** canals. Then, the fluid flows into the **lacrimal sac** and passes through the **nasolacrimal duct** into the nasal cavity.

The small, red, conical body in the medial corner of the eye is the **caruncula.** It secretes a whitish substance that collects in this region. Lateral to the caruncula is a fold of the conjunctiva, the **plica semilunaris** (not labeled in Figure 25.1), a remnant of a third eyelid that is well developed in many other vertebrates.

Figure 25.1 The lacrimal apparatus of the eye.

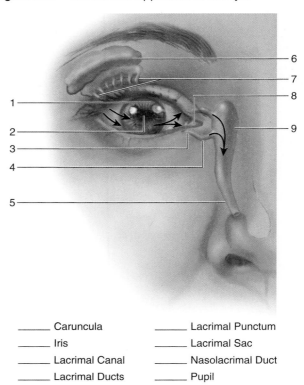

_____ Caruncula	_____ Lacrimal Punctum
_____ Iris	_____ Lacrimal Sac
_____ Lacrimal Canal	_____ Nasolacrimal Duct
_____ Lacrimal Ducts	_____ Pupil
_____ Lacrimal Gland	

Figure 25.2 The extrinsic muscles of the eye.

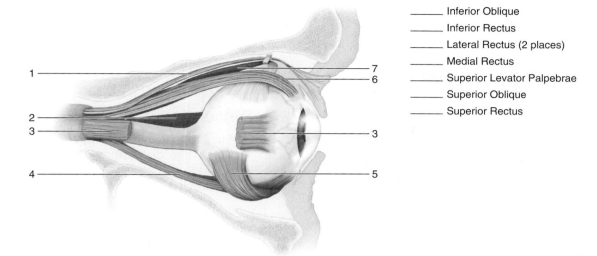

_____ Inferior Oblique
_____ Inferior Rectus
_____ Lateral Rectus (2 places)
_____ Medial Rectus
_____ Superior Levator Palpebrae
_____ Superior Oblique
_____ Superior Rectus

Extrinsic Eye Muscles

The eye orbit contains seven muscles: six extrinsic muscles that move the eye and the **superior levator palpebrae** muscle, which raises the upper eyelid. The insertions of these muscles are shown in Figure 25.2. Their origins are on bones at the back of the orbit.

The **lateral rectus** has been partially removed in the illustration. Its opposing muscle is the **medial rectus** (label 2). The **superior rectus** inserts on the upper surface of the eyeball and is opposed by the **inferior rectus** (label 4) inserted on the lower surface. The **superior oblique** passes through a cartilaginous loop allowing it to exert an oblique pull on the eyeball. The **inferior oblique** is oriented in a similar manner, but only a portion (label 5) is shown. The interaction of these muscles enables movement of the eye in all directions.

Note in Figure 25.2 how the optic nerve (yellow) exits the posterior surface of the eye.

Internal Anatomy

A transverse section of the right eye as seen looking down on it is shown in Figure 25.3. The wall of the eyeball consists of three layers: an outer, fibrous,

Figure 25.3 Internal anatomy of the eye.

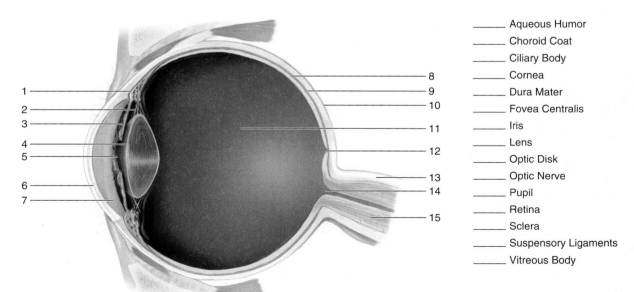

_____ Aqueous Humor
_____ Choroid Coat
_____ Ciliary Body
_____ Cornea
_____ Dura Mater
_____ Fovea Centralis
_____ Iris
_____ Lens
_____ Optic Disk
_____ Optic Nerve
_____ Pupil
_____ Retina
_____ Sclera
_____ Suspensory Ligaments
_____ Vitreous Body

scleroid coat (sclera); a middle, pigmented **choroid coat** containing blood vessels; and an inner **retina** containing the light receptors.

The **optic nerve** (label 15) contains fibers leading from the retina to the brain. The point where the optic nerve leaves the retina lacks light receptors and is known as the **optic disk** or **blind spot.** An artery and vein (not shown in Figure 25.3), which supply the retina and interior of the eye, enter and leave the eye at the optic disc and continue down the center of the optic nerve. Note that the optic nerve is wrapped in an extension of the **dura mater** (label 13).

Lateral to the optic disk is a small pit, the **fovea centralis,** the area of sharpest vision that is located in the center of a round yellow spot, the **macula lutea.** The macula is 0.5 mm in diameter and contains only cones.

The **cornea** covers the anterior portion of the eye as an extension of the scleroid coat. It is transparent, and its greater curvature bends the light rays that pass through it. Light rays are precisely focused on the retina by the transparent, crystalline **lens** that is supported by **suspensory ligaments** attached to the **ciliary body.** Contraction or relaxation of smooth muscle in the ciliary body changes the shape of the lens to allow focusing of objects at different distances. Just anterior to the lens is the colored portion of the eye, the circular **iris,** that regulates the amount of light entering the eye through the **pupil.**

The large cavity between the lens and retina is filled with a transparent, jellylike material, the **vitreous body.** The cavity in front of the lens is filled with a watery liquid, the **aqueous humor** (label 7). Aqueous humor is constantly produced from capillaries near the posterior attachment of the iris to the ciliary body and is absorbed into small veins near the anterior attachment of the iris to the sclera.

ASSIGNMENT

1. Label Figures 25.1, 25.2, and 25.3.
2. Complete Sections A through D of the laboratory report.

Eye Dissection

An understanding of eye anatomy can best be obtained from an eye dissection. The procedures are described below. Figure 25.4 illustrates the major steps to be followed.

1. Examine the external surface of the eyeball. Locate the optic nerve and any remnants of extrinsic muscles. Identify the cornea, iris, and pupil. Is the shape of the pupil the same as in humans?
2. Hold the eyeball as shown in illustration 1 of Figure 25.4 and make a small incision in the sclera about 0.5 cm from the cornea. Insert scissors into the incision and cut around the eye while holding the cornea upward, as in illustration 2. Note the watery aqueous humor that exudes.
3. Gently lift off the anterior portion of the eye and place it on a tray with the inner surface upward. The lens usually remains with the vitreous body in fresh eyes, but it may remain with the cornea in preserved eyes.
4. In the anterior portion, identify the thickened black ring and the ciliary body, and locate the iris. Are radial and circular muscle fibers evident?
5. Gently pour the vitreous body and lens from the posterior portion of the eye. Use a

---

Figure 25.4 Three steps in beef eye dissection.

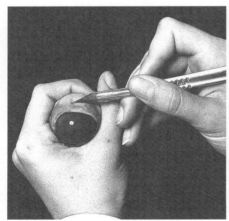

1. First cut through sclera of eyeball with scalpel.

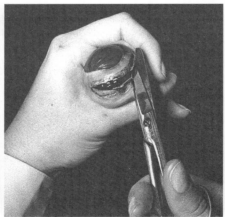

2. Second cut through wall with sharp scissors.

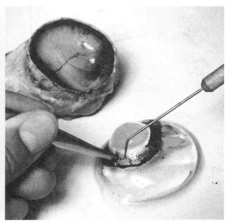

3. Separating the lens from the vitreous body with needle.

dissecting needle to carefully separate the lens from the vitreous body as shown in illustration 3 in Figure 25.4.

6. Hold the lens by the edges and look through it at a distant object. What is unusual about the image? Place it on printed matter. Does it magnify the letters? Compare the consistency of the center and periphery of the lens using a probe or forceps.
7. Observe the thin, pale retina in the posterior part of the eye. It separates easily from the choroid coat and forms wrinkles once the vitreous body is removed. Note the optic disk.
8. Observe the black choroid coat and the iridescent nature of part of it. This reflective surface causes animals' eyes to reflect light and appears to enhance night vision by reflecting some light back into the retina.

ASSIGNMENT

Complete Section E on the laboratory report.

Physiology of Vision

When light rays are focused on the retina by the cornea and crystalline lens, they stimulate light-sensitive cells to form impulses that are carried to the brain. The visual center in the brain interprets these impulses as vision. The retina contains two types of photosensitive cells—rods and cones—that are named for their shape. **Rods** are about twenty times more abundant than cones. They are very sensitive to light and enable us to see in very dim light. However, they only provide black and white vision. **Cones** require much more light to function but enable us to see in color. The fovea centralis, the area of the retina where light rays are focused for the sharpest, direct vision, contains only cones. The density of cones decreases and the density of rods increases with distance from the fovea centralis.

Rods contain **rhodopsin,** a photosensitive pigment composed of **opsin** and **retinal** in a loose chemical combination. When light strikes a rod cell, rhodopsin is broken down into opsin and retinal, and impulses are formed and sent to the brain. Rhodopsin must be reformed in order for the rod cell to respond to another light stimulus. The reformation of retinal requires vitamin A, and the recombination of opsin and retinal requires energy from ATP. The reformation of rhodopsin may take a few seconds, especially after exposure to very bright light.

Light rays stimulate impulse formation in cone cells in similar photochemical reactions. There are three types of cone cells, each with an opsin that is sensitive to a different color of light: blue, green, or red. The degree of stimulation of the three types of cones enables the wide range of color vision that we perceive.

ASSIGNMENT

1. Examine a prepared slide of retina sectioned through the fovea centralis. Compare your observations with Figure HA-13 in the Histology Atlas. The rods and cones are located in the outermost layer of the retina next to the choroid coat. Thus, light must pass through neuron processes, the ganglion cell layer, the bipolar cell layer, and the nuclear layer of rods and cones to reach their photosensitive tips.
2. Complete Section F on the laboratory report.

Visual Tests

The tests that follow will enable you to understand more about the functioning of the eye. Record the data for your eyes on your laboratory report.

The Blind Spot

Neither rods nor cones are present in the optic disk. Thus, light rays focused on this area produce no image. Determine the presence of the blind spot as follows.

1. Close your left eye and hold Figure 25.5 about 50 cm (20″) from your face while staring at the cross with the right eye. Note that you can see both the cross and the dot.
2. While staring at the cross, slowly move the figure closer to the eye and watch for the dot to disappear.
3. At the point where the dot is not visible, have your partner measure and record the distance from your eye to the figure.
4. Repeat the process for the left eye but stare at the dot and watch for the cross to disappear. Record the distance for your left eye.

Note that when the dot or cross disappear, they are not replaced with a black blotch, as might be expected. This is because the brain uses the image of

Figure 25.5 The blind spot test.

the surrounding area to fill the blind spot with a similar, but an imaginary, image.

Near Point Determination

The shortest distance from your eye that an object is in focus is known as the **near point.** To focus on close objects, the contraction of the ciliary body relaxes the suspensory ligaments allowing the lens to take a more spherical shape. The ability for the lens to do this depends upon its elasticity, which deteriorates with age. At age 20, the near point is about 3.5"; at age 40, it is 6.8"; at age 60, it is 33". After age 60, elasticity is minimal, a condition called **presbyopia.** Determine your near point as follows.

1. Close one eye and focus on this letter: T; gradually move the page closer to your face until the letter is blurred. Then move it away until you get a clear image. At this point, have your partner measure the distance from the page to your eye. Record this distance as the near point.
2. Determine and record the near point for the other eye.

Visual Acuity

The size of the cones and the distance between them in the fovea centralis determine the maximum acuity. Thus, it is possible for a "perfect" human eye to differentiate two points that are only one millimeter apart from a distance of 10 meters. Usually, poor acuity is due to incorrect focusing of light rays on the retina.

The Snellen eye chart (Figure 25.6) was developed to measure visual acuity. It is printed with letters of various sizes. When you stand 20 feet from the chart and can read the letters designated to be read at 20 feet, you have 20/20 vision. If you can read only the letters designated to be read at 200 feet, you have 20/200 vision. Determine your visual acuity as follows.

1. Stand 20 feet from the Snellen eye chart posted in the laboratory while your partner stands by the chart, indicates the lines to be read, and evaluates your accuracy.
2. Test each eye independently while covering the other eye. If you use corrective lenses, test your acuity with and without them. Record your results.

Astigmatism

An uneven curvature of the lens or cornea prevents some light rays from being sharply focused on the retina. The result is **astigmatism,** a condition in which an image will be blurred in one axis and sharp in other axes. Proceed with the test as follows.

Figure 25.6 The Snellen eye chart.

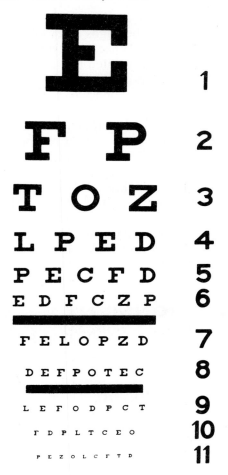

Using one eye at a time, stare at the center of Figure 25.7 and determine if all radiating lines are in focus and have the same intensity of blackness. If so, no astigmatism is present. If not, record the number(s) of the lines that are blurred or less dark.

Accommodation to Light Intensity

The diameter of the pupil may vary from 10 mm in darkness to only 1.5 mm in bright light. Sudden exposure of the retina to bright light causes an immediate and proportionate constriction of the pupil. In this reflex, impulses from the retina are carried via the optic nerve to the brain and return to the iris to cause its contraction.

Use a subject with light-colored eyes to test this reflex and perform it in dim light. Hold an unlighted desk lamp about 6 inches from the left eye. Have the subject look at the wall across the room, not at the lamp. While watching the pupil of the left eye, turn the light on for one second and then turn it off. Did the pupil constrict?

Figure 25.7 Astigmatism test chart.

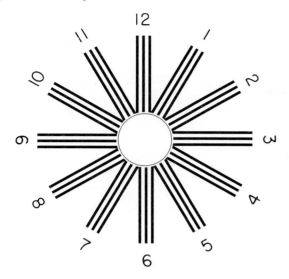

Now have the subject hold the edge of this manual against the forehead and extending down along the nose to exclude light from one side of the face. Repeat the test watching the right eye while shining light on the left eye. Record the results. What does this test tell you about the pathway of the impulses?

Color Blindness

The perception of color by the brain results from its receiving impulses from color-sensitive photoreceptors in the retina called **cones.** There are three types of cones, and each type responds maximally to a different color of light: red, blue, or green. The degree of stimulation of each type of cone determines the color that is perceived by the brain.

When the retina is exposed to *red* monochromatic light, the red-sensitive cones are stimulated 75%, the green-sensitive cones are stimulated 13%, and the blue-sensitive cones not at all. This ratio of stimulation, 75:13:0, is interpreted by the brain as red color.

When *blue* monochromatic light strikes the retina, red-sensitive cones are not stimulated, green-sensitive cones are activated at 14%, and blue-sensitive cones are stimulated 86%. The ratio of 0:14:86 is interpreted by the brain as blue light.

When *green* monochromatic light strikes the retina, the ratio of 50:85:15 is interpreted by the brain as green color. White light, composed of all colors of the spectrum, stimulates the cones equally.

Color blindness is a sex-linked hereditary trait affecting 8% of the male population and 0.5% of females. The most common type is red-green color blindness, in which either red-sensitive cones or green-sensitive cones are absent. People with these conditions have difficulty distinguishing reds and greens.

Test your color vision by viewing the Ishihara color-blindness test plates (*Ishihara's Tests for Colour Blindness,* Concise Edition). Start with plate 1 and progress through plate 14. Examine the samples of these test plates shown in Figure 25.8 so you will know what to expect. Your partner is to hold the test plates in good light about 30 inches from your eyes. Your partner is to record your responses on *your* laboratory report. Then, reverse the roles and repeat the process.

 ASSIGNMENT

Complete the laboratory report.

Figure 25.8 Ishihara color test plates.

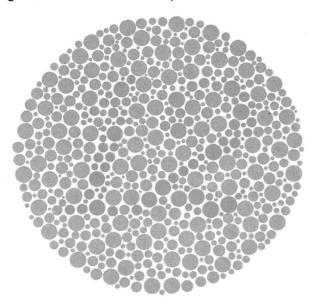

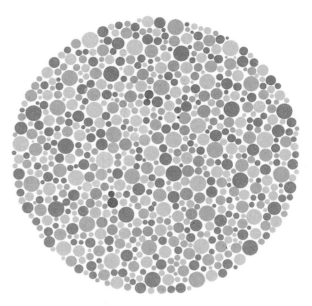

1. This color plate is nonselective. All individuals, normal or color-blind, will read this plate as 12.

2. Normal individuals read this plate as 8. Red/green color-blind individuals see a 3 here.

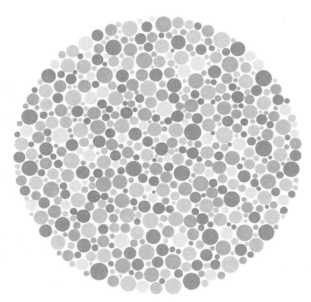

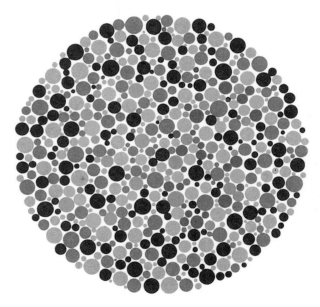

3. A normal person sees a 16 here. Color-deficient individuals read this plate incorrectly or not at all.

4. Normal individuals read a 96 here. Persons lacking red-sensitive cones see only a 6. Those lacking green-sensitive cones see only a 9.

These patterns have been reproduced from *Ishihara's Tests for Colour Blindness* published by Kanehara & Co., Ltd., Tokyo, Japan. Due to reproduction inaccuracies, these color replicas should not be used for accurate color-blind testing.

Exercise 26

THE EAR

After completing this exercise, you should be able to:

1. Identify the parts of the ear on charts and models, and describe their functions.
2. Perform and interpret simple tests of ear function.
3. Define all terms in bold print.

Materials
Tuning fork
Cotton earplugs
Prepared slides of cochlea, x.s., and crista ampullaris

In this exercise, you will study the anatomy of the human ear and its involvement in both hearing and equilibrium. The ear is anatomically divided into three parts: the external ear, the middle ear, and the inner ear. Figure 26.1 illustrates the basic anatomy of the ear.

The External Ear

The external ear consists of two parts: the auricle and the external auditory meatus. The **auricle,** sometimes called the **pinna,** is the outer shell-like structure of skin and elastic cartilage that is attached to the side of the head. The **external auditory meatus** is the canal that extends from the auricle about an inch into the temporal bone and terminates at the **tympanic membrane,** or **tympanum.** The skin lining the canal contains some small hairs and modified apocrine sweat glands that produce a waxy secretion called cerumen. The auricle collects sound waves and directs them through the external auditory meatus to the tympanum.

The Middle Ear

The middle ear consists of a small cavity within the temporal bone. This air-filled cavity lies between the tympanum and the inner ear, and it contains three small bones, the ear **ossicles.** These tiny bones form a sensitive lever system that transmits the vibrations of the tympanum to the fluids within the inner ear.

The outermost ossicle, the **malleus,** is a hammer-shaped or club-shaped bone that is attached to the tympanum by the end of its "handle." The middle ossicle is an anvil-shaped bone called the **incus.** The innermost ossicle, the **stapes,** is a stirrup-shaped bone whose "footplate" fits into the **oval window** of the inner ear. Because of the linkage of the ossicles, vibrations of the tympanum cause the stapes to move in and out of the oval window.

The **auditory tube** leads downward from the middle ear to the nasopharynx. This slender duct allows the air pressure in the middle ear to be equalized with that of the outside atmosphere. A valve at the nasopharynx end of the tube normally keeps the tube closed. However, yawning or swallowing temporarily opens the valve to allow air pressure equalization.

The Inner Ear

The inner ear consists of a **bony labyrinth,** shown in illustration B, Figure 26.1, that is the hollowed-out portion of the temporal bone. Within the bony labyrinth is a **membranous labyrinth** illustrated in Figure 26.2. The space between the bony and membranous labyrinths is filled with a fluid, the **perilymph.** The fluid within the membranous labyrinth is the **endolymph.** These fluids play important roles in hearing and equilibrium.

Note that the bony labyrinth consists of three semicircular canals, a vestibule, and the cochlea. The **vestibule** (label 13) is the portion that contains the oval window. The **semicircular canals** branch off one end of the vestibule, and the **cochlea** is the coiled extension at the other end.

Two branches of the vestibulocochlear nerve, the eighth cranial nerve, carry impulses to the brain from the inner ear. The **vestibular nerve** leads from the vestibule, and the **cochlear nerve** emanates from the cochlea.

The Cochlea

The bony labyrinth of the cochlea contains three chambers extending along its full length. See the cross section in illustration C, Figure 26.1. The upper chamber, the **scala vestibuli** (label 17), is continuous with the vestibule. The lower chamber, the **scala tympani,** is so-named because it contains the membranous

Figure 26.1 Anatomy of the ear.

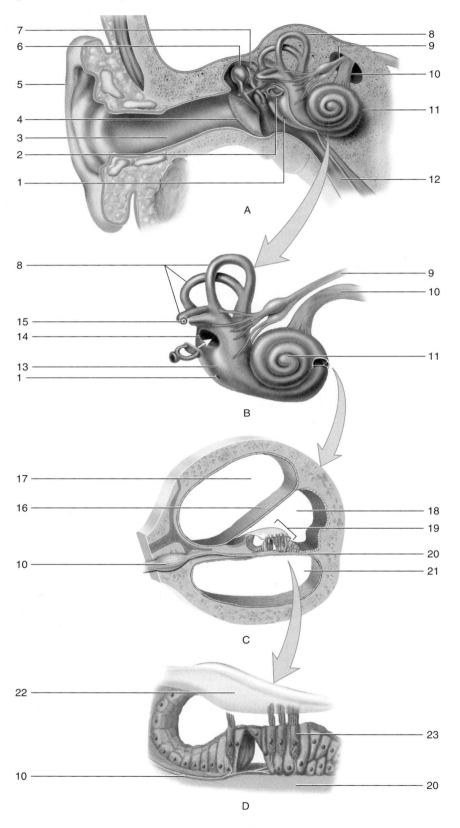

	Auditory Tube
	Auricle (Pinna)
	Basilar Membrane (2 places)
	Cochlea (2 places)
	Cochlear Duct
	Cochlear Nerve Fibers (4 places)
	External Auditory Meatus
	Hair Cells
	Incus
	Malleus
	Membranous Labyrinth
	Oval Window
	Round Window (2 places)
	Scala Tympani
	Scala Vestibuli
	Semicircular Canals (2 places)
	Spiral Organ (Organ of Corti)
	Stapes
	Tectorial Membrane
	Tympanic Membrane
	Vestibular Membrane
	Vestibular Nerve (2 places)
	Vestibule

Figure 26.2 The membranous labyrinth.

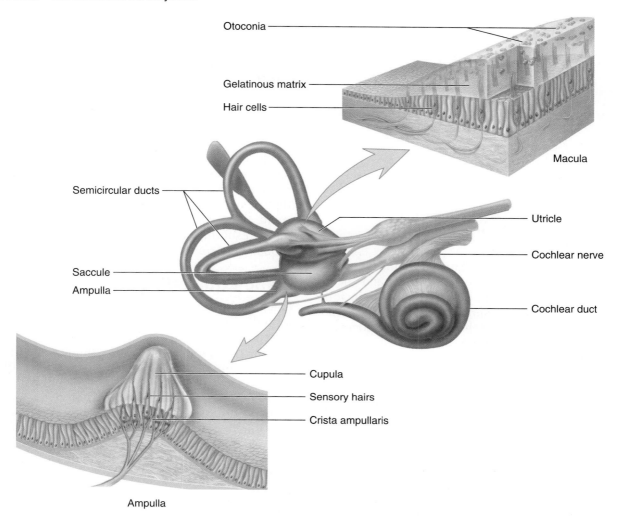

Otoconia

Gelatinous matrix

Hair cells

Macula

Semicircular ducts

Utricle

Cochlear nerve

Saccule

Ampulla

Cochlear duct

Cupula

Sensory hairs

Crista ampullaris

Ampulla

round window (label 1). Both of these chambers are filled with perilymph. The middle chamber, the **cochlear duct,** is bounded by the **vestibular membrane** above and the **basilar membrane** below. It is filled with endolymph. On the upper surface of the basilar membrane is the **spiral organ,** or **organ of Corti,** which contains **hair cells** that act as receptors for sound stimuli.

The detail of the organ of Corti is shown in illustration D. The hair cells sit on top of the basilar membrane with the tips of the hairs embedded in a gelatinous flap, the **tectorial membrane.** Neuron fibers leading from the hair cells become part of the cochlear nerve.

Vestibular Apparatus

Figure 26.2 shows the membranous labyrinth and the details of the vestibular apparatus that consists of the saccule, utricle, and semicircular canals. The **saccule** and **utricle** are located within the vestibule.

Sensory structures, the maculae, are located on the inner walls of these structures. Each **macula** contains hair cells that act as receptors for gravitational stimuli. The hairs are embedded in a gelatinous matrix that also contains crystals of calcium carbonate called **otoconia.** Neuron fibers leading from the hair cells become part of the vestibular nerve.

The hair cells that are receptors for dynamic equilibrium occur in the **ampullae,** the dilated basal portions of the semicircular canals. These hair cells are arranged in a crest, the **crista ampullaris,** within each ampulla. The hairs are embedded in a gelatinous mass called the **cupula.** Neuron fibers leading from the hair cells join the vestibular nerve.

ASSIGNMENT

Label Figure 26.1.

Figure 26.3 Pathway of sound wave transmission in the ear.

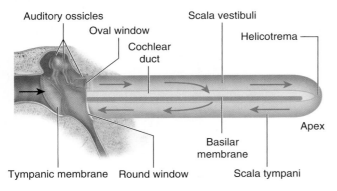

The Physiology of Hearing

Hearing is dependent upon (1) the conduction of sound waves from the tympanum to the inner ear by the ear ossicles, (2) the stimulation of hair cells in the organ of Corti, and (3) the formation of impulses and their transmission to the auditory centers of the brain for interpretation. Figure 26.3 shows the cochlea uncoiled to reveal the relationships of the inner ear components involved in sound wave transmission.

Transmission to the Inner Ear

Sound waves striking the tympanic membrane cause it to vibrate. These vibrations are transmitted to the perilymph of the inner ear by the stapes moving in and out of the oval window.

Activation of the Basilar Membrane

Note in Figure 26.3 that the scala vestibuli and scala tympani are contiguous at the apex of the cochlea. Since the perilymph cannot be compressed, the flexibility of the round window allows the fluid to move back and forth in these chambers in synchrony with the movements of the stapes. The pressure waves created by the sudden movements of the perilymph form bulges in the flexible cochlear duct, which activate the basilar membrane.

The basilar membrane contains about 20,000 transverse fibers of gradually increasing length, which are attached at one end to the bony center of the cochlea. Because they are unattached at the other end, they can vibrate like reeds of a harmonica to specific vibration frequencies. Fibers closest to the oval window are short (0.04 mm); they vibrate when stimulated by high-frequency sounds. Those at the apex of the cochlea are long (0.5 mm) and vibrate with low-frequency sounds. Because of the increasing lengths of these fibers, the human ear can detect sound frequencies from about 30 to 20,000 cycles per second.

Hair Cell Stimulation

When the fibers in a particular part of the basilar membrane vibrate in response to a specific sound frequency, the hairs of the hair cells above them are bent and stressed against the tectorial membrane. This action causes the hair cells to form impulses that are carried via the cochlear nerve to the brain, where they are interpreted as sound sensations.

Pitch and Loudness

The hair cells associated with each portion of the basilar membrane send impulses to slightly different portions of the brain. Thus, the detection of pitch is dependent upon the portion of the basilar membrane that is activated by a specific sound frequency and the part of the brain that receives the impulses.

Loudness is determined by the frequency of impulses received by the brain. The louder the sound, the greater the vibration of the basilar membrane, resulting in more hair cells sending a greater frequency of impulses to the brain.

Static Equilibrium

When the head is tilted in any direction, the otoconia are pulled by gravity, causing the gelatinous matrix to bend the hairs of the hair cells. Bending of the hairs increases the formation of impulses by the hair cells. The impulses are transmitted via the vestibular nerve to the brain.

Bending the head alone does not give a sense of malequilibrium because proprioceptors and the eyes send impulses to the brain to neutralize the effect of the vestibular impulses. Only when the whole body becomes disoriented are the vestibular impulses not inhibited.

Dynamic Equilibrium

Recall that the semicircular canals are oriented in the three planes of space. When the head moves in any direction, the semicircular canal in the direction of the movement moves in relation to the endolymph within it; i.e., the canal moves but the fluid is stationary. This movement causes the cupula to move in either direction because of the force of the endolymph against it. This movement bends the hairs of the hair cells and results in the generation of impulses that are carried via the vestibular nerve to the brain. Reception of these impulses by the brain initiates a reflex activating the appropriate muscles to maintain equilibrium.

ASSIGNMENT

1. Complete Sections A and B of the laboratory report.
2. Examine prepared slides of cochlea, x.s., and crista ampullaris, and compare your observations with Figures 26.1 and HA-14.

Figure 26.4 Tuning fork manipulations in hearing tests.

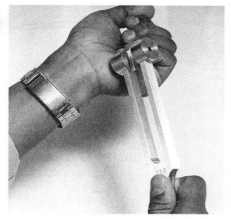

1. Tuning fork is activated by striking heel of hand.

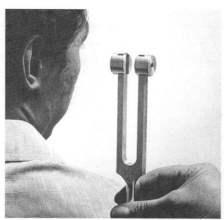

2. Rinne Test. Vibrating tuning fork is held six inches from ear.

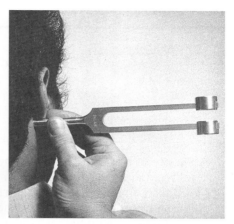

3. Rinne Test. Stem of tuning fork is placed on mastoid process.

Hearing Tests

There are two basic kinds of deafness. **Nerve deafness** results from damage to the cochlea, organ of Corti, cochlear nerve, or auditory center. It cannot be corrected. **Conduction deafness** results from damage to the tympanum or ear ossicles. It may be corrected by surgery or hearing aids. Perform the following hearing tests in a quiet room.

The Rinne Test

The Rinne test distinguishes between nerve deafness and conduction deafness.

1. Plug one of your ears with cotton.
2. Set the tuning fork in motion by striking it against the heel of the hand. Hold it about 6 inches from your unplugged ear with a tine facing the ear as shown in Figure 26.4. Persons with normal hearing and slight hearing loss will hear the sound, but those with a severe hearing loss will not hear the sound or will hear it only briefly.
3. As the sound fades to where it is no longer heard, place the stem of the tuning fork against the mastoid process behind your ear. If the sound reappears, conduction hearing loss is evident. If it does not, nerve hearing loss exists.
4. Test both ears and record the results.

The Weber Test

When the base of a vibrating tuning fork is placed against the center of the forehead of a person with normal hearing, the sound is heard with equal loudness in both ears. If conduction deafness exists in one ear, the sound will be louder in the defective ear than in the normal ear. This difference occurs because the defective ear is more attuned to sound waves conducted through the bone. If nerve deafness exists in one ear, the sound will be louder in the normal ear.

Place the base of a vibrating tuning fork against your forehead and compare the sounds in both ears. Record your results.

Static Balance Test

In addition to the vestibular apparatus, three other factors are important in maintaining equilibrium: (1) visual orientation, (2) proprioceptor sensations, and (3) cutaneous sensations. Impulses from these four sources are subconsciously integrated to provide correct reflex responses in maintaining equilibrium.

A simple way to evaluate the effectiveness of the static balance mechanism is to have the subject stand perfectly still with the eyes closed. If there is damage to the system, the subject will waiver and tend to fall.

1. Use a meterstick to draw a number of vertical lines on the chalkboard about 5 cm apart. Cover an area about one meter across. This chart will help detect body movements.
2. The subject is to remove his or her shoes and stand facing the examiner in front of the lined area.
3. Have the subject stand perfectly still for 30 seconds with arms at the sides and eyes open. Record the degree of swaying movement as small, moderate, or great.
4. Repeat the test with the subject's eyes closed. Record the results.
5. Repeat the test with the eyes closed while the subject stands on only one foot. What happens?

ASSIGNMENT

Complete the laboratory report.

Exercise 27

BLOOD TESTS

A variety of blood tests are routinely performed as part of a physical examination because they provide a significant amount of information about the health of a patient. This exercise contains four of the simpler blood tests that can easily be performed in a teaching laboratory. Although these tests have diagnostic value, our purpose is not to diagnose but to understand the purpose and nature of the tests. Your instructor will determine whether you are to use your own blood, simulated blood, or sterilized blood for these tests. If you perform these tests using your own blood, follow the precautions below and your instructor's directions precisely.

Precautions

In performing these tests, care must be taken to avoid (1) self-infection from environmental contaminants and (2) transmission of bloodborne infections, such as hepatitis and AIDS, from one person to another. Your instructor will determine whether you are to use your own blood or a blood substitute for each test. **These tests can be performed safely using your own blood if the following safeguards are rigidly applied.**

1. Wash your tabletop at the beginning of the laboratory period with a suitable disinfectant.
2. Wash your hands with soap and water before and after doing the blood tests.
3. **Avoid contact with blood of other students.** *Wear protective disposable gloves whenever there is a chance of such contact—for example, when piercing your partner's finger with a lancet.*
4. Before piercing the finger for a drop of blood, disinfect the skin with 70% alcohol.
5. Use disposable lancets only once and deposit them immediately into an autoclavable biohazard sharps container or a receptacle containing disinfectant as specified by your instructor. **Never** *place a used lancet on a tabletop or toss it into a wastebasket.*
6. Disposable microscope slides, glassware, paper towels, alcohol wipes, and similar items that may have been in contact with blood are to be placed *immediately* in the biohazard bag for autoclaving and disposal.
7. At the end of the laboratory period, wash the tabletop and equipment with a suitable disinfectant. Wash your hands with soap and water and rinse them with a disinfectant.

OBJECTIVES

After completing this exercise, you should be able to:

1. Identify the formed elements of blood when viewed microscopically or on charts.
2. Perform and interpret these tests: differential white cell count, packed red cell volume, coagulation (clotting) time, and blood typing.
3. Define all terms in bold print.

Materials

For each test:
  Alcohol pads
  Amphyl solution, 0.25%
  Biohazard bag
  Biohazard sharps container
  Household bleach solution, 10%
  Kimwipes
  Paper towels
  Sterile disposable lancets
  Protective disposable gloves
For white cell differential count:
  Bibulous paper
  Clean microscope slides
  Distilled water in dropping bottle
  Wax pencil
  Wright's blood stain in dropping bottle
  Prepared slides of a human blood smear
For volume of packed red cells:
  Heparinized capillary tubes
  Microhematocrit centrifuge
  Microhematocrit tube reader
  Seal-Ease (Clay Adams)

For clotting time:
  Capillary tubes, 0.5 mm diameter
  File, 3-cornered
For blood typing:
  Clean microscope slide
  Slide warming box with typing slide
  Toothpicks
  Typing sera: anti-A, anti-B, anti-D

Before You Proceed

Consult with your instructor about using protective disposable gloves when performing portions of this exercise.

Blood is the medium that transports substances to and from body cells. It consists of a fluid **plasma,** which forms 55% of the blood volume, and the **formed elements,** which constitute the remaining 45%. The formed elements consist of **erythrocytes** (red blood cells), **leukocytes** (white blood cells), and **thrombocytes** (platelets). Their relative abundance and functions are shown in Table 27.1. They are illustrated in Figure 27.1 and shown in photomicrographs in Figure 27.2 as they appear when stained with Wright's blood stain. Learn to recognize the characteristics of each type of formed element.

Caution: *Follow infection control procedures prescribed by your instructor when performing the blood tests that follow.* Record your results of the tests on the laboratory report.

TABLE 27.1
Formed Elements of the Blood

Formed Element	Concentration	Function
Erythrocytes	4,000,000–6,000,000 per cubic millimeter	Transport oxygen and carbon dioxide
Leukocytes	5,000–10,000 per cubic millimeter	Destroy pathogens; neutralize toxins
Granulocytes		
Neutrophils	50%–70% of leukocytes	Phagocytosis
Eosinophils	1%–3% of leukocytes	Neutralize products of allergic reactions; destroy parasitic worms
Basophils	0.5%–1% of leukocytes	Release heparin and histamine; intensify inflammation and allergic reactions
Agranulocytes		
Lymphocytes	20%–30% of leukocytes	B lymphocytes form antibodies
		T lymphocytes form chemicals that destroy antigens and/or stimulate phagocytosis
Monocytes	2%–6% of leukocytes	Phagocytosis
Thrombocytes	250,000–400,000 per cubic millimeter	Initiate clotting process

Figure 27.1 Formed elements of blood.

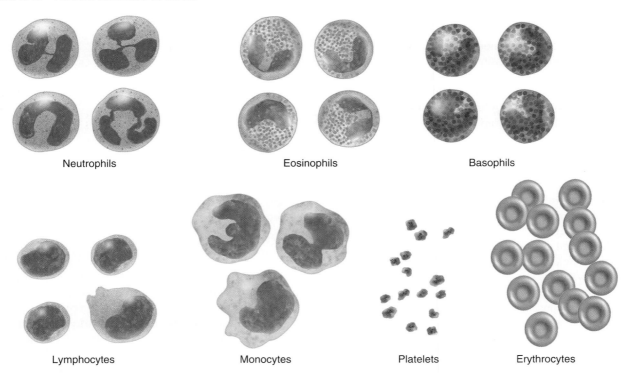

Neutrophils   Eosinophils   Basophils

Lymphocytes   Monocytes   Platelets   Erythrocytes

The Differential White Blood Cell Count

The differential white blood cell (WBC) count is performed to determine the relative percentages of the five types of white blood cells shown in Figure 27.1. Three of them, **neutrophils, eosinophils,** and **basophils,** have cytoplasmic granules and are classified as **granulocytes. Monocytes** and **lymphocytes** lack cytoplasmic granules and are classified as **agranulocytes.** High or low counts for certain types of white blood cells may be related to pathological conditions. High neutrophil counts usually indicate bacterial infection. High eosinophil counts occur in allergic reactions or parasitic worm infestations. High monocyte counts usually indicate a chronic infection. High lymphocyte counts usually indicate antigen-antibody reactions.

A differential WBC count is performed by preparing a stained blood slide and counting a total of 100 white cells to determine the percentage of each type of cell.

Slide Preparation

If your instructor wants you to use a prepared slide of a human blood smear, obtain the slide and skip down to the next section.

Keys to success in preparing a good slide are (1) placing the correct size drop of blood on the slide and (2) spreading the blood evenly over the slide.

1. Clean 3 or 4 slides with soap and water, and keep them free of fingerprints.
2. Cleanse a fingertip with an alcohol pad and pierce it from the side with a lancet. Place the lancet *immediately* in the biohazard bag. Place a drop of blood 2 cm from one end of a slide. *The drop should be 3–4 mm in diameter.*
3. Spread the blood on the slide using another clean slide as a spreader as shown in Figure 27.3. Note that the blood is *pulled* along by the slide. Hold the spreader slide at an angle of 50° to obtain a good smear. Place the spreader slide directly in the biohazard bag.
4. After the blood has dried, use a wax pencil to draw a line at each end of the smear to confine the stain.
5. Place the slide on a folded paper towel and add Wright's blood stain, *counting the drops* until the smear is covered.
6. Wait *4 minutes,* then add an equal number of drops of distilled water. Let it stand for *10 minutes* while blowing on the slide every few minutes to keep the solutions mixed.
7. Gently rinse the slide in slowly running water. Blot the slide dry with bibulous paper. When the slide is dry, make the count as described in the next section.

Figure 27.2   Photomicrographs of formed elements in blood (5000×).

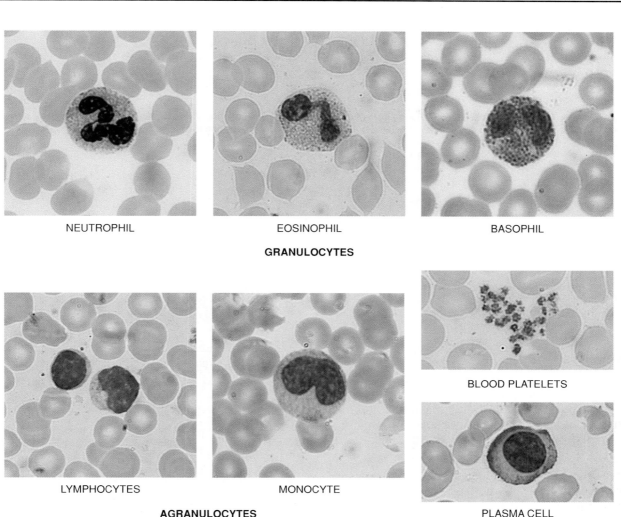

NEUTROPHIL   EOSINOPHIL   BASOPHIL

GRANULOCYTES

BLOOD PLATELETS

LYMPHOCYTES   MONOCYTE

AGRANULOCYTES

PLASMA CELL
(abnormal)

Neutrophils differ from the other two granulocytes in having small lavender granules in the cytoplasm. Nuclei of these leukocytes are characteristically lobulated with thin bridges between the lobules (see figure 27.1). During infections, when many of these are being produced in the bone marrow, the nuclei are horseshoe-shaped; these juvenile cells are often called "stab cells."

Eosinophils, generally, have bilobed nuclei as above; however, juveniles have nuclei that are horseshoe-shaped (figure 27.1). Note that the red eosinophilic granules in the cytoplasm are large and pronounced.

Basophils, which are few in number, are distinguished from the other granulocytes by the presence of dark blue basophilic granules in the cytoplasm. The nuclei of these cells are large and varied in shape. The fact that heparin has been identified in these cells indicates that they probably secrete this substance into the blood.

Two sizes of lymphocytes—large and small—are shown above and in figure 27.1. The small ones, which are more abundant than the large ones, have large, dense nuclei surrounded by a thin layer of basophilic cytoplasm. The large lymphocytes have indented nuclei and more cytoplasm than small lymphocytes.

Monocytes are the largest cells found in normal blood. The nucleus may be ovoid, indented, or horse-shoe shaped, as illustrated in figure 27.1. Monocytes have considerably more cytoplasm than lymphocytes and are voracious phagocytes.

Blood platelets, which form from megakaryocytes in the bone marrow, function in the formation of blood clots.

Plasma cells are only rarely seen in the blood. The photomicrograph above was made from the blood of a patient with plasma cell leukemia. Although they resemble lymphocytes, the plasma cells have more cytoplasm that is very basophilic.

Figure 27.3 Smear preparation technique for differential WBC count.

1. A small drop of blood is placed about 3/4" away from one end of slide. The drop should not exceed 1/8" diameter.

2. The spreader slide is moved in direction of arrow, allowing drop of blood to spread along slide's back edge.

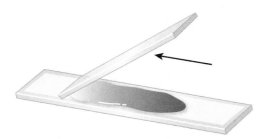

3. The spreader slide is moved along the slide, dragging the blood over the surface of the slide.

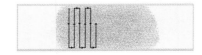

4. A china marking pencil is used to mark off both ends of the smear to retain the staining solution on the slide.

Making the Count

Whether you are using a prepared slide or one that you have just stained, the counting procedure is the same. Ideally, one should use an oil immersion lens for this procedure; high-dry optics can be used, but they are not as reliable for best results. Proceed as follows:

1. Scan the slide with the low power objective to find an area where cell distribution is best. A good area is one in which the cells are not jammed together or scattered too far apart.
2. Once an ideal area has been selected, place a drop of immersion oil on the slide near one edge and lower the oil immersion objective into the oil. If you are using a slide you have just prepared you needn't worry about oil being placed directly on the stained smear.

   If the high-dry objective is to be used, omit placing oil on the slide.
3. Systematically scan the slide, following the pathway indicated in Figure 27.4. As each leukocyte is encountered identify it, using Figures 27.1 and 27.2 for reference.
4. Record on the laboratory report the type of each white cell encountered until you have tabulated 100 cells.
5. When finished, either place the smear that you made in the biohazard bag or return the prepared slide to its tray.

Figure 27.4 Examination path for differential count.

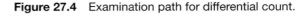

Volume of Packed Red Cells

Hemoglobin, the red blood pigment, is the oxygen-carrying component in erythrocytes. A person with a reduced concentration of hemoglobin is said to be anemic. One of the causes of anemia is a subnormal erythrocyte concentration.

One test used to determine the concentration of erythrocytes in the blood is called the **VPRC,** or **volume of packed red cells** (hematocrit). This test determines the volume percentage of the blood occupied by red blood cells. It is performed by using a microhematocrit centrifuge and heparinized microhematocrit capillary tubes.

In males, the normal range is 40%–54%; 47% is average. In females, the normal range is 37%–47%; 42% is average. Perform the test as described below and shown in Figure 27.5.

1. Cleanse and pierce the fingertip as before and place the lancet in the biohazard bag. Wipe away the first drop of blood with the alcohol pad.

Figure 27.5 Determination of the volume of packed red cells.

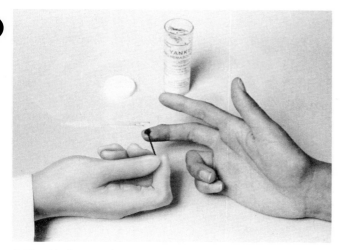

1. Blood is drawn up into heparinized capillary tube for hematocrit determination.

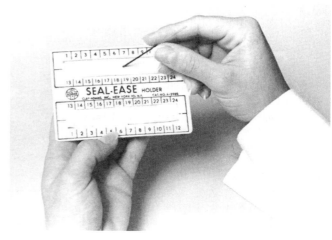

2. The end of the capillary tube is sealed with clay.

3. The capillary tubes are placed in the centrifuge with the sealed end toward the perimeter.

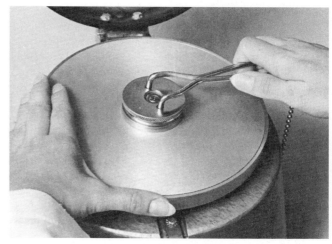

4. The safety lid is tightened with a lock wrench.

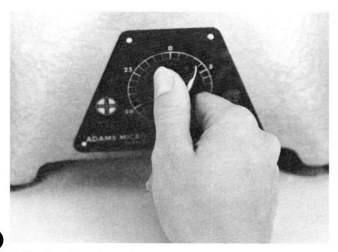

5. The timer is set for 4 minutes by turning the dial clockwise.

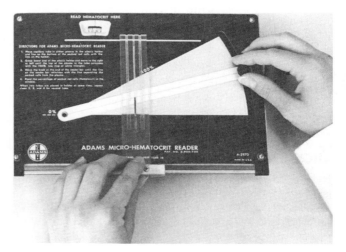

6. The capillary tube is placed in the tube reader to determine the hematocrit.

2. Place the red end of the heparinized capillary tube in the second drop of blood and lower the other end so the blood will fill about two-thirds of the tube.

3. Insert the blood end of the tube into Seal-Ease to plug it.

4. Place the tube in the centrifuge with the sealed end against the ring at the perimeter. Record your tube slot number. Load the centrifuge with an even number of tubes that are arranged opposite each other to balance the load.

5. Secure the inside cover and fasten the outside cover.

6. Turn on the centrifuge and set the timer for *4 minutes*.

7. Determine the VPRC by using the mechanical tube reader. Instructions are on the instrument. The VPRC may be read on the head of some centrifuges.

8. Place blood-contaminated materials in the biohazard bag.

Clotting Time

Coagulation of the blood is a complex process that culminates in the conversion of a soluble protein, **fibrinogen,** into an insoluble fibrous protein, **fibrin.** Normal clotting time is 3 to 6 minutes. Figure 27.6 shows the basic procedures. Determine your clotting time as follows.

1. Cleanse and pierce the fingertip and place the lancet in the biohazard bag. *Record the time.* Place one end of a capillary tube into the drop of blood and lower the other end, allowing blood to fill it completely.

2. At *1-minute intervals* break off small segments of the capillary tube by scratching the glass first with the file. Place the tube on a paper towel when scoring it with the file. *Separate the broken ends slowly while looking for strands of fibrin between them.* Record the clotting time as the time from the first appearance of blood on your finger to the formation of fibrin strands.

3. Place the tube fragments and paper towel in the biohazard bag.

Blood Typing

Red blood cells possess a variety of antigens on their surfaces. The presence or absence of some of these antigens is used to determine one's blood type, which is genetically controlled and never changes throughout the lifetime of an individual.

An **antigen** is a chemical molecule that stimulates the production of a specific **antibody,** which, in

Figure 27.6 Determination of clotting time.

1. Blood is drawn up into a nonheparinized capillary tube.

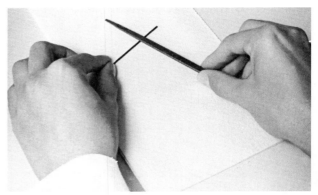

2. At 1-minute intervals, sections of the tube are filled and broken off for the test.

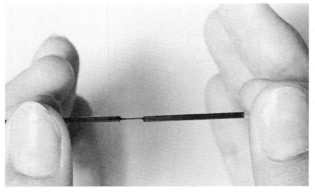

3. When the blood has coagulated, strands of fibrin will extend between broken ends of tube.

turn, will attach to the antigen. Although there are many antigens on the red blood cells, it is the **A, B,** and **D (Rh)** antigens that are most commonly used in blood typing. In this exercise, we will study blood typing using the A, B, and D antigens to determine your ABO and Rh blood type.

Normally, transfusions involve a donor and a recipient with the same blood type. However, it may be possible for a recipient to receive a transfusion of a different blood type in emergency situations.

Not all blood types are compatible, and a transfusion of incompatible blood may be fatal. Thus, typing the blood of both the potential donor and the recipient is essential. Because of differences in concentration, it is the *antigens of the donor* and the *antibodies of the recipient* that must be considered. Table 27.2 shows the relationship of antigens and antibodies in normal blood. By understanding this table and the antigen-antibody reaction, you can predict the compatibility of various blood type combinations.

In the clinical setting, blood is usually diluted with saline when doing ABO blood typing. However, in this exercise, you will use whole blood because you will be determining the ABO group and Rh type simultaneously. Rh typing requires whole blood and a slide warming box to yield a temperature of about 50° C.

You will type your blood by mixing a drop of blood with a drop of **antiserum** that contains specific, known antibodies. If the corresponding antigen

TABLE 27.2
Antigen and Antibody Associations in Blood

Blood Type	Antigen	Antibody
O	none	a,b
A	A	b
B	B	a
AB	A,B	none
Rh+	D	none
Rh−	none	d*

*Antibodies are formed by an Rh− person only after Rh+ erythrocytes enter his or her blood.

is present on the red blood cells, an antigen-antibody reaction occurs causing the erythrocytes to **agglutinate** (clump together). This clumping can be identified visually.

If you are to use simulated blood, follow the directions of your instructor. The procedure for determining your ABO and Rh blood type using your own blood is described below and shown in Figure 27.7. *Be sure to follow the safety precautions.*

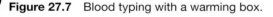

Figure 27.7 Blood typing with a warming box.

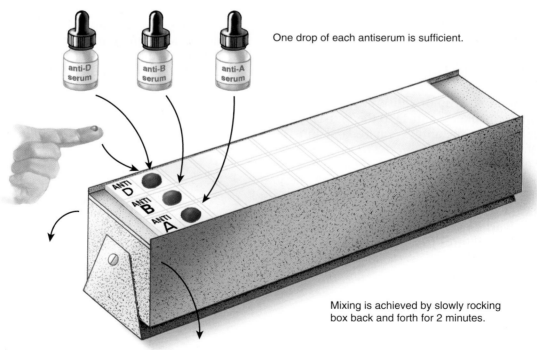

One drop of each antiserum is sufficient.

anti-D serum

anti-B serum

anti-A serum

ANTI D

ANTI B

ANTI A

Mixing is achieved by slowly rocking box back and forth for 2 minutes.

Figure 27.8 Interpretation of agglutination patterns.

Anti-A + Blood	Anti-B + Blood	Anti-D + Blood

Type O — Rh⁻ (above)

Type A — Rh⁺

Type B

Type AB

1. Place a clean microscope slide across the three typing squares on the typing slide of the warming box. Allow 5 minutes for temperature equilibration.
2. Cleanse and pierce a fingertip. Discard the lancet in the biohazard bag. Place a small drop of blood on the slide over each of the three squares.
3. Quickly add one drop of antiserum to the blood in each square: anti-D to the anti-D square, anti-A to the anti-A square, and anti-B to the anti-B square. *Record the time.*
4. Mix the antiserum and blood in each square *using a separate clean toothpick for each one.* Why? Place the toothpicks directly into the biohazard bag.
5. Rock the warming box for *2 minutes,* then examine the mixtures for agglutination,

starting with the anti-D square. Clumps in the anti-D square must appear within 2 minutes to be valid. These clumps will be very small and require close scrutiny for detection. Agglutination in any square indicates the presence of the antigen in question. For example, clumping in the anti-A square indicates presence of the A antigen. See Figure 27.8.
6. Place the slide in the biohazard bag. Clean your workstation with 0.25% Amphyl solution.

ASSIGNMENT

Complete the laboratory report.

THE HEART

OBJECTIVES

After completing this exercise, you should be able to:

1. Identify the parts of the heart on preserved specimens, charts, or models.
2. Describe the functions of each part of the heart.
3. Define all terms in bold print.

Materials

Sheep heart, fresh or preserved
Dissecting instruments and tray
Protective disposable gloves

 Before You Proceed

Consult with your instructor about using protective disposable gloves when performing portions of this exercise.

The heart is a double pump that circulates blood via arteries and veins throughout the body. The two pumps are located side by side, and each consists of two chambers: an atrium (receiving chamber) and a ventricle (pumping chamber). The **right atrium** and **right ventricle** compose one pump. It receives deoxygenated blood from the body and pumps it to the lungs. The **left atrium** and **left ventricle** form the other pump. It receives oxygenated blood from the lungs and pumps it to all other parts of the body. Like all pumps, the heart has valves that prevent a backflow of blood and keep blood flowing in the correct direction.

Internal Anatomy

With the basic function of the heart in mind, locate in Figure 28.1 the heart structures described below and correlate each structure with its function. Vessels colored red carry oxygenated blood; those colored blue carry deoxygenated blood.

Wall of the Heart

The wall of the heart consists of three layers. The thick **myocardium** is composed of cardiac muscle tissue. It is covered by two thin serous membranes attached to the muscle tissue: an inner **endocardium** and an outer **epicardium,** or **visceral pericardium.** External to the epicardium is the double-layered **parietal pericardium,** composed of an inner serous membrane and an outer fibrous membrane. The **pericardial cavity** (label 20) lies between the epicardium and parietal pericardium. Fluid secreted by the serous membranes reduces friction, enabling the heart to move freely within the pericardial sac.

Chambers of the Heart

The **left** and **right atria** are the upper, thin-walled chambers, which receive blood returning to the heart. Note that the right atrium is on the left side of the frontal section in the figure. There is no connection between the atria. A wall of tissue, the **interatrial septum** (not labeled) separates the two atria. During fetal life, an opening in the interatrial septum allows blood to flow between the atria. This opening is closed and sealed after birth, leaving only a slight depression, the **fossa ovalis** (not labeled) to mark the site. The lower, thick-walled chambers are the ventricles, which pump blood into the arteries leaving the heart. Note the **interventricular septum** separating the **left ventricle** from the **right ventricle,** and the thicker myocardium of the left ventricle.

The heart has two atrioventricular (AV) valves and two semilunar valves. The **tricuspid atrioventricular valve** has three flaps, or cusps, and it is located between the right atrium and ventricle. The **mitral,** or **bicuspid, atrioventricular valve** has only two cusps; it separates the left atrium and ventricle. The tricuspid and bicuspid atrioventricular valves are indicated in the superior sectional view, and they are shown but not labeled in the coronal section of the heart. Atrioventricular valves prevent a backflow of blood from the ventricles into the atria during **ventricular systole,** the contraction of the ventricles. The cusps have thin cords, the **chordae tendineae** (label 14) that anchor them to **papillary muscles** (label 15) on the walls of the ventricles and

Figure 28.1 Internal anatomy of the heart.

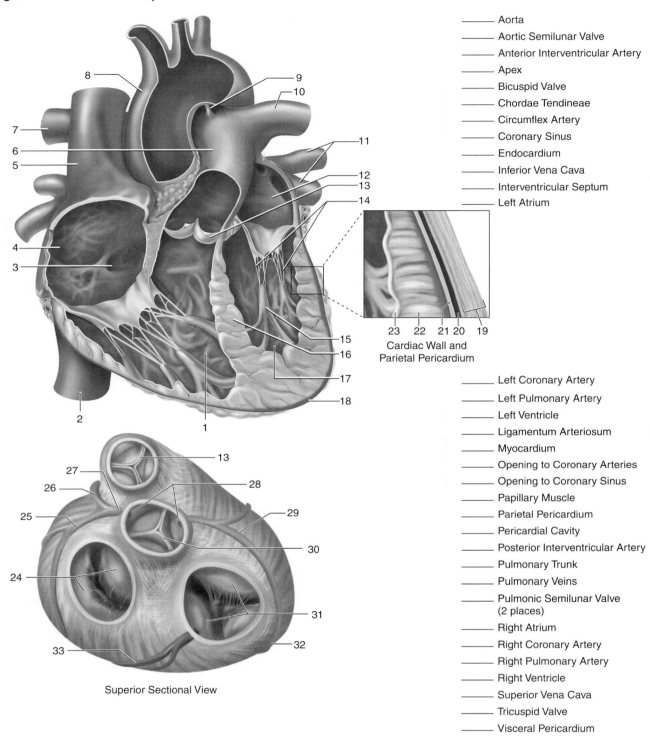

Cardiac Wall and
Parietal Pericardium

Superior Sectional View

_____ Aorta
_____ Aortic Semilunar Valve
_____ Anterior Interventricular Artery
_____ Apex
_____ Bicuspid Valve
_____ Chordae Tendineae
_____ Circumflex Artery
_____ Coronary Sinus
_____ Endocardium
_____ Inferior Vena Cava
_____ Interventricular Septum
_____ Left Atrium

_____ Left Coronary Artery
_____ Left Pulmonary Artery
_____ Left Ventricle
_____ Ligamentum Arteriosum
_____ Myocardium
_____ Opening to Coronary Arteries
_____ Opening to Coronary Sinus
_____ Papillary Muscle
_____ Parietal Pericardium
_____ Pericardial Cavity
_____ Posterior Interventricular Artery
_____ Pulmonary Trunk
_____ Pulmonary Veins
_____ Pulmonic Semilunar Valve
         (2 places)
_____ Right Atrium
_____ Right Coronary Artery
_____ Right Pulmonary Artery
_____ Right Ventricle
_____ Superior Vena Cava
_____ Tricuspid Valve
_____ Visceral Pericardium

prevent the valve cusps from being forced into the atria during contraction.

The **pulmonic semilunar valve** and the **aortic semilunar valve** are located at the base of the pulmonary trunk and aorta, respectively. They each have three cusps and prevent the backflow of blood from the arteries into the ventricles during **ventricular diastole,** the relaxation of the ventricles. The pulmonic semilunar valve is indicated in both the coronal section of the heart and the superior sectional view. The aortic semilunar valve is shown only in the superior sectional view.

Vessels of the Heart

The major vessels leading to and from the heart are the two venae cavae, four pulmonary veins, the coronary sinus, a pulmonary trunk, and the aorta. During diastole, blood is returned to the atria. Deoxygenated blood is returned to the right atrium from body regions above the heart by the **superior vena cava** (label 5) and from body regions below the heart by the **inferior vena cava.** Simultaneously, oxygenated blood is returned to the left atrium by the **left** and **right pulmonary veins** (label 11).

During systole, blood is pumped from the ventricles. Deoxygenated blood is carried from the right ventricle to the lungs via the **pulmonary trunk,** which branches to form the **left** and **right pulmonary arteries.** Simultaneously, the left ventricle pumps oxygenated blood to all parts of the body except the lungs via the **aorta** (label 8). The emergence of the aorta from the left ventricle is not visible in the figure.

The **ligamentum arteriosum** (label 9) is a nonfunctional remnant of the **ductus arteriosus** that allows blood to flow from the pulmonary artery to the aorta during fetal life.

Blood that flows through the heart does not provide oxygen and nutrients to the heart itself. This function is provided by the coronary circulation, which consists of coronary arteries and cardiac veins. In the superior sectional view, note that the **left coronary artery** (label 27) and **right coronary artery** branch from the aorta. Two **openings to the coronary arteries** (label 28) are located in the aorta just superior to the aortic semilunar valve. Each artery produces several branch arteries to serve the heart tissues. The superior sectional view shows that the left coronary artery branches to form the **circumflex artery** that encircles the superior portion of the left ventricle, and the **anterior interventricular artery** (label 26) that runs down the surface of the anterior interventricular septum as shown in Figure 28.2. The right coronary artery encircles the right ventricle giving off several branches. The

Figure 28.2   External anatomy of the heart.

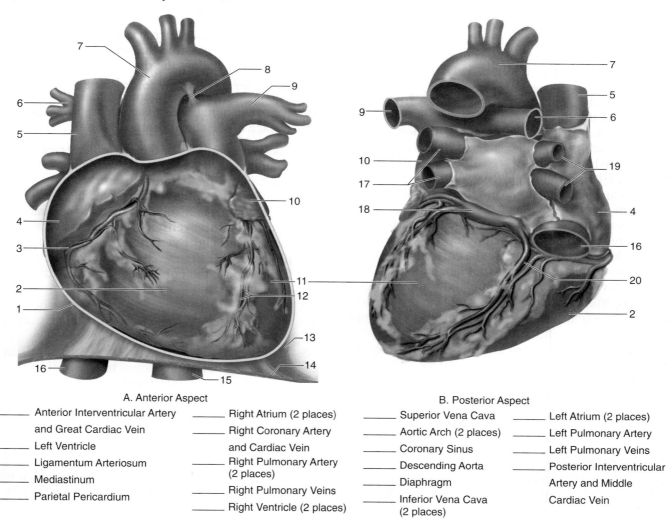

A. Anterior Aspect

_____ Anterior Interventricular Artery
        and Great Cardiac Vein
_____ Left Ventricle
_____ Ligamentum Arteriosum
_____ Mediastinum
_____ Parietal Pericardium

_____ Right Atrium (2 places)
_____ Right Coronary Artery
        and Cardiac Vein
_____ Right Pulmonary Artery
        (2 places)
_____ Right Pulmonary Veins
_____ Right Ventricle (2 places)

B. Posterior Aspect

_____ Superior Vena Cava
_____ Aortic Arch (2 places)
_____ Coronary Sinus
_____ Descending Aorta
_____ Diaphragm
_____ Inferior Vena Cava
        (2 places)

_____ Left Atrium (2 places)
_____ Left Pulmonary Artery
_____ Left Pulmonary Veins
_____ Posterior Interventricular
        Artery and Middle
        Cardiac Vein

Figure 29.3   Major veins of the body.

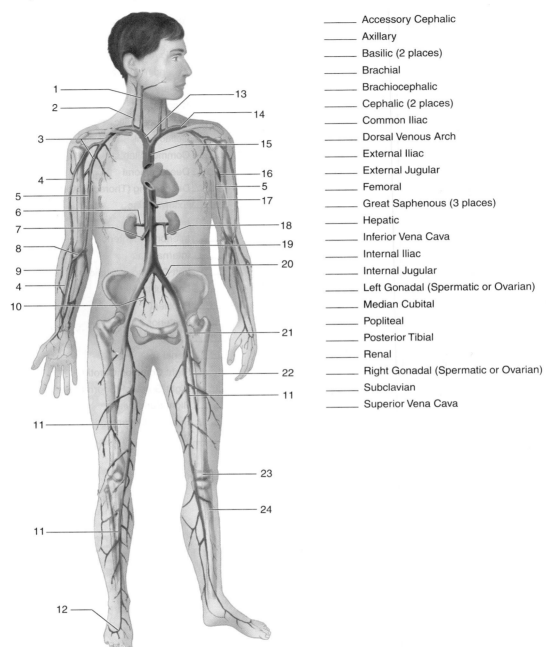

_____ Accessory Cephalic
_____ Axillary
_____ Basilic (2 places)
_____ Brachial
_____ Brachiocephalic
_____ Cephalic (2 places)
_____ Common Iliac
_____ Dorsal Venous Arch
_____ External Iliac
_____ External Jugular
_____ Femoral
_____ Great Saphenous (3 places)
_____ Hepatic
_____ Inferior Vena Cava
_____ Internal Iliac
_____ Internal Jugular
_____ Left Gonadal (Spermatic or Ovarian)
_____ Median Cubital
_____ Popliteal
_____ Posterior Tibial
_____ Renal
_____ Right Gonadal (Spermatic or Ovarian)
_____ Subclavian
_____ Superior Vena Cava

leg is the **great saphenous,** which originates at the **dorsal venous arch** (label 12) on the superior surface of the foot and ascends to join with the femoral vein to form the **external iliac** vein. The **internal iliac** (label 10) and external iliac unite to form the **common iliac** vein, which empties into the inferior end of the **inferior vena cava.**

Of the many small veins entering the inferior vena cava, only three are shown in Figure 29.3. The **hepatic** vein returns blood from the liver and is

located just below the heart. The two **renal** veins return blood from the kidneys. The branch from the left renal vein is the **left gonadal** vein, which carries blood from the left testis or ovary. The **right gonadal** enters the inferior vena cava just below the right renal vein.

The larger veins contain valves that prevent a backflow of blood and keep blood moving toward the heart. If the valves break down and the veins dilate, blood flow is reduced. This condition is

Figure 29.4 Hepatic portal circulation.

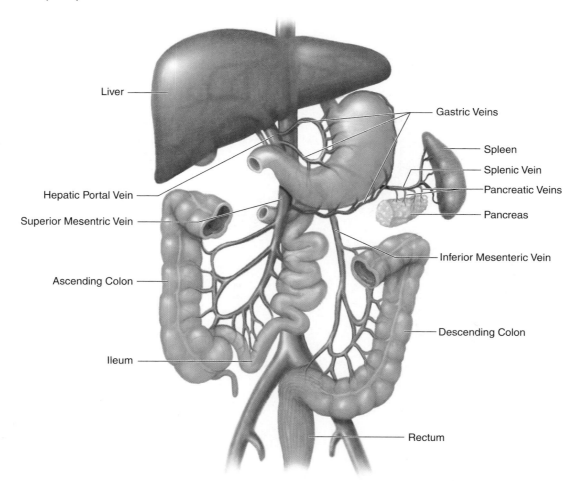

Liver

Gastric Veins

Spleen

Splenic Vein

Pancreatic Veins

Hepatic Portal Vein

Pancreas

Superior Mesentric Vein

Inferior Mesenteric Vein

Ascending Colon

Descending Colon

Ileum

Rectum

known as *varicose veins,* and it is promoted by aging and inactivity.

Hepatic Portal Circulation

A portal system consists of veins that carry blood from capillary beds in one or more organs to a capillary bed in another organ. Figure 29.4 shows that the **hepatic portal** vein receives venous blood from digestive organs and carries it to the liver. This pathway allows the liver to process nutrients absorbed from the digestive tract before they enter the general circulation. The hepatic portal vein receives blood from four major veins. Blood from the intestines is drained by the **superior mesenteric** and the **inferior mesenteric** veins. The **splenic** vein brings blood from the spleen, pancreas, and part of the stomach to join with the inferior mesenteric. The **gastric** veins serve most of the stomach. After flowing through capillary beds in the liver, blood from the liver enters the inferior vena cava via the **hepatic vein.**

ASSIGNMENT

1. Label Figures 29.2 and 29.3.
2. Demonstrate the presence of valves in veins by applying pressure with your fingertip to a prominent vein on the back of your hand near your wrist. Then, continue to apply pressure as you trace the vein toward your knuckles. Note that blood does not flow back to fill the vein until you release the pressure.
3. Examine a prepared slide of artery and vein, x.s. Note the thicker, more muscular wall of the artery.
4. Examine a prepared slide of an atherosclerotic artery, x.s. Locate the fatty plaque that partially plugs the vessel. High levels of blood cholesterol and genetic tendencies lead to the development of atherosclerosis. Make drawings of normal and atherosclerotic arteries and a normal vein in Section D on the laboratory report.
5. Complete Section B of the laboratory report.

Figure 29.5 Fetal circulation.

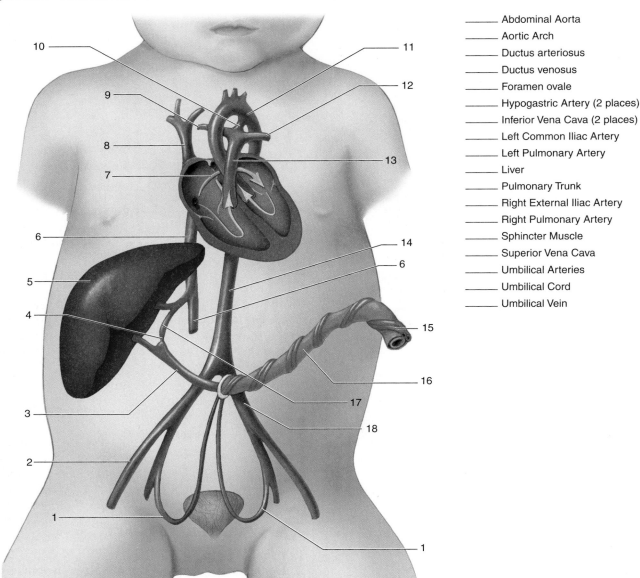

Abdominal Aorta
Aortic Arch
Ductus arteriosus
Ductus venosus
Foramen ovale
Hypogastric Artery (2 places)
Inferior Vena Cava (2 places)
Left Common Iliac Artery
Left Pulmonary Artery
Liver
Pulmonary Trunk
Right External Iliac Artery
Right Pulmonary Artery
Sphincter Muscle
Superior Vena Cava
Umbilical Arteries
Umbilical Cord
Umbilical Vein

Fetal Circulation

Circulation during fetal development is necessarily different from that of an adult. During fetal development, the placenta is the site of the exchange of nutrients and oxygen from maternal blood to fetal blood and, simultaneously, the transfer of wastes and carbon dioxide from fetal blood to maternal blood. Also, the lungs are nonfunctional and receive a reduced supply of blood.

Fetal circulation through the heart is shown in Figure 29.5. Vessels colored purple are carrying a mixture of oxygenated and deoxygenated blood.

Locate the red-colored **umbilical vein** (label 3), which brings oxygenated blood from the placenta to the fetus via the **umbilical cord** (label 16). The umbilical vein continues to the liver. Near the liver, the umbilical vein gives off a branch, the **ductus venosus,** that bypasses the liver to join a vein from the liver and merge with the inferior vena cava. This arrangement allows oxygen-rich blood to quickly enter the general circulation. Note that there is a small constriction at the beginning of the ductus venosus where a **sphincter muscle** is present in the vessel wall. The sphincter muscle remains open before birth to allow most of the blood to flow directly into the inferior vena cava, but it closes soon after birth.

Two **umbilical arteries** carrying blood to the placenta are wrapped around the umbilical vein in the umbilical cord. They are extensions of the **hypogastric arteries** (internal iliacs) that pass along each side of the urinary bladder.

Two additional fetal circulatory adaptations enable blood with relatively more oxygen to be delivered quickly to body cells. In the fetal heart, the **foramen ovale** is an opening between the right and left atria. It allows much of the blood entering the right atrium to flow into the left atrium and on into the left ventricle, which pumps blood to all parts of the body except the lungs. A smaller amount of blood from the right atrium enters the right ventricle, which pumps blood into the pulmonary artery. However, much of the blood in the pulmonary artery does not go to the lungs, but instead passes through the **ductus venosus** (label 10) into the aorta, increasing the amount of blood carried to body cells. Both of these fetal adaptations allow much of the blood entering the right ventricle to bypass the nonfunctional lungs, which aids the rapid transport of oxygen and nutrients to body cells.

When the fetus is separated from the placenta after birth, circulatory changes occur. The constriction and closure of the ductus venosus helps to decrease the loss of blood via the umbilical vein. It subsequently becomes the **ligamentum venosum** of the liver. The umbilical vein becomes nonfunctional and is converted into the **round ligament** extending from the umbilicus to the liver. The distal portions of the hypogastric arteries become fibrous cords, and the proximal portions remain functional and supply the urinary bladder.

As soon as the infant begins to breathe, more blood is channeled to the lungs. The decreased use of the ductus arteriosus leads to its closure and conversion into the **ligamentum arteriosum.** The resulting increase in blood volume and pressure in the left atrium closes the flap on the foramen ovale. Subsequently, connective tissue seals this opening to permanently separate the atria.

Failure of the foramen ovale or ductus arteriosus to close after birth results in a circulatory defect that allows continued mixing of oxygenated and deoxygenated blood. Such defects may deprive body tissues of adequate oxygen, seriously impairing the newborn. These defects are usually repaired surgically.

The Lymphatic System

The lymphatic system collects excess interstitial fluid (tissue fluid) from interstitial spaces in body tissues and returns it to the blood. On the way back to the blood, the fluid is cleansed of bacteria and cellular debris as it passes through lymph nodes. In addition, lymph nodes contain masses of lymphocytes that play a major role in immunity.

The Lymph Pathway

The smallest lymphatic vessels are the **lymphatic capillaries,** tiny closed-end vessels that are abundant among cells of body tissues. After entering a lymphatic capillary, interstitial fluid is called **lymph.** Lymphatic capillaries combine to form **lymphatic vessels** (lymphatics), which are shown (label 11) in Figure 29.6. Like veins, lymphatic vessels contain numerous valves that prevent a backflow of lymph, which is propelled by contractions of skeletal muscles depressing the walls of the vessels.

When the lymphatic vessels enter the trunk, lymph passes through one or more **lymph nodes** that are associated with the lymphatics. Lymph nodes are especially abundant in cervical, axillary, abdominal, pelvic, and inguinal areas.

The lymphatics from the right arm, right side of the thorax, and right side of the neck and head join to form the **right lymphatic duct** (label 5), which drains into the right subclavian vein.

Lymphatics from the legs, abdomen, left side of the thorax, left arm, and left side of the neck and head empty into the **thoracic duct.** At its inferior origin is a saclike enlargement, the **cisterna chyli,** that receives fat-rich lymph from the intestines. The thoracic duct empties into the left subclavian vein.

Lymph Node Structure

The enlarged lymph node in Figure 29.6 reveals its structure. The small indentation at its upper end is the **hilum,** the site where blood vessels and an **efferent lymphatic vessel** exit the node. Each node has only one hilum and one efferent lymphatic vessel, but it may have two or more **afferent lymphatic vessels** that carry lymph into the node.

A **capsule** of fibrous connective tissue envelopes the node. Just inside the capsule is the **cortex** (label 18) that contains many **germinal centers** (nodules) which contain masses of lymphocytes. The central portion of the node is called the **medulla.**

ASSIGNMENT

Label Figure 29.5 and add arrows to show the direction of blood flow.

ASSIGNMENT

Label Figure 29.6 and add arrows to show the direction of lymph flow.

Figure 29.6   The lymphatic system.

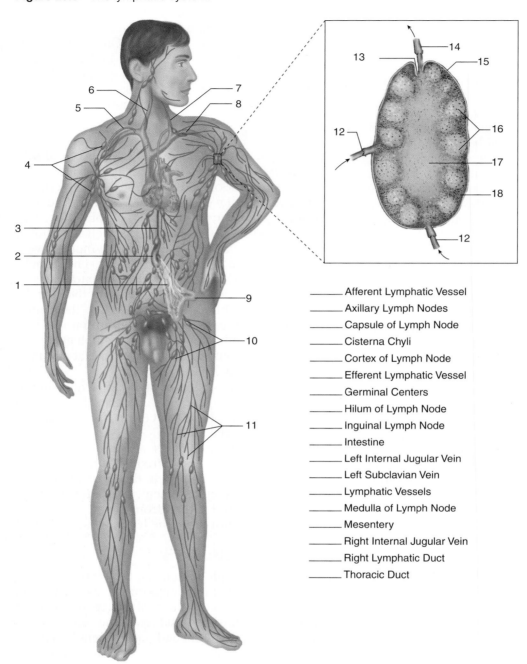

_____ Afferent Lymphatic Vessel
_____ Axillary Lymph Nodes
_____ Capsule of Lymph Node
_____ Cisterna Chyli
_____ Cortex of Lymph Node
_____ Efferent Lymphatic Vessel
_____ Germinal Centers
_____ Hilum of Lymph Node
_____ Inguinal Lymph Node
_____ Intestine
_____ Left Internal Jugular Vein
_____ Left Subclavian Vein
_____ Lymphatic Vessels
_____ Medulla of Lymph Node
_____ Mesentery
_____ Right Internal Jugular Vein
_____ Right Lymphatic Duct
_____ Thoracic Duct

Lymphocytes and Immunity

Although all types of white blood cells are involved in an immune response, it is lymphocytes that play a predominant role in immunity. They are responsible for *active acquired immunity,* immunity that develops only after exposure to specific **antigens,** certain proteins on the surface of pathogens.

There are two basic types of lymphocytes, **B lymphocytes (B cells)** and **T lymphocytes (T cells),** although they appear morphologically similar. There are three basic types of T cells: *helper T cells (T_H), cytotoxic T cells (T_C),* and *memory T cells (T_M).*

Like all blood cells, both types of lymphocytes are derived from **stem cells** in red bone marrow before and for a short time after birth. Lymphocytes go through a maturation process to become immuno-competent (Figure 29.7). The maturation of B lymphocytes occurs in red bone marrow. T lymphocytes mature in the thymus gland, an endocrine gland

Figure 29.7 Activation of a helper T cell or a cytotoxic T cell and cell-mediated immunity. (APC = antigen presenting cell; T_H = helper T cell; T_C = cytotoxic T cell; T_M = T memory cell)

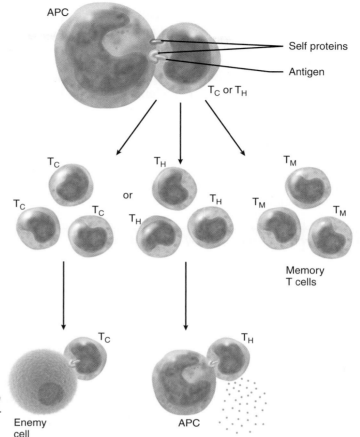

1. Activation of a T cell
T cell binds to nonself antigen and self proteins of APC and becomes activated.

2. Clone formation
Activated T cell undergoes repeated mitotic division producing a clone of identical T cells.

3. Action of clone cells
T_C cells bind to cells displaying the targeted antigen and inject chemicals to destroy them.

T_H cells bind to the targeted antigen and self proteins on APC, become activated, and release cytokines, which stimulate activity of T_C and B cells, inflammation, and phagocytosis by neutrophils and macrophages.

T_M cells launch a secondary response if the targeted antigen later reappears.

located in the thoracic cavity above the heart. The maturation process yields thousands of different varieties of B cells and T cells. Each variety is capable of recognizing a different specific antigen. Thus, any pathogen that may enter the body will be recognized by some (but not all) B cells and T cells. After maturation, both B cells and T cells are carried to lymph nodes, spleen, and other lymphatic tissue throughout the body, and some circulate in the blood.

There are two basic forms of immunity: T cells provide cell-mediated immunity, and B cells provide antibody-mediated immunity.

Cell-Mediated Immunity

In **cell-mediated immunity,** T cells directly attack foreign antigens or diseased body cells, such as cancer cells. It also destroys *intracellular* pathogens and the invaded cells.

Cell-mediated *primary immune response* begins when an *antigen-presenting cell (APC),* often a macrophage, engulfs a foreign antigen. A part of the antigen is presented on the cell surface along with the APC's self antigens. When a T cell, which has been programmed in the maturation process to recognize the antigen, binds to both the antigen and the APC's self proteins, it becomes activated. The activated T cell undergoes repeated cell divisions forming a clone of identical T cells (Figure 29.7).

If the activated T cell is a helper T cell (T_H), it forms a clone of mostly active T_H cells and fewer dormant memory T cells (T_M). When active T_H cells bind with the antigen, they secrete *cytokines* (chemical) that promote phagocytosis by neutrophils and macrophages and stimulate activity of T_C cells and B cells. After the foreign antigen has been destroyed, the dormant T_M cells will launch a quicker and stronger *secondary immune response* if the antigen should reappear.

If the activated T cell is a cytotoxic T cell (T_C), it forms a clone of mostly active T_C cells and fewer memory T cells (T_M). When an active T_C cell binds to the antigen, it injects lethal chemicals that kill the cell and any pathogen within. It then detaches and searches for more antigens to destroy. After the foreign antigen has been destroyed, the dormant T_M cells will launch a quicker and stronger secondary immune response if the antigen should reappear.

Antibody-Mediated Immunity

Antibody-mediated immunity involves both B cells and helper T cells, and it provides a defense against *extracellular* pathogens. Instead of directly attacking a foreign antigen, it involves B cells producing **antibodies** that bind to the targeted antigens, which makes it easier for them to be destroyed by other means, such as phagocytosis.

Antibody-mediated primary immune response begins when foreign antigens bind to the complementary receptors of B cells, which have been programmed to recognize these antigens. The antigen is engulfed by a B cell and displayed on the cell surface along with the B cell's self proteins. A helper T cell with receptors for this antigen binds to the antigen and self proteins and secretes cytokines that activate the B cell. The activated B cell forms a clone of identical B cells. Most cells of the clone become plasma cells but some become memory B cells. The plasma cells secrete antibodies that bind to the targeted antigens, tagging them for easier destruction. After the foreign antigen has been destroyed, the dormant memory B cells will launch a quicker and stronger secondary immune response if the antigen should reappear (Figure 29.8).

ASSIGNMENT

1. Examine a prepared slide of lymph node, comparing it with Figure 29.6. Using the high-dry objective, observe lymphocytes in a germinal center. Can you see any dividing lymphocytes?
2. Complete the laboratory report.

Figure 29.8 B-cell activation and antibody-mediated immunity.

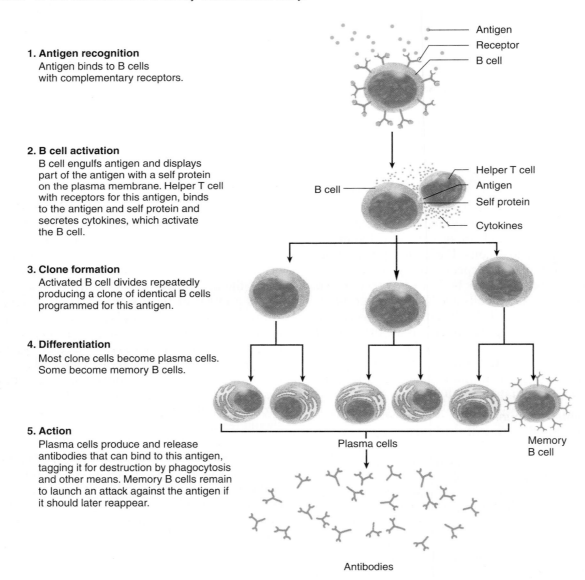

1. Antigen recognition
Antigen binds to B cells with complementary receptors.

2. B cell activation
B cell engulfs antigen and displays part of the antigen with a self protein on the plasma membrane. Helper T cell with receptors for this antigen, binds to the antigen and self protein and secretes cytokines, which activate the B cell.

3. Clone formation
Activated B cell divides repeatedly producing a clone of identical B cells programmed for this antigen.

4. Differentiation
Most clone cells become plasma cells. Some become memory B cells.

5. Action
Plasma cells produce and release antibodies that can bind to this antigen, tagging it for destruction by phagocytosis and other means. Memory B cells remain to launch an attack against the antigen if it should later reappear.

Antigen
Receptor
B cell

B cell
Helper T cell
Antigen
Self protein
Cytokines

Plasma cells
Memory B cell

Antibodies

Exercise 30

CARDIOVASCULAR PHENOMENA

OBJECTIVES

After completing this exercise, you should be able to:

1. Identify the heart sounds by stethoscope auscultation.
2. Determine the blood pressure and pulse of a subject.
3. Describe the control of capillary circulation.
4. Define all terms in bold print.

Materials

Stethoscope
Alcohol pads
Sphygmomanometer
Grass frog or goldfish
Frog board and wrapping cloth
Dissecting pins
Rubber bands
Epinephrine (1:1,000) in dropping bottles
Histamine (1:10,000) in dropping bottles

In this exercise, you will study heart sounds, blood pressure, pulse, and the control of capillary circulation.

Heart Sounds

There are three heart sounds that result from contraction and valvular movements. The **first sound** (lub) results from ventricular contraction and the simultaneous closure of the atrioventricular valves. The **second sound** (dub) results from the simultaneous closure of the aortic and pulmonic semilunar valves. The **third sound** seems to be caused by the vibration of the ventricular walls and the atrioventricular valve cusps during systole. It is difficult to detect unless the subject is lying down. Abnormal heart sounds are called **murmurs** and usually result from a damaged or defective atrioventricular valve that allows blood to leak back into an atrium. Many murmurs have no clinical significance.

Figure 30.1 Auscultatory areas.

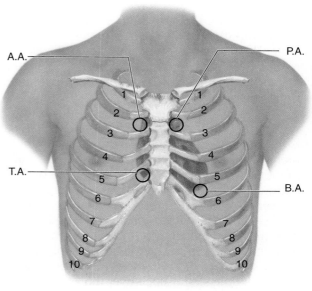

A.A. Aortic Area   P.A. Pulmonic Area
B.A. Bicuspid Area   T.A. Tricuspid Area

Auscultation of (listening to) the heart sounds provides important information for a physician and aids diagnosis of many heart conditions, including heart murmurs. For example, abnormal splitting of the heart sounds occurs in heart block, septal defects, and hypertension. For best results, you must use the **auscultatory areas** shown in Figure 30.1. These areas do not coincide with the anatomical locations of the valves, because heart sounds are projected to different spots on the rib cage.

Your instructor will demonstrate the heart sounds using an audio monitor. If a sensitive unit is used, such as the Grass AM7, it will not be necessary to bare the subject's chest. After you can distinguish the heart sounds, work in pairs to auscultate each other's heart with a stethoscope.

ASSIGNMENT

1. Clean the earpieces of the stethoscope with an alcohol pad. Fit them into your ears, directing them inward and forward.
2. With the subject in a sitting position, try to maximize the first sound by placing the stethoscope on the tricuspid and bicuspid auscultatory areas. Note that the bicuspid valve auscultatory area is at the apex of the heart (fifth rib) and that the tricuspid area is 2 to 3 inches medial from this site.
3. Now, move the stethoscope to the aortic and pulmonic areas to maximize the second sound. The aortic auscultatory area is on the right edge of the sternum at the level of the second rib, and the pulmonic area is on the left side of the sternum between the second and third ribs. Note that the aortic auscultatory area is the only one on the right side of the sternum.
4. Try to detect a **splitting of the second sound** while the subject is inhaling. This phenomenon is normal and occurs because during inspiration more venous blood is forced into the right side of the heart and causes a delayed closure of the pulmonary semilunar valve. Use hand signals, not verbalization, to inform the subject when to inhale. Have the subject inhale when you raise your hand and exhale when you lower it.
5. Now, have the subject lie down, face up, and repeat steps 2 to 4. Try to detect the **third sound** by placing the stethoscope over the apex of the heart. Do you detect any murmurs?
6. If time permits, compare the sounds before and after exercise.
7. Complete Sections A and B of the laboratory report.

Blood Pressure Determination

Blood pressure is usually taken when a person is at rest, and normal blood pressure values apply to a person at rest. However, blood pressure is not constant, and many factors affect it. For example, blood pressure is increased by an increase in heart rate, blood volume, and blood viscosity or a decrease in blood vessel diameter. These factors, in turn, are affected by a person's health, physical activity, and emotional state. The contractions of the ventricles of the heart pump blood into the pulmonary artery and aorta in pulses about 70 times a minute. This pumping action causes hydrostatic pressure in the arteries to fluctuate during the heart cycle. Blood pressure is greatest during **ventricular systole,** the contraction phase, and least during **ventricular diastole,** the relaxation phase.

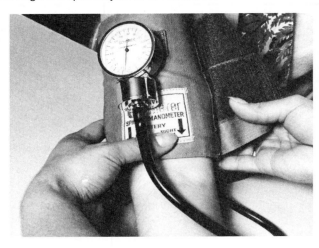

Figure 30.2 Sphygmomanometer cuff is wrapped around the upper arm, keeping lower margin of cuff above a line through the epicondyles of the humerus.

There are several ways to measure blood pressure, but we will use a standard **sphygmomanometer** and a **stethoscope** to measure blood pressure indirectly. In this method, a stethoscope is used to listen for the **Korotkoff sounds** that are produced by blood rushing through a partially occluded brachial artery.

Figure 30.2 shows how to apply the cuff of the sphygmomanometer to the upper arm with its lower margin above the epicondyles of the humerus. When the cuff is properly positioned, it is inflated to shut off the flow of blood in the brachial artery. As shown in Figure 30.3, the bell of the stethoscope is positioned over the brachial artery. Then, air is released from the cuff through the control valve on the hand pump while listening for the Korotkoff sounds.

Korotkoff sounds are caused by turbulent blood flow through a partially constricted artery. They are not produced by blood flow in a completely open

Figure 30.3 The inflated cuff shuts off the flow of blood in the brachial artery. Korotkoff sounds are listened for as pressure is gradually lowered by releasing air from the cuff.

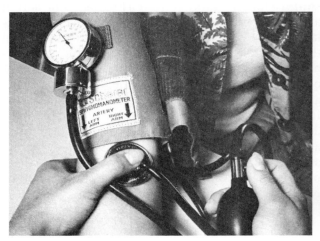

artery. Korotkoff sounds change as air pressure in the cuff is decreased. First, they sound like a clear tapping (at each heartbeat), and they get louder and deeper in tone as the pressure in the cuff is decreased. Then they become muffled and suddenly cease.

When the first Korotkoff sound is heard, the pressure on the gauge is read. This is the **systolic pressure.** At this point, the air pressure in the cuff is equal to the blood pressure that is just sufficient to force blood through the partially occluded artery. As air continues to be released, the Korotkoff sounds change in tone and then cease. The pressure where the sounds suddenly cease is the **diastolic pressure.** At this point, the artery is no longer partially occluded during ventricular diastole.

Normal systolic pressure in adults is 120 mm Hg; normal diastolic pressure is 80 mm Hg. Such blood pressures are usually designated as 120/80. Systolic and diastolic pressures may vary by ±10 mm Hg and still be considered normal.

Your instructor will demonstrate the procedure and Korotkoff sounds using a microphone and audio monitor. After you understand the procedure, work with a partner to determine each other's blood pressure. Record your results on the laboratory report.

ASSIGNMENT

1. Have the subject's arm resting on the table with the palm upward. Wrap the cuff of the sphygmomanometer around the upper arm. The method of attachment will vary with the type of instrument. Your instructor will indicate how to secure it.
2. Close the air valve on the neck of the rubber bulb *but not so tightly that you can't open it.* Attach the pressure gauge on the cuff so that you can easily read it.
3. Pump air into the cuff by squeezing the bulb. Stop when the pressure gauge reads 180 mm Hg. *Do not leave the cuff inflated for more than 30 seconds.*
4. Slip the diaphragm of the stethoscope under the cuff over the brachial artery—about midway between the epicondyles of the humerus. See Figure 30.3.
5. Loosen the valve control screw *slightly* to slowly release air from the cuff while listening for the Korotkoff sounds. When the first sound is heard, read the pressure on the gauge. This is the systolic pressure.
6. Continue listening to the Korotkoff sounds as air continues to escape from the cuff. At the point where the sounds cease, read the pressure on the gauge. This is the diastolic pressure.
7. Repeat several times until you can get consistent results. Wait several minutes between repetitions.

8. Measure the blood pressure after the subject has done some exercise.
9. Complete Section C of the laboratory report.

The Pulse

Ventricular contraction produces a surge of blood into the arterial tree causing a sudden pressure wave that produces expansion of the arteries. During ventricular relaxation the arterial walls recoil to their prior state. This alternating expansion and recoil of the elastic walls of arteries during each cardiac cycle is called the **pulse.** The pulse results from the difference between systolic and diastolic blood pressures— the **pulse pressure.**

The pulse is commonly taken by *palpation* (feeling of organs with the hands) of any artery that lies close to the body surface. The most common sites for taking the pulse are (1) the radial artery, which runs over the distal, anterior end of the radius at the wrist, and (2) a common carotid artery lying between the larynx and sternocleidomastoid muscle on each side of the neck. Figure 30.4 shows locations (pressure points) where the pulse may be felt.

The pulse rate reflects the heart rate, and the pulse strength reflects the difference between systolic and diastolic blood pressure. The pulse rate varies with body position, exercise, physical condition, and emotional state. In this section, you will palpate the pulse at various pressure points and note any rate changes.

ASSIGNMENT

1. Work in pairs during this assignment.
2. Place the tips of your index and middle fingers over your partner's radial artery to locate the pulse. Determine your partner's pulse rate (beats/minute) when lying supine, sitting, and standing. Allow 3 minutes at each position for equilibration before taking the pulse rate. Count the beats for 15 seconds and multiply by 4 to obtain the beats per minute. Record the results on your partner's laboratory report.
3. Determine your partner's pulse rate after jogging in place for 3 minutes and at 1-minute intervals until it has returned to normal. The faster the pulse returns to normal, the better is one's physical condition.
4. Locate your partner's pulse in the brachial artery, common carotid artery, temporal artery, and posterior tibial artery.
5. Complete Section D of the laboratory report.

Figure 30.4 Pressure points where the pulse may be detected.

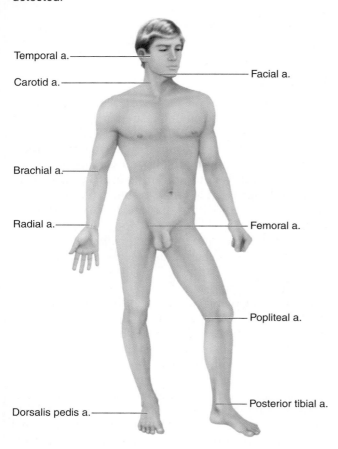

Figure 30.5 Capillary circulation.

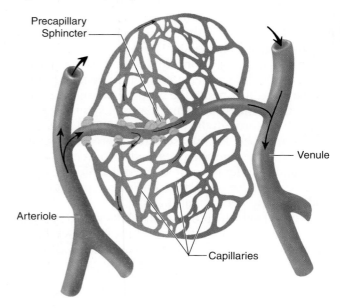

Peripheral Circulation

Arteries and veins are united by an intricate network of capillaries. **Capillaries** are short, thin-walled vessels composed of endothelial cells. They are so numerous that all body cells are within two or three cells of a capillary. The combined diameter of capillaries causes blood to slow down as it passes through them. This slowing allows time for the exchange of materials between the capillary blood and body cells. Interstitial fluid is an intermediate in the exchange.

capillary blood $\rightleftarrows$ interstitial fluid $\rightleftarrows$ body cells

There is insufficient blood to fill all capillaries of the body simultaneously, so many capillaries are intermittently filled and deprived in order to meet the needs of all tissues. Blood flow into a capillary is controlled by a **precapillary sphincter muscle** located at the junction of an arteriole and a capillary. See Figure 30.5.

The diameter of arterioles and precapillary sphincters is under the control of the **vasomotor center** in the medulla oblongata. Many factors affect this center, but carbon dioxide concentration of the blood is of major importance. At the localized tissue level, carbon dioxide, epinephrine, and histamine cause a direct influence on capillary blood flow. A high level of carbon dioxide dilates arterioles and precapillary sphincters (vasodilation), increasing capillary blood flow to the affected tissues. Increased levels of epinephrine cause vasodilation in some tissues (heart, lungs, and muscles) and vasoconstriction in others. Histamine, a product of allergic reactions, causes vasodilation.

In this section you will study the effect of histamine and epinephrine on capillary blood flow.

Figure 30.6 Blood flow setup.

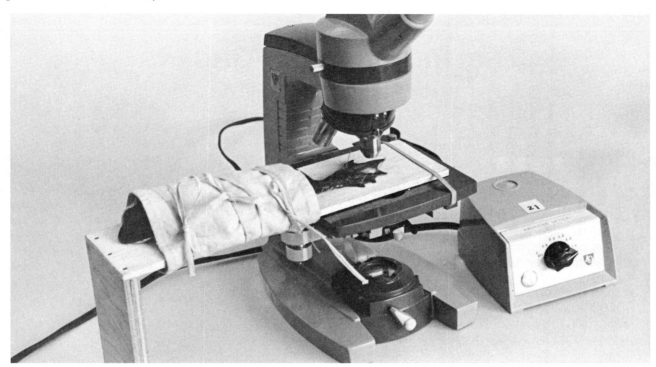

ASSIGNMENT

1. Obtain and wrap a small frog onto a frog board as shown in Figure 30.6. The cloth should be moist before wrapping. (If using a goldfish instead of a frog, wrap the fish in a *wet* paper towel. Secure the fish to the frog board with a rubber band.)
2. Spread the webbing of a hind foot over the hole in the board and use pins to secure the foot in that position. Attach the board to a microscope stage with rubber bands as shown. (If using a goldfish, spread the caudal [tail] fin over the opening in the frog board.)
3. Use a 10× objective to observe circulation in the webbing of the foot. Locate an arteriole with its pulsating, rapid flow, and a venule with its slower, steady flow.
4. Locate a capillary. Note its small diameter and the blood cells moving slowly in a single file through it. Try to find a site of a precapillary sphincter by observing the irregular flow of blood cells through a capillary. You will not be able to actually see the sphincter.
5. Blot the water from the webbing and add 2 drops of histamine (1:10,000). Observe the effect on capillary flow.
6. Rinse the foot with water and add 2 drops of epinephrine (1:1,000). Observe the effect on capillary flow.
7. Wash the foot and return the frog to the stock table. Clean the microscope.
8. Complete the laboratory report.

THE RESPIRATORY ORGANS

OBJECTIVES

After completing this exercise, you should be able to:

1. Identify the components of the respiratory system on charts, models, and a sheep pluck and describe their functions.
2. Trace the flow of air into and out of the lungs.
3. Identify lung and tracheal tissues when viewed microscopically.
4. Define all terms in bold print.

Materials

Model of a midsagittal section of the head
Torso model with removable organs
Corrosion preparation of the bronchial tree
Sheep plucks without livers
Dissecting instruments and trays
Prepared slides of human trachea, normal lung tissue, emphysematous lung tissue, and "smoker's lung" tissue
Protective disposable gloves

 Before You Proceed

Consult with your instructor about using protective disposable gloves when performing portions of this exercise.

Body cells must be continuously supplied with oxygen in order to carry out their metabolic activities. The metabolic activities of cells, in turn, produce carbon dioxide that must be removed from the body. The primary function of the respiratory system is to carry out the exchange of these respiratory gases between the body and the atmosphere. Secondary functions include warming and filtering the air that is inhaled.

The exchange of respiratory gases involves three related processes: (1) breathing air into and out of the lungs, (2) the exchange of oxygen and carbon dioxide between the air in the lungs and the blood, and (3) the transport of oxygen and carbon dioxide by the blood.

The respiratory organs include not only the lungs, the primary organs of gas exchange, but also the series of tubes and passageways that enable air to enter the lungs during inhalation and exit the lungs during exhalation. The respiratory organs are often subdivided into upper and lower respiratory tracts. Those respiratory organs located outside the thorax constitute the upper respiratory tract. Respiratory organs within the thorax form the lower respiratory tract.

In this exercise, you will study both macroscopic and microscopic aspects of the respiratory organs as well as the basic functions of each organ. The respiratory organs include the nose, pharynx, larynx, trachea, bronchi, and lungs.

The Upper Respiratory Tract

The midsagittal section of the head in Figure 31.1 shows the parts of the upper respiratory system. The **nasal cavity** is divided into left and right chambers by the **nasal septum** (not shown), which is composed of cartilage and bone. The surface area of the nasal cavity is increased by the superior, middle, and inferior **nasal conchae,** which project from the lateral walls. Air enters the nasal cavity via the **nares,** or nostrils.

The **palate** separates the nasal cavity from the **oral cavity.** The palate consists of an anterior portion supported by bone, the **hard palate,** and a posterior **soft palate,** which terminates in a median fingerlike projection, the **uvula.**

The **pharynx** consists of three parts. The **oropharynx** lies posterior to the oral cavity, and the **nasopharynx** is just above it and posterior to the nasal cavity. Inferior to the oropharynx is the **laryngopharynx,** where air enters the larynx and food enters the esophagus. The openings of the **auditory tubes** (label 10) are visible on the lateral walls of the nasopharynx.

Figure 31.1 Upper respiratory system. (a) Midsagittal section; (b) divisions of the pharynx.

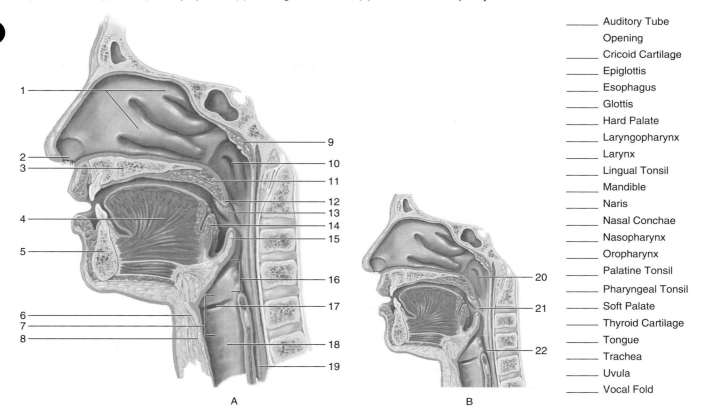

_____ Auditory Tube Opening
_____ Cricoid Cartilage
_____ Epiglottis
_____ Esophagus
_____ Glottis
_____ Hard Palate
_____ Laryngopharynx
_____ Larynx
_____ Lingual Tonsil
_____ Mandible
_____ Naris
_____ Nasal Conchae
_____ Nasopharynx
_____ Oropharynx
_____ Palatine Tonsil
_____ Pharyngeal Tonsil
_____ Soft Palate
_____ Thyroid Cartilage
_____ Tongue
_____ Trachea
_____ Uvula
_____ Vocal Fold

As air passes through the nasal cavity, it is warmed, filtered, and humidified by the mucous membrane lining. Airborne particles are trapped in mucus secreted by goblet cells. The mucus and entrapped particles are moved by beating cilia to the pharynx and swallowed.

From the nasal cavity, air passes through the pharynx, larynx, and trachea on its way to the lungs. The **larynx,** or voice box, has walls formed of an upper **thyroid cartilage** (label 6) and a lower **cricoid cartilage.** Within the larynx are the **vocal folds** (label 17), which vibrate to produce sounds when activated by passing air. The **epiglottis** is a flaplike structure diagonally situated over the **glottis** (label 16), the upper opening to the larynx. It prevents food from entering the larynx when swallowing. The opening into the **esophagus** (label 19), which carries food to the stomach, lies just posterior to the larynx.

Clumps of lymphoid tissue compose the tonsils. The **pharyngeal tonsils,** or **adenoids** (label 9), are located in the roof of the nasopharynx, the **palatine tonsils** (label 13) occur on each side of the oropharynx, and the **lingual tonsils** are on the posterior inferior portion of the tongue.

The Lower Respiratory Passages

Figure 31.2 illustrates the lower respiratory structures. The **trachea,** or windpipe, extends below the larynx to the center of the thorax, where it branches into the **primary bronchi,** which enter the lungs. Within the lungs, the primary bronchi divide to form the **secondary bronchi,** one for each lobe. Continued branching forms the **tertiary bronchi** and **bronchioles,** which end in a cluster of tiny sacs called **alveoli.** The exchange of gases occurs between the air in the alveoli and blood in capillaries surrounding the alveoli. Cartilaginous "rings" support the walls of both the trachea and bronchi. Bronchioles lack cartilaginous rings but contain smooth muscle in their walls. In an **asthma** attack, contraction of the muscles and an excessive production of mucus in the air passages restrict the air flow.

Each lung is covered by a tightly adhering membrane, the **pulmonary pleura** (label 4), and the **parietal pleura** lines the inner wall of the thoracic cavity. Fluid secreted into the **pleural cavity,** the potential space between these membranes, reduces friction during lung movements. The **diaphragm,** the

Figure 31.2 The respiratory system. (a) Gross anatomy; (b) alveoli.

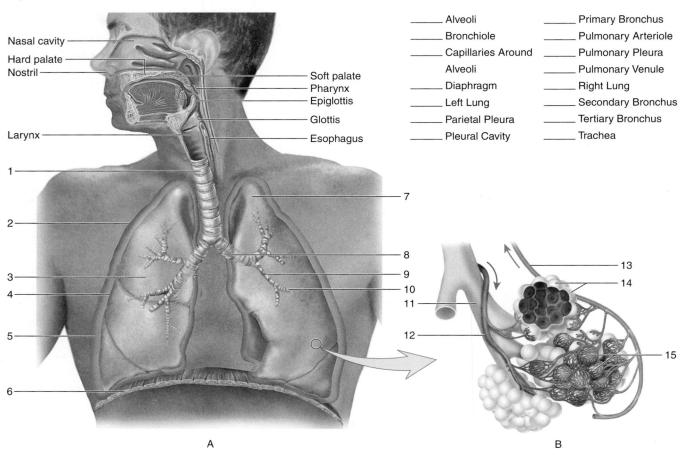

Labels list:

_____ Alveoli   _____ Primary Bronchus
_____ Bronchiole   _____ Pulmonary Arteriole
_____ Capillaries Around   _____ Pulmonary Pleura
      Alveoli   _____ Pulmonary Venule
_____ Diaphragm   _____ Right Lung
_____ Left Lung   _____ Secondary Bronchus
_____ Parietal Pleura   _____ Tertiary Bronchus
_____ Pleural Cavity   _____ Trachea

Diagram labels (A): Nasal cavity, Hard palate, Nostril, Larynx, Soft palate, Pharynx, Epiglottis, Glottis, Esophagus, 1, 2, 3, 4, 5, 6, 7, 8, 9, 10, 11, 12

Diagram labels (B): 13, 14, 15

A   B

primary muscle of respiration, separates the thoracic and abdominal cavities.

ASSIGNMENT

1. Label Figures 31.1 and 31.2.
2. Complete Sections A, B, and C of the laboratory report.
3. Locate the parts of the respiratory system on models available in the laboratory.
4. Examine a corrosion preparation of the bronchial tree and identify the air passages composing it.

Sheep Pluck Dissection

A sheep "pluck" consists of the larynx, trachea, lungs, and heart removed during routine slaughter. The study of a fresh sheep pluck is valuable because its components are similar to those in humans. Although it is as safe to handle as any material from a meat market, use sanitary precautions. Record your observations on the laboratory report.

1. Lay out the pluck on the dissecting tray. Locate the heart, lungs, trachea, larynx, and diaphragm.
2. Identify the epiglottis and the thyroid and cricoid cartilages of the larynx. Look into the larynx to find the vocal folds.
3. Use scissors to cut through the trachea below the larynx and examine the cut end. Note the shape of the cartilaginous rings and the smooth, slimy interior of the trachea.
4. Note that the lungs are divided into lobes. Are the same number of lobes found in both sheep and human lungs? Feel the surface of a lung. What membrane makes it so smooth?
5. Separate or remove the connective tissue around the pulmonary trunk to expose the pulmonary arteries. Locate the pulmonary veins. Can you locate the ligamentum arteriosum extending between the pulmonary trunk and the aorta?

6. Trace the trachea down to where it branches into the primary bronchi, removing surrounding tissue as necessary. Note that a bronchus leading to the upper right lobe branches off some distance above the primary bronchi.

7. Use scissors to cut through the trachea at the level of the top of the heart. Then, cut down the dorsal side (opposite the heart) of the trachea to the primary bronchi. Continue the cut along one bronchus into the center of the lung and note the extensive branching of the air passages.

8. Insert a plastic straw into a bronchiole and blow into the lung. Note the expansibility and elasticity of lung tissue.

9. Cut off a piece of a lung and examine the cut surface. Note the sponginess of lung tissue.

10. Dispose of the sheep pluck as directed by your instructor. **Important:** Wash the tray, instruments, and your hands with soap and water.

Microscopic Study

Compare your observations with Figure HA-15 and make drawings of your observations on the laboratory report.

1. Examine the photomicrographs of the nasal cavity in Figure HA-15. Note the nasal septum, concha, and pseudostratified ciliated epithelium.

2. Examine a prepared slide of trachea. Locate the hyaline cartilage composing cartilaginous rings and the ciliated cells and goblet cells of the epithelial lining.

3. Examine a prepared slide of normal lung tissue. Note the porous nature of lung tissue. Look for the thin-walled alveoli, a bronchiole, and blood vessels. Why are thin alveolar walls advantageous?

4. Examine a prepared slide of emphysematous lung tissue. Note that many walls of the alveoli have been destroyed, decreasing the respiratory surface of the lung. Emphysema also reduces the elasticity of lung tissue so that it is difficult to force air out of the lungs.

5. Examine a prepared slide of "smoker's lung." Note the breakdown of alveoli and the deposits of carbon and tar particles, which can cause cancer.

 ASSIGNMENT

Complete the laboratory report.

Exercise 32

RESPIRATORY PHYSIOLOGY

OBJECTIVES

After completing this exercise, you should be able to:

1. Describe the factors that stimulate breathing.
2. Describe the mechanics of breathing.
3. Determine the various lung volumes using a spirometer.
4. Define all terms in bold print.

In this exercise you will (1) investigate the control of breathing and (2) determine your lung capacities using either nonrecording Propper spirometers or the computer-based Intelitool Spirograph.

The Control of Breathing

The rhythmic cycle of respiration is controlled by the **respiratory control center** in the medulla oblongata. The primary stimulus on the respiratory control center for inspiration is an increase in the hydrogen ion concentration of the blood and cerebrospinal fluid. Recall that an increase in carbon dioxide concentration increases the hydrogen ion concentration.

carbon dioxide + water $\leftrightarrows$ carbonic acid
carbonic acid $\leftrightarrows$ hydrogen ion + bicarbonate ion

Both carbon dioxide and hydrogen ions act directly on the respiratory control center. A decreased level of oxygen acts indirectly and is detected by chemoreceptors in the carotid arteries and aortic arch. Impulses from these receptors are then transmitted to the respiratory control center. Usually, oxygen concentration is of little significance in stimulating breathing.

Materials
Paper bag
Paper cup
Drinking straw

Hyperventilation

Hyperventilation reduces the concentrations of carbon dioxide and hydrogen ions in the blood so that the respiratory center is depressed and, in turn, the desire to breathe is reduced. Demonstrate this effect as follows. Record your results on the laboratory report.

1. Breathe deeply at the rate of 15 cycles per minute for 1–2 minutes. It will be increasingly difficult to breathe. Stop if you *start* to get dizzy.
2. Place a paper bag over your nose and mouth, and breathe into it for about 3 minutes. Note how much easier it is to breathe into the bag. Why?
3. After breathing normally for 5 minutes, determine how long you can hold your breath.
4. Repeat step 1 and determine how long you can hold your breath.

Deglutition Apnea

Obviously, breathing and swallowing food or drink cannot occur simultaneously. A reflex called **deglutition apnea** prevents one from wanting or attempting to breathe while swallowing. Demonstrate this reflex and record your results on the laboratory report.

1. Hold your breath until you have a strong need to breathe.
2. Then, sip water through a straw and note whether the desire to breathe decreases.

ASSIGNMENT

Complete sections A and B on the laboratory report.

Spirometry and Lung Capacities

A **spirometer** is used to determine the volume of air moving into and out of the lungs during breathing. A recording spirometer may produce a **spirogram** similar to the one shown in Figure 32.1, which indicates the various lung capacities. Note their relationships.

Figure 32.1 Spirogram of lung capacities.

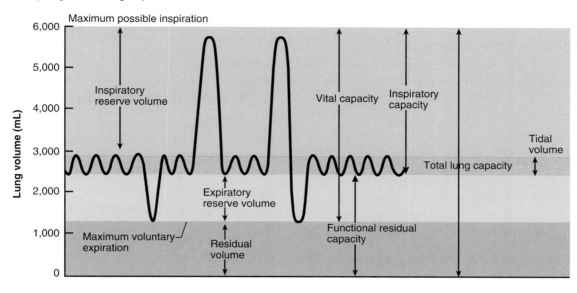

Lung capacities vary with sex, age, and weight of the subject and are useful in diagnosing certain pulmonary disorders. The measurement of lung volumes is essential to determine how well the lungs are functioning and in diagnosing lung disorders. For example, the breakdown of alveoli and loss of lung elasticity in emphysema cause an increase in the residual volume, which decreases the expiratory reserve volume and vital capacity.

You will determine lung volumes using one of two methods as determined by your instructor: Propper spirometers or Intelitool Spirocomp™.

Using Propper Spirometers

In this section, you will use a nonrecording spirometer to determine your lung capacities as described below. Record your results on the laboratory report. In these experiments you are to *exhale only* through the spirometer and inhale through your nose. **Never inhale through the spirometer because you may inhale pathogens from a prior user.**

Materials

Propper spirometers with sterile, disposable mouthpieces
Alcohol pads
Biohazard bag

Tidal Volume (TV)

The amount of air that moves into and out of the lungs during a normal respiratory cycle is the **tidal volume.** Normal tidal volume averages about 500 ml.

1. Cleanse the stem of a Propper spirometer with an alcohol pad, slip on a sterile mouthpiece,

and rotate the dial of the spirometer to zero as shown in Figure 32.2. When using the spirometer always hold the dial upward.
2. Take a normal breath through your nose and exhale through the spirometer. Repeat for a total of three expirations. *Do not inhale or exhale forcibly.*
3. Record the total volume from the spirometer dial. Divide this number by 3 to determine your tidal volume.

Minute Respiratory Volume (MRV)

The amount of tidal air that moves into and out of your lungs in one minute is your **minute respiratory volume.** To determine this, count the number of

Figure 32.2 Dial face of spirometer is rotated to zero prior to measuring exhalations.

respirations in one minute and multiply this number by your tidal volume.

Expiratory Reserve Volume (ERV)

The amount of air that you can expire beyond the tidal volume is your **expiratory reserve volume.** It is usually about 1,100 ml.

1. Set the spirometer dial on 1,000.
2. After 2–3 normal tidal expirations through your nose, exhale forcibly all of the *additional* air that you can through the spirometer.
3. Subtract 1,000 from the dial reading to determine your expiratory reserve volume.

Vital Capacity (VC)

As shown in Figure 32.1, the **vital capacity** of the lungs is the total of the tidal, expiratory reserve, and inspiratory reserve volumes. Vital capacity varies with age, sex, and size, so you must obtain your predicted normal vital capacity from one of the tables in Appendix C. VC averages about 4,300 ml but may vary as much as 20% from established norms and still be considered normal.

1. Set the spirometer dial on zero.
2. After inhaling as deeply as possible and exhaling completely, take a second deep breath and exhale as much air as possible through the spirometer. An even, forced exhalation is best.
3. Repeat three times and record the value for each exhalation. The values should not vary by more than 100 ml. Divide the total by 3 to determine your vital capacity. Compare your vital capacity with the predicted normal values in Appendix C.
4. Place the mouthpiece in the biohazard bag. Wipe the spirometer with an alcohol pad.

Inspiratory Capacity (IC)

The **inspiratory capacity** is the maximum volume of a deep inhalation after expiring the tidal air. It is usually about 3,000 ml. This volume must be calculated since the Propper spirometer cannot measure inspirations. Determine your inspiratory capacity as follows:

$$IC = VC - ERV$$

Inspiratory Reserve Volume (IRV)

The **inspiratory reserve volume** is the volume of air that can be inspired after filling the lungs with tidal air. Calculate your IRV as follows:

$$IRV = IC - TV$$

Residual Volume (RV)

The **residual volume** is air that cannot be expelled from the lungs—about 1,200 ml. It cannot be determined by usual spirometric methods.

ASSIGNMENT

Complete those portions of the laboratory report pertaining to the tests that you have done.

Using Intelitool Spirocomp™

The principal advantage of the Intelitool Spirocomp computerized setup is that all of the calculations are made *automatically* from data that are fed into the program. In addition, the program allows one to compute and average the values of up to 30 subjects. Respiratory efficiency comparisons of smokers with nonsmokers, athletes with nonathletes, and males with females can also be made easily.

Spirocomp can be run on four computer platforms: (1) IBM or compatible PC (MS-DOS 3.3 or higher), (2) Macintosh, (3) Windows 95 or higher, and (4) Apple II (ProDOS). For whichever platform you are using there will be separate instructions. It is assumed here that you have a basic knowledge of the platform you are using. Note also that in the materials list two Spirocomp manuals are available for reference.

In performing these experiments you will be working with a laboratory partner. While one acts as the subject, breathing into the spirometer, the other person can run the computer and prompt the subject as to the procedures.

Materials

computer with monitor and printer
Phipps & Bird wet spirometer
Spirocomp scale arm and interface box
computer game port to transducer cable
Spirocomp program diskette
Spirocomp User Manual
Spirocomp Lab Manual
blank diskette for saving data
valve assembly (Spirocomp)
4 reusable mouthpieces
noseclip
stopwatch or watch with second hand
Lysol disinfectant (or household bleach)
1 beaker (1 liter size) of 70% alcohol
container of mild soap solution large enough to immerse one-way valve assembly

Equipment Hookup

The Spirocomp scale arm and interface box will already be attached to the spirometer post

Figure 32.3 The Spirocomp™ setup for monitoring respiratory volumes.

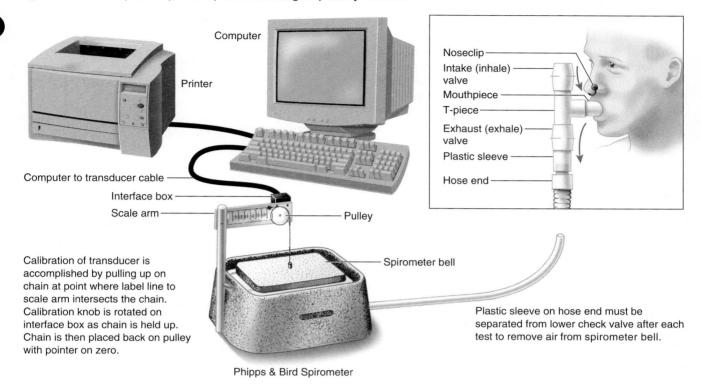

Computer

Printer

Computer to transducer cable

Interface box

Scale arm

Pulley

Noseclip

Intake (inhale) valve

Mouthpiece

T-piece

Exhaust (exhale) valve

Plastic sleeve

Hose end

Calibration of transducer is accomplished by pulling up on chain at point where label line to scale arm intersects the chain. Calibration knob is rotated on interface box as chain is held up. Chain is then placed back on pulley with pointer on zero.

Spirometer bell

Plastic sleeve on hose end must be separated from lower check valve after each test to remove air from spirometer bell.

Phipps & Bird Spirometer

(Figure 32.3), so all you have to do is hook up the components, fill the spirometer with water, and calibrate the Spirometer before starting any tests. Proceed as follows:

1. Fill the spirometer to a little over the fill line with fresh tap water, and add Lysol or bleach to the water for disinfection.
2. Plug in the computer and printer power cords to *grounded* 110-volt electrical outlets.
3. *With the computer turned off,* connect the transducer to computer cable between the interface box and the game port of the computer.
4. Insert the program diskette and turn on the computer.

Preliminary Preparations
Before you start the experiments make the following preparations:

1. Place the four reusable mouthpieces, valve assembly, and noseclip in a beaker of 70% alcohol. Between subjects, always swish the valve assembly through the mild soap solution and a water rinse and immerse it in 70% alcohol before reusing.
2. Seat the subject in a straight high-backed chair and allow him or her to rest for 5 minutes

before starting the test. Unless the subject is relaxed, calm, and motionless, normal respiratory volumes are unattainable. (Even slight shifting of the body during data acquisition can alter results.)

In addition, the subject should not face his or her lab partner or the computer screen. Watching the results on the monitor can cause the subject to, consciously or unconsciously, alter the breathing pattern. (This is a problem only when measuring the TV and ERV.)

3. When breathing into the mouthpiece, the subject's nose must be securely closed with the noseclip. In addition, the lips must form a tight seal around the mouthpiece to avoid any air leakage.
4. Calibrate the transducer by following the procedures outlined by the software. If, during data acquisition, the chain slips on the scale arm, recalibrate the spirometer before continuing.

Groups and Data Files
If there will be an attempt to compare nonsmokers with smokers, nonathletes with athletes, or pre- with postexercise, your instructor will find it necessary to make specific group assignments so that separate

data files can be made. If the class is quite small, the entire class may be a group.

You will be able to add records to any group data file, even if it has been saved to disk; so if there are two or more different lab sections, students from each section may be put into the same group data file. *One thing you don't have to do is separate the groups by gender, because the software will do that automatically.*

Experimental Procedure

Now that all components are active, proceed as follows, using the instructions that apply to your particular computer platform.

1. Enter subject data.

MS-DOS: Select "Experimental Menu" from the main SPIROCOMP MENU and choose "New" group file. Enter the subject data as prompted. The Spirocomp data acquisition screen will then appear.

Macintosh and Windows: Select **New Database** from the **File** menu and name the file appropriately. Then select **Add New Record** from the **Acquire** menu. The Subject Data Entry window will appear. Enter the subject data and then click the **Volumes** button. The Spirocomp data acquisition window will then appear.

Apple II: Select "Run Spirocomp" from the MAIN MENU, and then choose "Run Spirocomp" again. Enter the subject data as prompted. The Spirocomp data acquisition screen will then appear.

Note: If you do not know your height in centimeters, enter it in inches, followed by a quote mark ("). The software will automatically convert the value to centimeters. For example, a value of 69" will be converted to 175 cm.

2. Attach the valve assembly.

Remove the valve assembly from the beaker of alcohol, shake it dry, and attach it to the spirometer hose. Next, remove a reusable mouthpiece from the alcohol, shake it dry, and insert it on the valve assembly. Refer to Figure 32.4 to make sure that your assembly is correct.

3. Put the noseclip on the subject.

Remove the noseclip from the alcohol, shake dry, and place it in position on the subject.

4. Measure the *tidal volume (TV).*

Note in Figure 32.1 that the tidal volume is the amount of air that moves into and out of the lungs during a single normal respiratory cycle. Although sex, age, and weight determine this value, the average normal tidal volume is around **500 ml.** To get reliable results, make certain that the subject is rested, not excited, and not moving.

Have the subject breathe normally into the spirometer for one or two cycles, and then

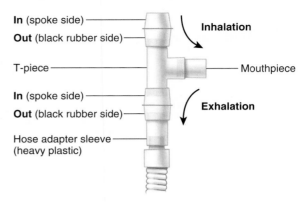

Figure 32.4 Correct method of assembling the valve assembly to the hose and mouthpiece.

press the T key during an inhalation. Follow the instructions on the screen.

The software will read the changes in the spirometer for three cycles, automatically stop, and then graph the average Tidal Volume. Record the TV here:

TV:_____

5. Count the *resting respiratory rate.*

This value pertains to the number of breaths per minute. To determine the *resting respiratory rate (RRR),* simply use a watch to count the number of tidal cycles in one minute. A typical respiratory rate is between 12 and 15 cycles per minute. Record this rate here:

RRR: _____

6. Determine the *resting minute respiratory volume.*

This value (MRV) is the amount of air moved into and out of the lungs in one minute. At rest, it is simply the product of the tidal volume and respiratory rate, or RRR × TV. Record here:

MRV: _____

7. Empty air from the bell float.

Empty the air from the bell float of the spirometer by pulling the valve assembly apart as shown in Figure 32.5. You can speed up the emptying process by pressing *gently* down on the bell float.

8. Determine the expiratory reserve volume (ERV).

As indicated in Figure 32.1, the ERV is the amount of air that one can expire beyond the tidal volume.

To measure this volume have the subject breathe normally into the spirometer for one or two cycles, and then *press the "E" key during*

Figure 32.5 When emptying the bell float, separate the valve assembly from the hose as shown.

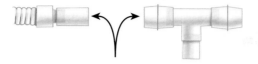

Separate components at this point to empty air from bell float.

an inhalation and follow the instructions on the screen.

After reading two normal tidal cycles, the computer will instruct you to stop after the next (third) normal exhalation and wait for the next prompt. (This means that after exhaling the third time, the subject should stop all air movement and wait for the next prompt.)

The computer will then prompt you to *exhale maximally.* At this point the subject should empty his or her lungs as forcefully and as completely as possible. ***Absolutely no inhaling should take place at this time.***

When the subject is no longer capable of expelling more air, the software will calculate the ERV and graph it on the screen.

If you felt that you made a mistake or did not get good results for some reason, repeat the reading by emptying the bell float and pressing the "E" key.

9. **Determine the *vital capacity* and *FEV*T.** Note in Figure 32.1 that the vital capacity (VC) is the sum of the tidal, expiratory reserve, and inspiratory reserve volumes. This value is determined by directing an individual to take as deep a breath as possible and exhale all the air possible. The average vital capacity for men and women is around **4,500 ml.** Because vital capacity varies with age, sex, and size, you must obtain your predicted normal vital capacity from one of the tables in Appendix C.

While measuring the vital capacity you will also be able to determine *timed vital capacity,* also referred to as the *forced expiratory volume*T (FEV$_T$).

To determine the FEV$_T$ of an individual, the subject takes a deep breath and expels it as fast as possible. An individual with no respiratory impairment should be able to expire 95% of his or her vital capacity within 3 seconds; however, it is the percentage of vital capacity that is expelled within the first second that is of paramount importance. **An individual with no pulmonary impairment should be able to expel 75% of his or her vital capacity within one second.** Individuals with emphysema and asthma will have a much lower percentage due to air entrapment.

In doing these two tests, both the **volume** and **rate** of air flow will be measured. It is for this reason that the subject must exhale as completely and rapidly as possible. The subject must keep trying to squeeze out every little bit of air until the software stops and graphs the values. Proceed as follows to do these two tests:

MS-DOS, Macintosh, and Windows: Empty the air from the spirometer's bell as described in step 7 and then press the "V" key. Follow the instructions on the screen. You will be prompted to *inhale maximally* and then *exhale as forcefully, rapidly, and completely as possible* into the spirometer.

The computer will automatically detect the onset of exhalation and compute the forced expiratory volumes FEV$_1$, FEV$_2$, and FEV$_3$.

Apple II: Empty the spirometer's bell as described in step 7 and then press the "V" key. Follow the instructions on the screen.

At this point let the subject operate the computer because the computer needs to know *exactly* when the subject begins to exhale.

The subject will be prompted to inhale maximally and then *press the "V" key as he or she is exhaling maximally. Emphasis must be on exhaling as forcefully, rapidly, and completely as possible.*

10. **Repeat for each subject.**

MS-DOS and Apple II: When you are satisfied with all the measurements for this subject, press the "N" key and repeat steps 1 through 9 above for each subject in the group.

Macintosh and Windows: When you are satisfied with all the measurements for this subject select **"Add New Record"** from the **Acquire** menu, enter the subject data, click the **Volumes** button, and repeat steps 1 through 9 above for each subject.

11. **View the group averages.**

MS-DOS: Press ESC to return to the main SPIRO-COMP MENU. Select **"Review/Analyze Menu"** and then choose **"Graph Averages."** A stacked bar graph representing the average volumes for all members of the group is displayed. Press the **"M"** or **"F"** key to view the averages for just the males or females, respectively. A single record can be viewed by pressing the **"S"** key.

Macintosh and Windows: View the average volumes for all subjects, only males, or only females by choosing the appropriate items from the **Analyze** menu.

A stacked bar graph representing the average volumes for all members of the group is displayed. An option for viewing any single record is also available from the **Analyze** menu.

Apple II: From the MAIN MENU, select "Data Review (Graph)." You can then choose to view the average volumes for all males, all females, or the entire group. An option for viewing any single record is also available.

12. View the spreadsheet.

MS-DOS: From the Spirocomp Review/Analysis Menu, select "View Spreadsheet." All the records from the entire group are displayed in row/column format. Averages and other summary information are displayed at the end of the listing. If there are more records than will fit on the screen, press the "N" key to display subsequent pages.

Macintosh and Windows: From the **Analyze** menu select **Summary Table.** A new window appears with the data from the entire file displayed in row/column format. Averages are displayed at the top of the listing. Resize the window or use the scroll bars to view any portion of the data not visible.

Apple II: From the MAIN MENU, select "Data Review (Spreadsheet)." You can then choose to view the average volumes for all males, all females, or the entire group.

An option for viewing any single record is also available. Averages and other summary information are displayed at the end of the listing. If there are more records than will fit on the screen, pressing the "N" key will display subsequent pages.

13. Save the data to disk.

MS-DOS: Press ESC to return to the SPIROCOMP MENU and select "Data File/Disk Commands." Then choose "SAVE" File, give the file an appropriate name, and press the Return (Enter) key.

Macintosh and Windows: Each record is automatically saved before a new one is added. To manually save the current record select **Save Record** from the **File** menu.

Apple II: From the MAIN MENU select "Save Current Data" and the select "Catalog/Save Data File." On single floppy drive systems, you will be prompted to insert a data disk. Give the file an appropriate name and press the Return key.

14. Print out.
Press the "P" key to print out any screens needed by individuals.

 ASSIGNMENT

Complete the laboratory report and attach your spirograms to the laboratory report.

THE DIGESTIVE ORGANS

OBJECTIVES

After completing this exercise, you should be able to:

1. Identify the parts of the digestive system on charts and models and describe the function of each.
2. Identify tissues studied microscopically.
3. Define all terms in bold print.

Materials

Torso model with removable organs
Prepared slides of:
    Vallate papillae with taste buds
    Stomach wall, x.s.
    Ileum wall, x.s.
    Jejunum wall, x.s.
    Liver

The digestive system consists of the alimentary canal and the associated glands. The **alimentary canal** is the tube through which food passes, and it consists of the mouth, pharynx, esophagus, stomach, small intestine, and large intestine. The associated glands are the salivary glands, liver, and pancreas. As you study the descriptions that follow, locate the structures on Figures 33.1 through 33.5.

The Oral Cavity

The teeth, tongue, and salivary glands play significant roles in the digestive functions of the oral cavity. When food is taken into the mouth and chewed, it is mixed with saliva secreted by the salivary glands. The food mass is then pushed posteriorly by the tongue into the **oropharynx,** where the swallowing reflex is initiated. The oropharynx is the portion of the pharynx that is posterior to the oral cavity.

The basic structure of the oral cavity is shown in Figure 33.1. Note that the cheeks have been cut and the lips retracted to expose pertinent structures. A mucous membrane lines the lips and cheeks. It forms thickened folds, the **labial frenula,** at the midline on the inner surface of the lips. The **oral vestibule** (label 11) is the space between the cheeks or lips and the teeth of each jaw.

Figure 33.1 The oral cavity.

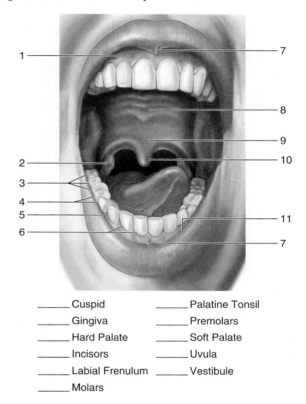

_____ Cuspid	_____ Palatine Tonsil
_____ Gingiva	_____ Premolars
_____ Hard Palate	_____ Soft Palate
_____ Incisors	_____ Uvula
_____ Labial Frenulum	_____ Vestibule
_____ Molars	

The roof of the mouth is formed by the anterior **hard palate,** which is reinforced by bone, and the posterior **soft palate.** The **uvula** is the fingerlike downward projection from the posterior margin of the soft palate. Note that portions of the **palatine tonsils** are visible on the sides of the oropharynx.

The Teeth

All twenty deciduous teeth usually have erupted by the time a child is two years old. Five teeth are present in each quadrant of the jaws. Starting from the median line, they are **central incisor, lateral incisor, cuspid, first molar,** and **second molar.**

There are thirty-two permanent teeth, which begin appearing when a child is about six years of age. The adult dentition is shown in Figure 33.1.

Figure 33.2 Tooth anatomy.

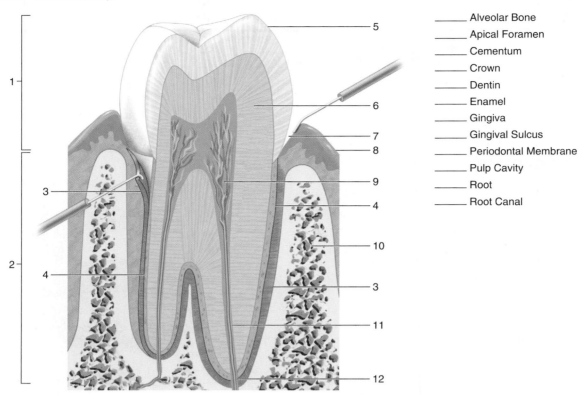

_____	Alveolar Bone
_____	Apical Foramen
_____	Cementum
_____	Crown
_____	Dentin
_____	Enamel
_____	Gingiva
_____	Gingival Sulcus
_____	Periodontal Membrane
_____	Pulp Cavity
_____	Root
_____	Root Canal

Note that there are eight teeth in each quadrant of the jaws. Starting at the median line, they are **central incisor, lateral incisor, cuspid** (or **canine**), **first premolar** (or **bicuspid**), **second premolar** (or **bicuspid**), **first molar, second molar,** and **third molar.** The chisel-like incisors are effective in biting off bits of food. The pointed cuspids are used to puncture and tear food. Premolars and molars have low, rounded cusps suitable for grinding food.

Figure 33.2 illustrates the basic anatomy of a tooth. Note the two major subdivisions of a tooth: a **root** that is embedded in alveolar bone, and a **crown** that is not embedded in bone. The **neck** of the tooth is at the junction of the crown and root.

Four substances compose a tooth. **Enamel** is the hardest substance in the body and forms an ideal chewing surface for the crown. **Dentin** forms most of a tooth and lies under the enamel in the crown. It is similar to bone in composition and hardness. **Cementum** (label 4) is modified bone tissue that covers the exterior of the root and attaches the tooth to the **periodontal membrane.** Fibers of the periodontal membrane penetrate into the bone, holding the tooth firmly in place.

The **pulp cavity** occupies the central part of a tooth. It extends down into the roots in the **root canals.** Pulp consists of loose connective tissues and contains blood vessels, lymphatic vessels, and nerves that enter via the **apical foramina** (label 12) of the roots.

The mucosa covering the jawbones and the neck of each tooth is the **gingiva.** The upper margin of the gingiva is pulled away from the tooth in Figure 33.2 to reveal the **gingival sulcus,** a common site for bacterial putrefaction that may cause periodontal disease.

The Tongue

The tongue consists of skeletal muscle covered by a mucous membrane. The **root** (label 1 in Figure 33.3) of the tongue is covered by the **lingual tonsils.** Note in Figure 33.3 the **epiglottis** extending upward near the root's posterior margin and the **palatine tonsils** located laterally in the oropharynx. The **body** of the tongue extends from the root to the tip. The **lingual frenulum** (not shown) extends between the lower surface of the tongue and the floor of the oral cavity.

The upper surface of the tongue is covered by tiny projections called **papillae.** The enlarged cutout in Figure 33.3 shows the different types. **Filiform papillae** have tapered points, are most numerous,

Figure 33.3 Tongue anatomy.

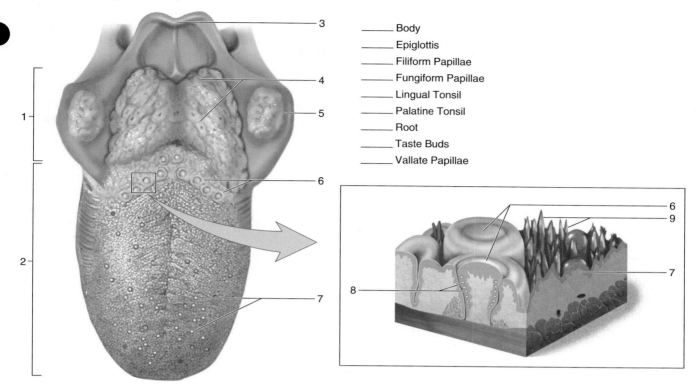

_____ Body
_____ Epiglottis
_____ Filiform Papillae
_____ Fungiform Papillae
_____ Lingual Tonsil
_____ Palatine Tonsil
_____ Root
_____ Taste Buds
_____ Vallate Papillae

and provide a rough texture for handling the food. **Fungiform papillae** (label 7) are low, rounded projections scattered over the surface. Eight to twelve large **vallate** or **circumvallate papillae** occur in a V-shaped pattern on the posterior portion of the body. Both fungiform and vallate papillae contain **taste buds,** which are shown in the circular furrow of a vallate papilla.

The Salivary Glands

Saliva is secreted from three pairs of salivary glands as shown in Figure 33.4. The **parotid gland** is located

Figure 33.4 Major salivary glands.

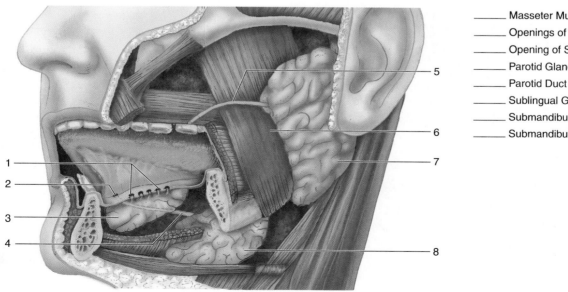

_____ Masseter Muscle
_____ Openings of Sublingual Ducts
_____ Opening of Submandibular Duct
_____ Parotid Gland
_____ Parotid Duct
_____ Sublingual Gland
_____ Submandibular Duct
_____ Submandibular Gland

anterior to the ear and exterior to the masseter muscle. Its watery secretion contains salivary amylase, which catalyzes the digestion of starch, and enters the mouth through a duct that opens near the upper second molar.

The **submandibular gland** (submaxillary) lies within the posterior portion of the mandibular arch. Its secretion contains mucin, which aids in holding food together and imparts a slippery nature to the saliva, facilitating swallowing food. The submandibular duct empties into the anterior floor of the mouth.

The **sublingual gland** is located under the tongue in the anterior floor of the oral cavity. Several small ducts carry its secretion into the floor of the mouth under the tongue. Its secretion contains a high level of mucin.

The Esophagus and Stomach

Refer to Figure 33.5 as you study the discussion of the alimentary canal. When swallowed, food passes into the **esophagus,** a long tube extending from the pharynx to the stomach. Wavelike **peristaltic contractions** begin at the top of the esophagus and carry the food to the **stomach.**

The upper opening of the stomach is guarded by a sphincter muscle, the **cardiac (lower esophageal) sphincter.** It is usually closed but opens to allow the entrance of food. The rounded bulge at the upper end of the stomach is the **fundus,** where most of the food to be digested is held. The lower portion, the **pyloric region,** is smaller in diameter and is the primary site of digestion in the stomach. The stomach region between the pylorus and the fundus is called the **body.**

Figure 33.5   The alimentary canal.

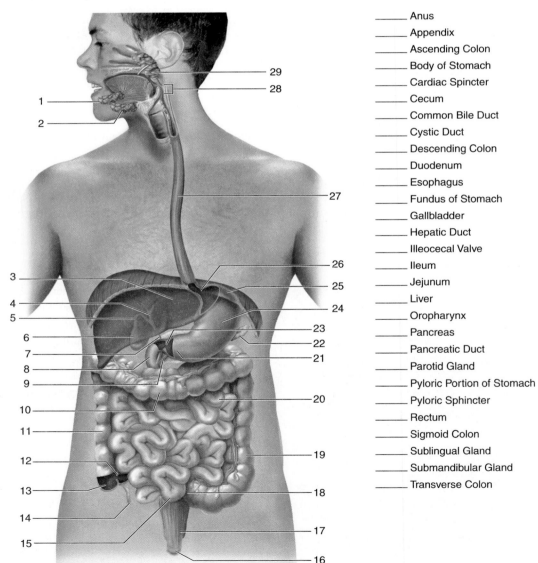

_____ Anus
_____ Appendix
_____ Ascending Colon
_____ Body of Stomach
_____ Cardiac Spincter
_____ Cecum
_____ Common Bile Duct
_____ Cystic Duct
_____ Descending Colon
_____ Duodenum
_____ Esophagus
_____ Fundus of Stomach
_____ Gallbladder
_____ Hepatic Duct
_____ Illeocecal Valve
_____ Ileum
_____ Jejunum
_____ Liver
_____ Oropharynx
_____ Pancreas
_____ Pancreatic Duct
_____ Parotid Gland
_____ Pyloric Portion of Stomach
_____ Pyloric Sphincter
_____ Rectum
_____ Sigmoid Colon
_____ Sublingual Gland
_____ Submandibular Gland
_____ Transverse Colon

Gastric juice, secreted by the stomach mucosa, is highly acidic and contains enzymes to catalyze the digestion of proteins and milk fats. Food is mixed with gastric juice and is converted into a semiliquid substance called **chyme.** The chyme enters the small intestine through another sphincter, the **pyloric sphincter** (label 23) of the stomach.

The Small Intestine

The small intestine is about 23 feet long and consists of three parts: the duodenum, jejunum, and ileum. In Figure 33.5, the small intestine has been shortened for clarity. The first 10–12 inches is the **duodenum** (label 8), which receives bile from the **liver** and digestive secretions from the **pancreas.**

The **hepatic duct** from the liver joins the **cystic duct** from the gallbladder to form the **common bile duct.** Pancreatic juice is carried from the pancreas by the **pancreatic duct** (label 9), which empties into the common bile duct near the duodenum. This union forms a short, dilated duct, the **hepatopancreatic ampulla** (not labeled), that opens into the duodenum. A sphincter muscle, located at the distal end of the ampulla, controls the passage of bile and pancreatic juice into the duodenum. When no food is in the duodenum, the sphincter constricts so that bile from the liver is forced up the cystic duct to the **gallbladder** for temporary storage. When food is present in the duodenum, the sphincter opens, allowing the passage of bile, and the gallbladder contracts, expelling stored bile into the cystic and common bile ducts. **Gallstones** may form in the gallbladder if the bile contains too much cholesterol or is too concentrated.

The middle section of the small intestine is the **jejunum** (label 20), which is 7–8 feet long; the remainder is the **ileum.**

In the small intestine, chyme is mixed with bile and pancreatic juice, which enter via the common bile duct, and with intestinal juice that is secreted by the intestinal mucosa. Digestion of food and absorption of the end products of digestion are completed in the small intestine. Absorption occurs through **villi,** tiny, fingerlike projections of the intestinal mucosa. The villi are extremely numerous and greatly increase the absorptive surface of the small intestine.

The Large Intestine

Indigestible food and water pass from the small intestine into the **large intestine** through the **ileocecal valve** (label 12 in Figure 33.5). The large intestine is about 3½–4 feet in length. There are three continuous sections of the large intestine: cecum, colon, and rectum. The ileum opens into the inferior end of the **ascending colon** at its junction with the pouchlike **cecum** from which the narrow **appendix** extends. The ascending colon runs up the right side of the abdominal cavity and is continuous with the **transverse colon,** which extends across the abdominal cavity below the diaphragm. The **descending colon** runs downward on the left side of the abdomen and continues as the S-shaped **sigmoid colon.** The final 5–6 inches is the **rectum,** which terminates with an opening, the **anus.**

The principal functions of the large intestine are the decomposition of undigested food by resident bacteria, absorption of certain vitamins formed by the bacteria, and reabsorption of water. These actions bring about the formation of the feces, which are passed from the body by defecation.

ASSIGNMENT

1. Label Figures 33.1 through 33.5.
2. Complete Sections A–C of the laboratory report.
3. Locate the digestive organs on the human torso model. Note their relative positions.

Microscopic Study

Compare your observations of the following prepared slides with Figures HA-16 through HA-19 of the Histology Atlas.

1. *Vallate papillae with taste buds.* Note the structure of the papillae and the location of the taste buds. Only substances in solution can reach the taste buds.
2. *Stomach wall.* Locate the columnar epithelium and gastric pits. Try to find chief and parietal cells near the base of a gastric pit. Chief cells produce pepsinogen, a precursor of pepsin, and parietal cells produce hydrochloric acid.
3. *Ileum wall.* Note how the villi increase the surface area of the mucosa. Try to find a **lacteal,** a small lymph capillary, in a villus. Colloidal fats are absorbed into lacteals.
4. *Jejunum wall.* Locate the four layers of the intestinal wall: serosa (visceral peritoneum), muscularis, submucosa, and mucosa. Observe the columnar epithelium with goblet cells on the surface of the villi. Nutrients are absorbed in the villi.
5. *Liver.* Locate a lobule with its **central vein** and **sinusoids** radiating from it between rows of liver cells. Find branches of the **portal vein, hepatic artery,** and **bile duct** at a "corner" of the lobule. On its way to the central vein,

blood from the portal vein and hepatic artery flows through the sinusoids, bathing the liver cells. All central veins merge to form the hepatic vein. Bile flows from the liver cells through small bile channels into the bile duct.

ASSIGNMENT

Make drawings of your observations in Section D on the laboratory report.

Exercise 34

DIGESTION

OBJECTIVES

After completing this exercise, you should be able to:

1. Describe the role of digestive enzymes in the process of digestion.
2. Describe the effect of temperature and pH on the action of salivary amylase.
3. Define all terms in bold print.

Materials

Beaker, 250 ml
Dropping bottles of:
    Iodine (IKI) solution
    10% saliva solution
    pH 5 buffer solution
    pH 7 buffer solution
    pH 9 buffer solution
0.1% soluble starch solution in beaker
Medicine droppers
Test tubes, 13 mm diam. $\times$ 100 mm long
Test tube rack, Wasserman type
Electric hot plate
Water baths at 20° C and 37° C
Thermometers, Celsius
Glass-marking pen

Organic nutrients are required for the normal metabolic activities of body cells. The processes involved in providing organic nutrients to body cells are (1) ingestion of food, (2) digestion of food, (3) absorption of nutrients into the blood, and (4) distribution of nutrients to body cells. In this exercise you will investigate the process of digestion using starch as a food and amylase as the enzyme that catalyzes starch digestion.

Digestion and Digestive Enzymes

A normal diet contains carbohydrates, lipids, and proteins plus small quantities of vitamins and minerals. Although vitamins and minerals play a vital role in metabolic processes, it is the carbohydrates, lipids, and proteins that provide the necessary raw materials for tissue repair and growth. In addition, they are the sources of energy required to power body functions.

However, the large molecules of carbohydrates, lipids, and proteins cannot be absorbed by the mucosa of the digestive tract and enter the blood. They must be reduced to absorbable **nutrients** by digestion.

Digestion is a complex process that involves both mechanical and chemical components. Food is mechanically broken down into smaller pieces by mastication (chewing), which increases the surface area of the food. The conversion of food into absorbable nutrients occurs by a chemical process called **enzymatic hydrolysis.** This process may be expressed by the following:

$$\begin{array}{ccc} \text{large} & & \text{small} \\ \text{nonabsorbable} & \xrightarrow[\text{(+ H}_2\text{O)}]{\textit{(digestive enzymes)}} & \text{absorbable} \\ \text{molecules} & & \text{molecules} \end{array}$$

In hydrolysis, water combines with large molecules and splits them into smaller molecules. Hydrolysis without the catalytic action of enzymes is too slow to sustain life. It is the **digestive enzymes** found in saliva and gastric, pancreatic, and intestinal juices that greatly increase the rate of the hydrolysis reactions and enable us to derive nutrients from the food we eat. Relatively few enzymes can catalyze a great number of reactions since the enzymes are not altered in the reaction and are simply recycled. A summary of the major digestive enzymes, their sources, and their actions are shown in Table 34.1.

Enzymes are proteins. Each type of enzyme interacts with a specific type of substrate molecule. This specificity results from the three-dimensional shape of the enzyme, which allows it to fit onto a particular type of substrate molecule. The shape of an enzyme is largely determined by hydrogen bonds that form between certain amino acids in the molecule. Temperature and pH changes may alter the hydrogen bonding and change the shape of the enzyme so that it is no longer able to fit onto the substrate molecule. When this happens, the enzyme is inactivated. Most enzymes function optimally within rather narrow temperature and pH ranges.

TABLE 34.1
Summary of the Major Digestive Enzymes

Enzyme	Digestive Action
Saliva	
Amylase	Starch to maltose
Gastric Juice	
Lipase	Milk fat to fatty acids** and monoglycerides**
Pepsin	Proteins to peptides
Pancreatic Juice	
Amylase	Starch to maltose
Trypsin	Proteins to peptides
Chymotrypsin	Proteins to peptides
Carboxypeptidase	Peptides to peptides and amino acids
Lipase	Lipids to fatty acids** and monoglycerides**
Intestinal Juice	
Maltase	Maltose to glucose*
Sucrase	Sucrose to glucose* and fructose*
Lactase	Lactose to glucose* and galactose*
Lipase	Lipids to fatty acids** and monoglycerides**
Peptidase	Peptides to amino acids***

*End products of carbohydrate digestion
**End products of lipid digestion
***End products of protein digestion

 ASSIGNMENT

Complete Sections A, B, and C of the laboratory report.

Starch Digestion Experiments

The enzyme **amylase** catalyzes the hydrolysis of starch, a polysaccharide, into maltose, a disaccharide. Starch digestion begins in the mouth through the action of salivary amylase and is completed in the small intestine by the action of pancreatic amylase.

In the experiments that follow, you will investigate the effect of temperature and pH on the digestive action of amylase on starch dissolved in an aqueous solution.

The experiments are best performed by students working in pairs. If time is limited, it may be necessary to divide the class into three groups, with each performing part of the experiments as shown in Table 34.2. Results are to be shared with the entire class.

Solutions will be provided in dropping bottles. You will dispense the solutions with medicine droppers in quantities of "drops" or "droppers." A

TABLE 34.2
Division of Labor

Group	Assignment
A (1/3 of class)	Controls, 20° C, boiling
B (1/3 of class)	Controls, 37° C, boiling
C (1/3 of class)	Controls, pH effects

dropper means *one dropper full of solution,* approximately 1 ml. You must maintain cleanliness and use good laboratory procedures in dispensing the solutions in order to achieve good results. Contamination must be avoided. Do not allow a dropper to touch any solution except the one it is used to dispense. You can avoid problems in performing the experiments if you understand the procedures thoroughly and establish a division of labor among your partners before you start.

Amylase Preparation

Prior to the laboratory session, your instructor prepared either a 10% saliva solution or a 1% commercial amylase solution. The directions that follow assume the use of a 10% saliva solution. Glassware in contact with saliva is to be placed in a 0.25% Amphyl solution after use.

Controls

In the experiments, the activity of the amylase will be determined by using a color test: the iodine (IKI) test for starch. Prepare a set of control tubes as follows to compare with your experimental results:

1. Label two clean test tubes 1 and 2, and add to them the ingredients as follows:
   Tube 1: 1 dropper starch, 2 drops iodine solution
   Tube 2: 1 dropper distilled water, 2 drops iodine solution
2. Your results are interpreted as follows:
   Tube 1: A blue-black coloration is positive for starch.
   Tube 2: An amber coloration is negative for starch.

Effect of Temperature

In this section, you will compare the activity of amylase at 20° C (room temperature) and 37° C (body temperature). A water bath is provided at each temperature. The general protocol for the experiment is shown in Figure 34.1. Proceed as follows for each temperature to be tested:

1. Label 10 test tubes 1 through 10 and arrange them sequentially in the test tube rack.
2. Add 1 dropper of starch solution to each of the 10 test tubes.

Figure 34.1 Amylase activity at 20° C and 37° C.

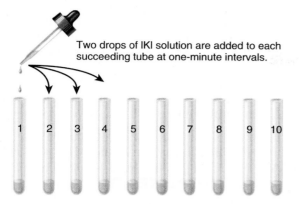

Two drops of IKI solution are added to each succeeding tube at one-minute intervals.

Ten Tubes: Each contains 1 ml of starch solution and 1 drop of 10% saliva. Keep in water baths at 20° C and 37° C.

Figure 34.2 Effect of boiling on amylase.

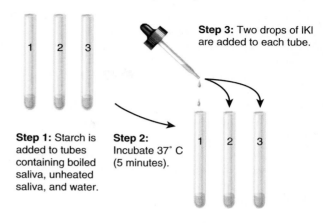

Step 1: Starch is added to tubes containing boiled saliva, unheated saliva, and water.

Step 2: Incubate 37° C (5 minutes).

Step 3: Two drops of IKI are added to each tube.

3. Place the rack of test tubes and the dropping bottle of saliva in the water bath. Allow 5 minutes for temperature equilibration.
4. *Record the time,* and place 1 drop of saliva solution into each tube starting with tube 1. Be sure that each drop lands directly in the starch solution and does not run down the side of the tube. Mix by agitation.
5. *Exactly 1 minute* after tube 1 received the saliva solution, add 2 drops of IKI to tube 1.
6. *One minute later,* add 2 drops of IKI to tube 2. Continue to add IKI sequentially to the tubes at 1-minute intervals until all 10 tubes have received IKI solution.
7. Remove the tubes from the water bath after adding the iodine solution and compare the color of the solutions with the controls to determine the time required to digest the starch. Record your results on the laboratory report.
8. Discard the tube contents in the sink and place the tubes in the pan containing Amphyl solution.

Effect of Boiling

In this section, you will compare the activity of amylase that has been heated to boiling with unheated amylase. Figure 34.2 depicts the general procedure.

1. Label three test tubes 1, 2, and 3.
2. Add 5 drops of saliva solution to tubes 1 and 2, and 5 drops of distilled water to tube 3.
3. Place about 100 ml water in a beaker and heat it to boiling on a hot plate. Place tube 1 in the boiling water bath for 3 minutes.
4. Add 1 dropper of starch solution to each tube and mix by agitation. Place the tubes in a water bath at 37° C for 5 minutes.

Figure 34.3 Effect of pH on amylase activity.

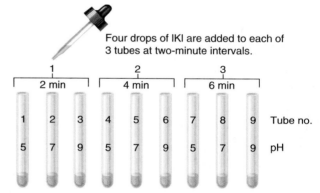

Four drops of IKI are added to each of 3 tubes at two-minute intervals.

1			2			3			
2 min			4 min			6 min			
1	2	3	4	5	6	7	8	9	Tube no.
5	7	9	5	7	9	5	7	9	pH

Nine Tubes: Each contains 1 ml buffered solution, two drops of 10% saliva, and 1 ml of starch solution in 37° C water bath.

5. Remove the tubes from the water bath and add 2 drops of IKI to each tube. Record the results on the laboratory report.
6. Discard the tube contents in the sink and place the tubes in the pan of Amphyl solution.

Effect of pH

In this section, you will determine the effect of three different hydrogen ion concentrations on amylase activity. The general procedure is shown in Figure 34.3.

1. Label nine test tubes 1 through 9. Arrange them in sequence in the test tube rack and write the pH designation on each tube as

shown in Figure 34.3. Note that there are three tubes at each pH.

2. Add 1 dropper of buffer solution as follows:
   pH 5 buffer to tubes 1, 4, and 7
   pH 7 buffer to tubes 2, 5, and 8
   pH 9 buffer to tubes 3, 6, and 9

3. Add 2 drops of saliva solution to each tube, mix by agitation, and place the rack with the tubes in the water bath at 37° C. Also, place the dropping bottle of starch in the water bath. Allow 5 minutes for temperature equilibration.

4. *Record the time* and, starting with tube 1, add 1 dropper of starch solution to each tube. Mix by agitation.

5. *Two minutes* after the recorded time, add 4 drops of IKI to tubes 1, 2, and 3. *Four minutes* after the recorded time, add 4 drops of IKI to tubes 4, 5, and 6. *Six minutes* after the recorded time, add 4 drops of IKI to tubes 7, 8, and 9.

6. Remove the tubes from the water bath after adding the iodine solution and compare them with the controls. Record the results on the laboratory report.

7. Discard the contents and place the tubes in the Amphyl solution. Wash your workstation with Amphyl solution.

 **ASSIGNMENT**

Complete the laboratory report.

Exercise 35

THE URINARY ORGANS

OBJECTIVES

After completing this exercise, you should be able to:

1. Identify the components of the urinary system on charts or models and describe their functions.
2. Identify the parts of the kidney on charts, models, or specimens and describe their functions.
3. Identify a glomerulus and glomerular capsule when viewed microscopically.
4. Define all terms in bold print.

Materials

Models of the urinary system and a kidney
Sheep kidneys, fresh or preserved
Sheep kidneys, triple injected, sectioned
Dissecting kits and trays
Long, sharp knife
Prepared slides of kidney cortex and medulla
Protective disposable gloves

Before You Proceed

Consult with your instructor about using protective disposable gloves when performing portions of this exercise.

Organs of the Urinary System

The components of the urinary system are shown in Figure 35.1: two kidneys, two ureters, the urinary bladder, and the urethra. The **kidneys** are bean-shaped, reddish brown organs located on either side of the vertebral column and posterior to the parietal peritoneum (retroperitoneal). Each kidney receives blood from a **renal artery,** which branches from the **abdominal aorta.** Blood leaves each kidney through a **renal vein,** which empties into the **inferior vena cava.**

Urine, formed by the kidneys, is carried from each kidney to the **urinary bladder** through a slender tube, a **ureter.** Peristaltic contractions of the ureter wall propel the urine to the bladder. Each ureter originates as a funnel-like **renal pelvis** in the kidney and

Figure 35.1 The urinary system.

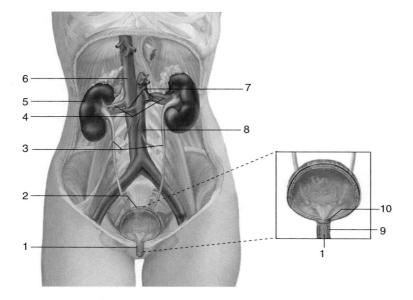

_____ Abdominal Aorta
_____ Inferior Vena Cava
_____ Kidney
_____ Renal Arteries
_____ Renal Veins
_____ Sphincter, External
_____ Sphincter, Internal
_____ Ureters
_____ Urethra (2 places)
_____ Urinary Bladder

descends parallel to the vertebral column and posterior to the peritoneum. The lower end enters the posterior surface of the urinary bladder.

Urine is temporarily stored in the distensible urinary bladder and then voided from the bladder via a short tube, the **urethra.** The male urethra is about 20 cm in length; the female urethra is approximately 4 cm in length. *Cystitis,* inflammation of the urinary bladder, is more common in females than males because the shorter length of the female urethra provides an easier entrance for pathogens.

The passage of urine from the bladder is called **micturition** and is controlled by two sphincter muscles. The **internal sphincter** is located at the junction of the bladder and urethra. It is formed of smooth muscle and is under parasympathetic control. The **external sphincter** is located in the urethra about 2 cm from the bladder. It consists of skeletal muscle and is under voluntary control.

When about 300 ml of urine has accumulated in the urinary bladder, the stretching of the bladder walls initiates an urge to urinate and a subconscious reflex, which causes the walls to contract. This contraction forces urine past the internal sphincter to the external sphincter, creating a sensation of urgency. When the external sphincter is consciously relaxed, micturition occurs.

ASSIGNMENT

1. Label Figure 35.1.
2. Locate the parts of the urinary system on the model and note their relationships to the adjacent structures.

Kidney Anatomy

Figure 35.2 shows the structure of a kidney in frontal section. The kidney is enveloped in a thin fibrous **renal capsule** (label 3) and encased in a thick layer of fat (not shown), which provides protection and support. The **cortex,** the outer portion of the kidney, lies just under the renal capsule. Its reddish brown color results from an extensive blood supply. The inner portion of the kidney is the **medulla** (label 2).

The medulla contains the cone-shaped **renal pyramids** (label 7), which are separated by inward extensions of the cortex known as **renal columns.** The **renal papilla,** the tip of a renal pyramid, projects into a short tube called a **minor calyx** (label 4). Each minor calyx opens into the centrally located **major calyx** (label 5). The major calyx, in turn, is continuous with the funnel-like **renal pelvis** at the upper end of the **ureter.** The indentation in the

Figure 35.2 Anatomy of the kidney.

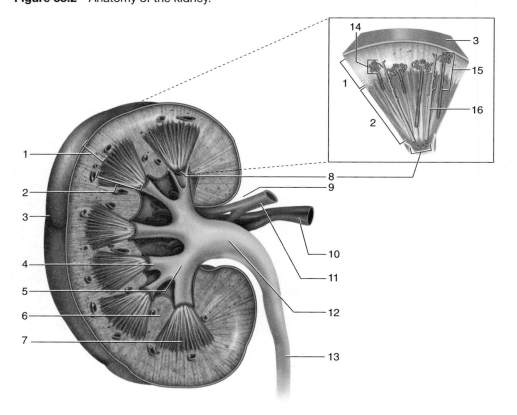

_____ Collecting Duct
_____ Cortical Nephron
_____ Hilum
_____ Juxtamedullary Nephron
_____ Major Calyx
_____ Minor Calyx
_____ Renal Artery
_____ Renal Capsule
_____ Renal Column
_____ Renal Cortex
_____ Renal Medulla
_____ Renal Papilla
_____ Renal Pelvis
_____ Renal Pyramid
_____ Renal Vein
_____ Ureter

Figure 35.3 The nephron.

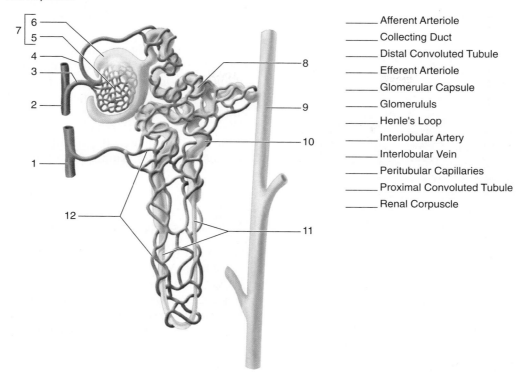

_____ Afferent Arteriole
_____ Collecting Duct
_____ Distal Convoluted Tubule
_____ Efferent Arteriole
_____ Glomerular Capsule
_____ Glomerululs
_____ Henle's Loop
_____ Interlobular Artery
_____ Interlobular Vein
_____ Peritubular Capillaries
_____ Proximal Convoluted Tubule
_____ Renal Corpuscle

kidney where renal blood vessels and the ureter are joined to the kidney is known as the **hilum** (label 9).

The Nephrons

The basic functional unit of the kidney is the **nephron.** There are about a million nephrons in each kidney. The enlargement in Figure 35.2 shows four nephrons, and Figure 35.3 shows a single nephron in greater detail. About 80% of the nephrons are located in the cortex. These are called **cortical nephrons** (label 14 in Figure 35.2). The remaining nephrons, the **juxtamedullary nephrons,** are located partially in the cortex and partially in the medulla.

Refer to Figure 35.3 as you study this section. Each nephron consists of two major parts: a renal corpuscle and a renal tubule. A **renal corpuscle** consists of an inner tuft of capillaries, the **glomerulus** (label 5), and an outer, double-walled **glomerular capsule** that envelopes the glomerulus. A **renal tubule** consists of three sequential segments: (1) the **proximal convoluted tubule,** which leads from the glomerular capsule, (2) **Henle's loop**—the downward U-shaped portion, and (3) the **distal convoluted tubule** (label 10)—the terminal segment that empties into a collecting duct. Several tubules empty into a single **collecting duct.** There are thousands of collecting ducts in each renal pyramid. Their presence produces the faint lines extending from the base to the apex of each pyramid.

Urine Formation

Nephrons form urine by three processes: (1) filtration, (2) reabsorption, and (3) secretion. In this way, water and essential substances in the blood are conserved while the concentrations of surplus substances, including **nitrogenous wastes** (urea, uric acid, and creatinine), are reduced. It is the process of urine formation that maintains the normal concentration of substances in blood plasma.

The blood supply to the kidney is shown in Figure 35.4. Note that the arteries and veins are essentially parallel to each other. The **renal artery** branches to form **interlobar arteries** that extend through the renal columns to form **arcuate arteries** that run along the base of the renal pyramids. The arcuate arteries give off small and numerous branches, the **interlobular arteries** (label 2, Figure 35.3), which permeate the cortex and give off extremely numerous branches, the afferent arterioles. Each **afferent arteriole** (label 3, Figure 35.3) carries blood to a **glomerulus.** An **efferent arteriole** (label 4, Figure 35.3) carries blood from a glomerulus to the **peritubular capillaries** (label 12, Figure 35.3) that enmesh the nephron tubule. From the capillaries, blood drains into an **interlobular vein,** which opens into an **arcuate vein.** Arcuate veins empty into **interlobar** veins, which carry blood to the **renal vein.**

The efferent arteriole has a smaller diameter than the afferent arteriole. This difference elevates the

Figure 35.4 Renal circulation.

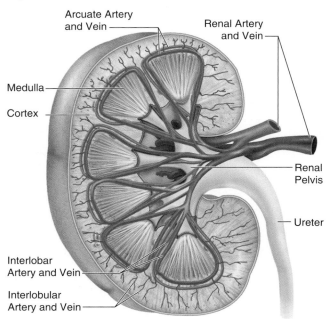

Arcuate Artery and Vein

Renal Artery and Vein

Medulla

Cortex

Renal Pelvis

Ureter

Interlobar Artery and Vein

Interlobular Artery and Vein

blood pressure within the glomerulus and forces the **glomerular filtrate,** a dilute fluid derived from blood plasma, into the glomerular capsule. About 600 ml of blood plasma flows through the glomeruli each minute: 860 liters in 24 hours. The kidneys form about 125 ml of filtrate per minute: 180 liters per day. The filtrate consists of all substances present in the blood except the formed elements. However, very few plasma proteins enter the filtrate because their molecules are too large.

As the filtrate passes through the renal tubule, various substances are reabsorbed into the peritubular capillaries. For example, almost all of the proteins that enter the filtrate are reabsorbed by tubule cells. A few substances are secreted from the

capillaries into the filtrate. This selective reabsorption and secretion of substances by the renal tubule plays a major role in maintaining the constancy of the body fluids.

The remaining filtrate passes into a collecting duct, where it may be further concentrated or diluted by the reabsorption of water or mineral ions. The filtrate flows from the renal papilla into a minor calyx and on into the major calyx, renal pelvis, and ureter. When it reaches the renal pelvis, it is called urine. Table 35.1 compares the quantity of selected substances in blood plasma, tubule filtrate, and urine.

ASSIGNMENT

1. Label Figures 35.2 and 35.3 and complete Section A of the laboratory report.
2. Examine a sectioned, triple-injected kidney under a demonstration dissecting microscope. Note the many renal corpuscles. Can you distinguish the glomerulus, glomerular capsule, and tubules?
3. Examine a prepared slide of kidney cortex and medulla. Compare your observations with Figure HA-20 and locate the labeled structures. Note that the walls of the glomerular capsule and the renal tubule are only one cell thick. Make diagrams of your observations in Section B of the laboratory report.
4. Complete Section C of the laboratory report.

Sheep Kidney Dissection

1. Obtain a sheep kidney for study. If it is still encased in fat, remove the fat carefully with your hands. Look for the adrenal gland embedded in the fat near one end of the kidney. Cut the gland in half and note that it has a distinct outer cortex and an inner medulla.
2. Insert a dissecting needle into the kidney to distinguish the tougher renal capsule from the softer underlying tissue.
3. With a long, sharp knife, cut the kidney longitudinally to make a frontal section similar to Figure 35.2. Wash the cut surfaces.
4. Locate all of the macroscopic structures shown in Figure 35.2.

ASSIGNMENT

Complete the laboratory report.

TABLE 35.1
Quantity of Selected Substances in a 24-hour Period in (1) Blood Plasma Flowing through the Kidneys, (2) Filtrate Passing through the Glomerular Capsule, and (3) Urine

Substance	Plasma	Filtrate	Urine
Total volume	806 l	180 l	1 l
Proteins	7,500 g	10 g	0
Chloride ions	3,180 g	667 g	7 g
Potassium ions	170 g	36 g	3 g
Sodium ions	2,924 g	612 g	5 g
Glucose	860 g	180 g	0
Creatinine	8.6 g	1.5 g	1.5 g
Urea	215 g	45 g	25 g
Uric acid	43 g	9 g	0.8 g

Exercise 36

URINE AND URINALYSIS

OBJECTIVES

After completing this exercise, you should be able to:

1. Identify the normal components and characteristics of urine.
2. List the common abnormal components and characteristics of urine, and identify possible clinical conditions associated with them.
3. Perform a simple urinalysis.
4. Define all terms in bold print.

Materials

Urine "unknowns"
10SG-Multistix reagent strips and analysis charts (Ames)
Urine collection cups, plastic
Test tubes and racks
Amphyl, 0.25%
Biohazard bag
Glass marking pen
Protective disposable gloves

 Before You Proceed

Consult with your instructor about using protective disposable gloves when performing portions of this exercise.

Urine is a complex aqueous solution of organic and inorganic substances. Most of the solutes are either waste products of cellular metabolism or products formed from certain foods.

The volume, pH, and solute concentration of urine may vary within normal limits in a healthy person as the constancy of the internal environment is maintained. In contrast, the physical and chemical characteristics of urine may change dramatically in certain pathological conditions. Thus, a **urinalysis** is a commonly used clinical test because it provides much information about the condition of the body.

The physical characteristics of normal urine are shown in Table 36.1, and the major organic and inorganic solutes in normal urine are found in Table 36.2.

TABLE 36.1
Physical Characteristics of Normal Urine

Characteristic	Description*
Volume	0.7 to 2.5 l in 24 hr. Varies with diet, water intake, concentration of plasma solutes, temperature, physical activity, emotional state, and other factors.
Color	Pale yellow to amber due to normal amounts of urochrome. The greater the solute concentration, the darker the color. Certain foods and drugs affect the color.
Turbidity	Clear when fresh but may become turbid after standing.
Odor	Characteristic aromatic odor, but becomes ammonia-like after standing due to bacterial breakdown of urea.
pH	4.8 to 7.5, average of 6.0. Acid in high protein diets; alkaline in vegetable diets.
Specific gravity	1.015 to 1.025 in 24-hr specimens. Single specimens from 1.002 to 1.030. The greater the concentration of solutes, the higher the specific gravity.

*Based on 24-hour urine specimen.

TABLE 36.2
Principal Solutes of a Normal 24-Hour Urine Specimen

Solute	Grams	Comments
Organic		
Urea	25.0	Forms up to 90% of nitrogenous wastes. Produced by liver after deamination of amino acids.
Creatinine	1.5	By-product of muscle metabolism.
Uric acid	0.8	Product of nucleic acid catabolism. Tends to form crystals; a common constituent of kidney stones (renal calculi).
Inorganic		
NaCl	15.0	Most abundant mineral salt. Amount varies with intake.
K^+	3.3	As chloride, phosphate, and sulfate salts.
Sulfates	2.5	As sodium, potassium, magnesium, or calcium compounds.
Phosphates	2.5	As sodium, potassium, magnesium, or calcium compounds.

Some variations from these values are normal, but marked variations are considered abnormal.

Abnormal Characteristics of Urine

In this section, you will study the abnormal characteristics of urine, which may have clinical implications.

Physical Characteristics

Volume   Normal volume ranges from 0.7 to 2.5 liters per day. An excessive production of urine, **polyuria,** occurs in diabetes insipidus and diabetes mellitus; very low or no urine production, **anuria,** occurs in renal failure.

Color   Urochrome gives the color to normal urine. It is the end product of hemoglobin breakdown. Certain foods may cause a color change without having pathological significance. Carrots may produce a yellow color, beets a reddish color, and rhubarb a brownish color.

Certain colors of urine may indicate pathological conditions.

> **Red to smoky brown:** presence of hemoglobin or red blood cells. One or both are found in urine in kidney or urinary tract injury, certain kidney or urinary tract infections, and hemolytic anemia.
> **Brownish yellow or green:** presence of bile pigments. Bile pigments are found in urine in jaundice.
> **Red-amber:** may indicate the presence of porphyrin. Porphyrin occurs in urine in liver damage, jaundice, Addison's disease, and other conditions.

Turbidity   Fresh normal urine usually is clear, but it may become cloudy after standing. Abnormal turbidity usually is caused by the presence of phosphates, urates, blood, pus, or bacteria.

Odor   Variations in odor may be due to diet or drugs, but only odors related to diseases are considered abnormal. For example, the urine may have a fruity odor in diabetes mellitus.

Hydrogen Ion Concentration   The average normal pH of urine is 6.0, but the normal range is from 4.8 to 7.5. Excessive acid reactions occur with fever, **nephritis** (inflammation of the kidney), **glomerulonephritis** (inflammation of the kidney involving glomeruli), and acute **cystitis** (inflammation of the urinary bladder). Excessive alkaline reactions occur in chronic cystitis and prostatic obstruction.

Specific Gravity   Specific gravity is a measure of the concentration of solids in the urine. Pure water has a specific gravity of 1.000. The specific gravity of a 24-hour specimen of normal urine ranges from 1.015 to 1.025. Single specimens may range from 1.002 to 1.030. The greater the concentration of solutes, the higher is the specific gravity. A low specific gravity (dilute urine) may result from excessive water intake, diabetes insipidus, or chronic nephritis. A high specific gravity (concentrated urine) may result from low water intake, diabetes mellitus, fever, or acute nephritis.

Constituents

Proteins   Proteins are normally absent in the urine because their large size usually prevents their passage into the glomerular capsule. However, under certain conditions, albumin, the smallest and most abundant of the plasma proteins, is present in the urine—a condition known as **albuminuria.** Temporary albuminuria may be caused by excessive physical exertion. Chronic albuminuria results from excessive permeability of the glomerular membrane and may be caused by bacterial toxins and glomerulonephritis.

Traces of mucin, a glycoprotein, may be normally present due to secretion from the urinary epithelium. Excessive mucin suggests irritation of the urinary tract.

Glucose   Only trace amounts of glucose are present in normal urine. The presence of larger amounts of glucose in the urine, **glycosuria,** occurs when the glucose level of the blood plasma exceeds normal limits (hyperglycemia), and glucose is not completely reabsorbed from the tubular filtrate. Persistent glycosuria is usually diagnostic of diabetes mellitus. Temporary glycosuria may result from other causes such as the excessive ingestion of carbohydrates.

Ketones   When the body uses an excessive amount of fatty acids in cellular respiration, the incomplete metabolism of fatty acids produces ketones, which are released into the blood and excreted in urine. The presence of ketones in urine, **ketonuria,** may result from an excessively low carbohydrate diet or diabetes mellitus. Diabetic ketosis may lead to acidosis, which can culminate in coma and death.

Hemoglobin   The presence of hemoglobin in urine, **hemoglobinuria,** occurs when hemoglobin is released into blood plasma by disintegrating red blood cells. This condition may result from hemolytic anemia, hepatitis, transfusion reactions, burns, malaria, and other causes.

Figure 36.1 Microscopic elements in urine.

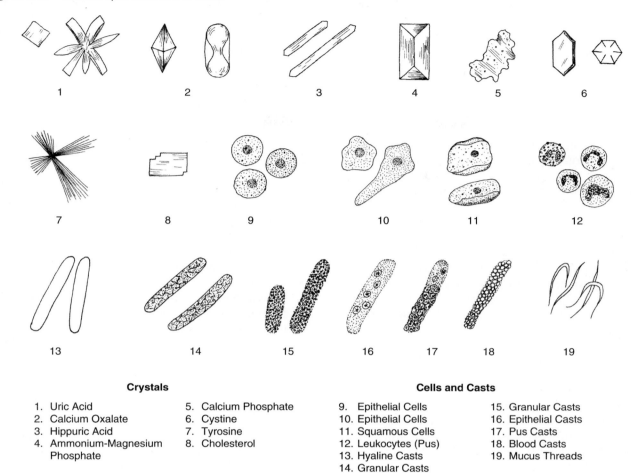

Crystals		Cells and Casts	
1. Uric Acid	5. Calcium Phosphate	9. Epithelial Cells	15. Granular Casts
2. Calcium Oxalate	6. Cystine	10. Epithelial Cells	16. Epithelial Casts
3. Hippuric Acid	7. Tyrosine	11. Squamous Cells	17. Pus Casts
4. Ammonium-Magnesium Phosphate	8. Cholesterol	12. Leukocytes (Pus)	18. Blood Casts
		13. Hyaline Casts	19. Mucus Threads
		14. Granular Casts	

Erythrocytes   The presence of red blood cells in urine, **hematuria,** indicates bleeding in the urinary tract, resulting from disease or trauma. It is often the first symptom of malignant kidney tumors.

Leukocytes   The presence of leukocytes or other components of pus in the urine, **pyuria,** indicates inflammation in the urinary tract, such as glomerulonephritis, cystitis, or urethritis (inflammation of the urethra).

Bilirubin   Bilirubin, a bile pigment, is formed by the breakdown of hemoglobin. Normally, only trace amounts are present in urine. The presence of larger concentrations, **bilirubinuria,** usually indicates liver pathology, such as hepatitis or cirrhosis, in which bilirubin is not removed from the blood by liver cells.

Urobilinogen   Bacteria in the large intestine convert bilirubin to urobilinogen. Some of this compound is absorbed into the blood and removed by liver cells in the formation of bile. Excessive amounts in the urine, **urobilinogenuria,** usually indicate an excessive destruction of hemoglobin or liver pathology.

Casts   Casts are concentrations of cells or cellular debris that have hardened in the tubules before being forced from the tubules by the buildup of filtrate behind them. See Figure 36.1. Casts always indicate kidney pathology, such as glomerulonephritis.

Renal Calculi   Calculi (kidney stones) are usually formed of crystals of uric acid, calcium oxalate, or calcium phosphate. They usually form in the pelvis of the kidney. They may lodge in the ureters, bladder, or urethra, where they may cause pain, hematuria, and pyuria.

Microbes   Urine normally contains low concentrations of bacteria ($<$1,000 bacteria/ml) derived from the urethral flora. An excessive concentration of bacteria in the urine, ***bacteriuria,*** or the presence of yeasts or protozoans indicates a urinary infection.

ASSIGNMENT

1. Study Tables 36.1 and 36.2.
2. Complete Sections A and B of the laboratory report.

Urinalysis

A routine urinalysis includes a description of the physical characteristics and a determination of the presence or absence of abnormal components. Microscopic examination of urine sediments and tests for the presence, identification, and concentration of microbes may also be performed; in some cases these tests are the most important parts of the analysis. Figure 36.1 illustrates the types of crystals, cells, and casts that may occur in urine. However, microscopic examination of urine sediments is beyond the scope of this course. As an alternative, your instructor may set up slides of urine sediments under demonstration microscopes for your examination.

Your instructor has prepared several simulated "unknown" urine specimens. You will be assigned two "unknowns" to analyze in addition to your own urine specimen. You are to determine the physical characteristics of the urine specimens by visual examination and to test for abnormal urine constituents using 10SG-Multistix reagent strips. After the analysis, you are to determine if any of the specimens indicate pathological conditions.

Each Multistix dipstick contains squares of chemical reagents that react with specific urine components. The reagent squares change color to indicate the presence of, or the degree of concentration of, these urine components. From top to bottom, a 10SG-Multistix dipstick has reagent squares that test for leukocytes (pus), nitrite, urobilinogen, protein (albumin), pH, blood (erythrocytes or hemoglobin), specific gravity, ketone, bilirubin, and glucose.

Each Multistix container has a color chart illustrating positive and negative results. After dipping a Multistix into a urine specimen, you must wait for the minimum number of seconds, as noted on the color chart, before reading your results. *Be sure that you know how to read the results before using a Multistix.*

Collect a sample of your own urine as directed by your instructor using the plastic collection cups available. Analyze it just as you do the unknowns. Follow the procedures below and record your findings on the laboratory report.

ASSIGNMENT

1. Swirl the specimen in the container to resuspend any sediments. Record the color and turbidity of the specimen.
2. Fill a clean test tube about three-fourths full with each urine sample to be analyzed. Use a glass-marking pen to label each tube.
3. Remove a 10SG-Multistix reagent strip from the jar while being careful not to touch the colored reagent squares. Compare the color of the reagent squares with the color chart provided, and note the color of negative and positive tests. Note that the resulting color is to be read only after a minimum number of seconds to obtain accurate results. *Be sure that you know how to read the tests before proceeding.*
4. Dip a 10SG-Multistix reagent strip into the urine sample so that all reagent squares are immersed. Remove the strip, tap off the excess urine on the side of the test tube, and lay it on a paper towel with the reagent squares up. Record the time.
5. Observing the minimum reading times, determine and record the results for each reagent square.
6. Discard the urine as directed by your instructor. Place the reagent strips, urine collecting cup, and paper towels in the biohazard bag.
7. Place the test tubes in the Amphyl solution provided. Wash your workstation with Amphyl.
8. Complete the laboratory report.

THE ENDOCRINE GLANDS

OBJECTIVES

After completing this exercise, you should be able to:

1. Identify the location of each endocrine gland, the hormones it produces, and their functions.
2. Indicate known effects of hyposecretion and hypersecretion.
3. Identify each endocrine gland when viewing its tissue microscopically.
4. Define all terms in bold print.

Materials

Prepared slides of endocrine glands

Endocrine glands are ductless glands whose secretions are absorbed into the blood and distributed throughout the body. The secretions of endocrine glands are **hormones,** chemical messengers that alter cellular metabolism. Although hormones are transported to all cells by blood, each hormone acts only on specific target cells that have receptors for that hormone. Other cells are not affected. In this way, the endocrine glands provide a chemical control of body functions.

In this exercise, you will study the location and histology of the endocrine glands and the action of their hormones. You will also learn of the disorders of the endocrine glands that result from either *hyposecretion* (deficient secretion) or *hypersecretion* (excessive secretion) of their hormones.

As you study the endocrine glands, locate each gland and label it in Figure 37.1. Then, correlate the description of its histology with the appropriate photomicrograph.

The Pituitary Gland

The pituitary gland, or **hypophysis,** consists of **anterior** and **posterior lobes** that have different embryological origins and functions. It is attached to the hypothalamus by a slender stalk, the **infundibulum.** The anterior lobe has secretory cells; the posterior lobe does not. The histological section in Figure 37.2 shows the cleft separating the lobes and the two portions of the posterior lobe.

The hypothalamus plays a major role in controlling pituitary functions and, in this way, provides an important link between the nervous and endocrine systems. Special hypothalmic neurosecretory neurons secrete stimulatory (releasing) and inhibitory hormones that are carried by blood directly to the anterior lobe, where they control the release of anterior lobe hormones. The hypothalmic hormones and their actions are shown in Table 37.1. Posterior lobe hormones are produced by hypothalmic neurosecretory neurons whose axons release the hormones within the posterior lobe of the pituitary, where they are absorbed into the blood.

The Anterior Lobe

Three types of secretory cells in the anterior lobe are distinguished by their colors after staining. **Chromophobes** do not take up common dyes used to stain tissues, so they appear pale or colorless. **Acidophils** are stained pink, and **basophils** are stained blue to purple. See Figure 37.3.

Six hormones are produced by various cells in the **anterior lobe,** or **adenohypophysis.** See Figure 37.4. Except for prolactin and growth hormone, the hormones are **tropic hormones** that stimulate other endocrine glands to produce their hormones. Production is regulated by releasing factors (hormones) formed by neurosecretory cells in the hypothalamus and carried by blood to the anterior lobe via the **hypophyseal portal system.**

Growth Hormone (GH) or Somatotropin (SH)

Growth hormone, secreted by a type of acidophil, increases the growth rate of body cells by enhancing protein synthesis and utilization of carbohydrates and lipids. Hypersecretion of GH during growing years produces **gigantism** due to excessive growth in the length of long bones. Hypersecretion in an adult causes **acromegaly,** which is characterized by an enlargement of the small bones in the hands and feet, the mandible, and frontal bones. Hyposecretion in the growing years causes **hypopituitary dwarfism.**

The secretion of GH is regulated by the antagonistic actions of two hypothalmic hormones: **growth hormone-releasing hormone (GHRH)** stimulates GH secretion, and **somatostatin (SS)** inhibits GH secretion.

Figure 37.1 The endocrine glands.

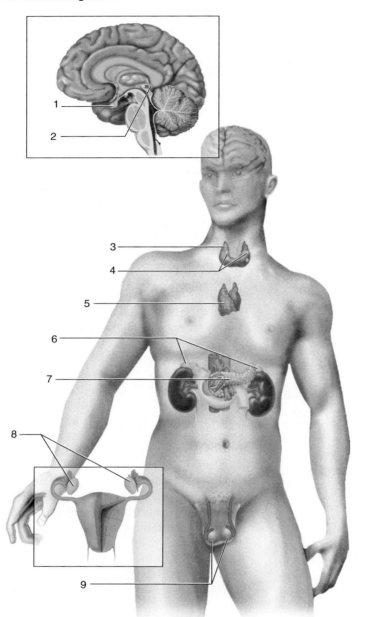

_____ Adrenal Glands
_____ Ovaries
_____ Pancreas
_____ Parathyroid Glands
_____ Pineal Gland
_____ Pituitary Gland
_____ Testis
_____ Thymus Gland
_____ Thyroid Gland

TABLE 37.1

Regulation of the Anterior Pituitary by Hypothalamus-Releasing and -Inhibiting Hormones

Hypothalmic Hormone	Action
GHRH: Growth hormone-releasing hormone	Promotes GH (somatoptropin) secretion
Somatostatin (SS)	Inhibits GH and TSH secretion
PRH: Prolactin-releasing hormone	Promotes PRL secretion
PIH: Prolactin-inhibiting hormone	Inhibits PRL secretion
TRH: Thyrotropin-releasing hormone	Promotes TSH secretion
GnRH: Gonadotropin-releasing hormone	Promotes FSH and LH secretion
CRH: Corticotropin-releasing hormone	Promotes ACTH secretion

Figure 37.2   The pituitary gland.

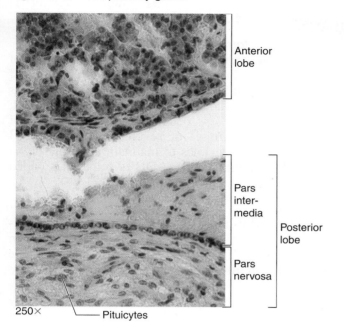

Anterior lobe

Pars intermedia

Posterior lobe

Pars nervosa

250×

Pituicytes

Figure 37.3   Cells of the anterior lobe of the pituitary.

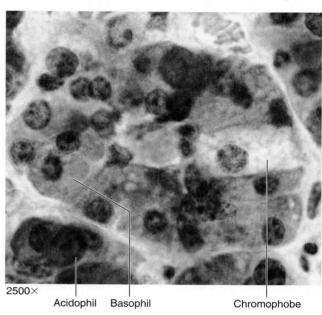

2500×

Acidophil   Basophil   Chromophobe

Figure 37.4   Target tissues of anterior pituitary hormones.

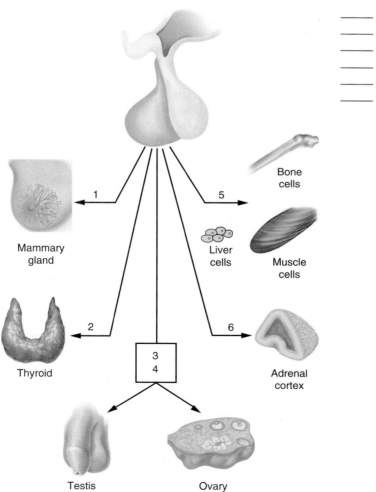

_____ Adrenocorticotropin

_____ Follicle Stimulating Hormone

_____ Growth Hormone

_____ Luteninizing Hormone

_____ Prolactin

_____ Thyroid Stimulating Hormone

Mammary gland

1

5

Bone cells

Liver cells

Muscle cells

Thyroid

2

3
4

6

Adrenal cortex

Testis

Ovary

Prolactin (PRL)

PRL, secreted by another type of acidophil, stimulates the production of milk by mammary glands after childbirth. Its secretion is governed by two antagonistic hypothalmic hormones: prolactin releasing hormone (PRH) and prolactin inhibiting hormone (PIH). Usually, PIH predominates. Near the end of pregnancy, prolactin secretion rises sharply, due to an increase in PRH, and milk production becomes possible. Suckling by an infant inhibits PIH production, continuing prolactin secretion that, in turn, maintains milk production.

Hypersecretion of prolactin causes the absence of menstrual cycles in females and impotence in males.

Thyroid-Stimulating Hormone (TSH) or Thyrotropin

TSH, secreted by basophils, stimulates the production of thyroid hormone by the thyroid gland. TSH secretion is triggered by thyrotropin releasing hormone (TRH) formed by the hypothalamus, which is inhibited by feedback inhibition exerted by an increased level of thyroid hormone. An excessive production of TSH causes hyperthyroidism, and a deficient production results in hypothyroidism.

Adrenocorticotropin (ACTH)

ACTH, secreted by chromophobes, stimulates glucocorticoid production by the adrenal cortex. ACTH secretion is promoted by the corticotropin releasing hormone (CRH) formed by the hypothalamus. ACTH regulation results from glucocorticoids exerting a negative feedback control on CRH production.

Follicle Stimulating Hormone (FSH)

FSH is secreted by basophilic cells that tend to be peripherally located. In females, FSH promotes development and maturation of ovarian follicles and the eggs they contain. In males, it promotes sperm formation by the testes. FSH production is promoted by gonadotropin releasing hormone (GnRH) from the hypothalamus.

Luteinizing Hormone (LH)

LH is secreted by large basophilic cells scattered throughout the anterior lobe. In females, a rapid rise in LH, stimulated by GnRH, causes ovulation and development of a corpus luteum from the empty follicle. High levels of estrogen and progesterone exert a negative feedback control on GnRH.

In males, LH is usually called **interstitial cell stimulating hormone (ICSH)** because it stimulates interstitial cells of the testes to secrete testosterone. Testosterone exerts a negative feedback effect on GnRH production.

The Posterior Lobe

The two hormones from the **posterior lobe** are formed in the hypothalamus by neurosecretory neurons and released within the posterior lobe, which is also called the **neurohypophysis.** Their production is controlled by the hypothalamus.

Antidiuretic Hormone (ADH)

Antidiuretic hormone promotes the reabsorption of water by the kidneys by increasing the permeability of the collecting tubules to water. Thus, **ADH** controls the osmotic balance of body fluids. Without ADH, excessive water loss occurs by the production of dilute urine. At times of severe blood loss, ADH also constricts arterioles to maintain adequate blood pressure. For this reason, ADH is sometimes called **vasopressin.**

Oxytocin

During childbirth, pressure on the cervix initiates a neural reflex that stimulates the secretion of oxytocin. In turn, **oxytocin** increases the strength of uterine contractions to facilitate childbirth.

The secretion of most hormones is controlled by a negative-feedback control mechanism, but oxytocin secretion is controlled by a positive-feedback control mechanism. Pressure on the cervix increases oxytocin secretion, which increases uterine contractions, increasing pressure on the cervix, which stimulates the secretion of more oxytocin. This cycle continually strengthens uterine contractions until childbirth is completed.

After childbirth, stimulation of a nipple by a nursing infant initiates a neural reflex that also releases oxytocin. In this case, oxytocin causes contraction of smooth muscles associated with the milk glands, which forces milk into the milk ducts, leading to release of milk from the breast.

The Thyroid Gland

The thyroid gland is located just below the larynx. It is formed of two lobes, one on each side of the trachea, that are joined by an isthmus of tissue across the anterior tracheal surface. Histologically, the gland contains numerous secretory **follicles** that secrete the thyroid hormone. The follicles are filled with **thyroglobulin,** a colloidal material that holds excess thyroid hormone until it is released. The parafollicular cells, interstitial cells between the follicles, produce a different hormone, calcitonin. See Figure 37.5.

The Thyroid Hormone

The **thyroid hormone (TH)** consists of several related compounds containing iodine in their molecules. Thyroxine (T_4) and triiodothyronine (T_3) are the most important, and they contain four iodine

Figure 37.5 Thyroid and parathyroid glands.

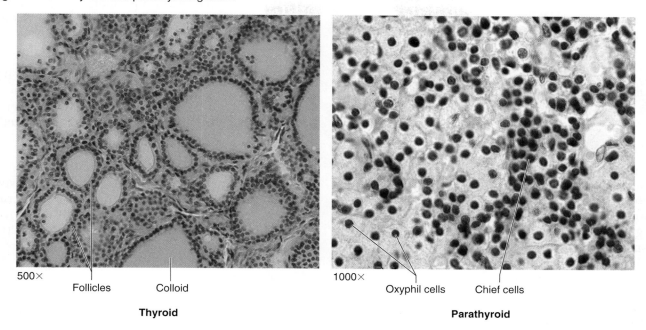

500× Follicles   Colloid

Thyroid

1000× Oxyphil cells   Chief cells

Parathyroid

atoms and three iodine atoms, respectively, in each molecule. Iodine is essential for the physiological activity of thyroid hormone. Thyroid hormone increases the rate of metabolism of most body tissues. For example, it stimulates cellular respiration, heart rate, protein synthesis, mental activity, and growth.

Like most hormones, the secretion of TH is regulated by a negative-feedback system. TRH from the hypothalamus stimulates the secretion of thyroid stimulating hormone (TSH) by the anterior pituitary. TSH promotes the secretion of TH, which, in turn, decreases the production of TRH by the hypothalamus and TSH by the anterior pituitary.

Simple goiter is enlargement of the thyroid gland caused by a deficiency of iodine in the diet. Without sufficient iodine, inadequate amounts of T_4 and T_3 are produced, and the thyroid gland enlarges in an attempt to produce more TH. Severe hyposecretion (hypothyroidism) in children may result in **cretinism,** a condition characterized by physical and mental retardation. In adults, it may cause **myxedema,** which is characterized by obesity, depression, slow heart rate, and sluggishness.

Severe hypersecretion (hyperthyroidism) results in weight loss, nervousness, sleeplessness, rapid heart rate, and sometimes exophthalmos, the protrusion of the eyeballs.

Calcitonin

Calcitonin reduces the level of calcium in the blood by stimulating bone deposition by osteoblasts and inhibiting calcium removal by osteoclasts. A high level of blood calcium stimulates calcitonin production; a low level inhibits production. The action of calcitonin is antagonistic to the function of parathyroid hormone, which is discussed below.

The Parathyroid Glands

The four parathyroid glands are small glands embedded on the posterior surface of the thyroid gland. Figure 37.5 illustrates the histology of a parathyroid gland. The small, numerous **chief cells** produce the parathyroid hormone. The function of the larger **oxyphil cells** is unknown.

Parathyroid Hormone

The **parathyroid hormone (PTH)** raises the calcium concentration and lowers the phosphorus concentration of the blood. This is done by promoting reabsorption of bone tissue by osteoclasts and stimulating calcium reabsorption and phosphorus excretion by the kidneys. The opposing actions of PTH and calcitonin maintain a normal concentration of calcium in the blood. A deficiency of PTH results in **hypocalcemia,** which causes hyperexcitability of the nervous system. When severe, it leads to muscle tremors, spasms, tetany, and even death. Excess PTH results in **hypercalcemia,** which depresses nervous system function leading to muscle weakness, sluggishness, emotional disturbances, and sometimes cardiac arrest.

Production of PTH is regulated by negative feedback control. High levels of blood calcium inhibit production; low levels promote production.

The Pancreas

The pancreas is a pennant-shaped gland that has both exocrine and endocrine functions. It is located between the stomach and duodenum. Each of the three pancreatic hormones (insulin, glucagon, and somatostatin) are produced by a different type of cell found in the **islets of Langerhans.** Cells surrounding the islets of Langerhans are acinar cells, which produce pancreatic juice. See Figure 37.6.

Insulin

Insulin (produced by beta cells) decreases blood glucose by (1) facilitating the passage of glucose through cell membranes into body cells where it can be metabolized and (2) stimulating the conversion of glucose into glycogen in the liver. A deficiency of insulin results in **diabetes mellitus,** a condition in which **hyperglycemia,** an abnormally high blood glucose level, is present.

The concentration of blood glucose regulates insulin production. A high level stimulates production; a low level inhibits production.

Glucagon

Glucagon (produced by alpha cells) increases the blood glucose level by stimulating the liver to (1) convert glycogen into glucose and (2) synthesize glucose from noncarbohydrate sources. Thus, insulin and glucagon have opposite effects on the concentration of blood glucose. A deficiency of this hormone results in **hypoglycemia,** a subnormal level of blood glucose.

Figure 37.6 Pancreas histology.

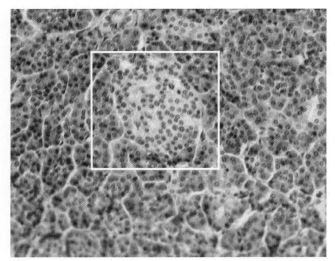

500×

Islet of Langerhans (boxed area)

Regulation is by negative feedback control. A low level of blood glucose promotes glucagon production; a high level decreases glucagon production.

Somatostatin

Somatostatin (produced by delta cells) is produced in the hypothalamus as well as the islets of Langerhans. It inhibits secretion of growth hormone by the anterior pituitary gland and insulin and glucagon by islet cells.

The Adrenal Glands

An adrenal gland sits atop each kidney and consists of two parts: the central **medulla** and the external **cortex.** See Figure 37.7. These two parts arise from different embryological origins and have distinctive functions.

The Adrenal Cortex

The adrenal cortex secretes many different hormones that are collectively called *corticosteroids* (corticoids). These hormones may be subdivided into three different groups. *Mineralocorticoids* control the concentrations of water and electrolytes in the blood, specifically Na^+ and K^+. *Glucocorticoids* control reactions to stress and the metabolism of glucose and proteins. *Gonadocorticoids,* or sex hormones, occur in such small concentrations that their effect is normally minimal. Of the thirty or more corticoids, the two most important are **aldosterone** and **cortisol.**

Figure 37.7 shows the three zones of the adrenal cortex. The outer **zona glomerulosa** lies just under the adrenal capsule. The **zona fasciculata** is the intermediate zone, and the **zona reticularis** is the innermost zone that is adjacent to the adrenal medulla.

Aldosterone

Aldosterone is a mineralocorticoid secreted by cells in the zona glomerulosa. It increases the rate of Na^+ reabsorption and K^+ excretion by the kidneys, which, in turn, increases water reabsorption and decreases urine output. In this manner, the concentrations of Na^+ and K^+ in the blood are maintained within normal levels. A deficiency of aldosterone results in an increase of K^+ in body fluids, a decrease of Na^+, and a reduction in water concentration. When severe, these conditions cause reduced cardiac output, shock, and, possibly, death.

The control of aldosterone secretion is complex. The primary stimulus is an increase in K^+ concentration and a decrease in Na^+ concentration in body fluids.

Cortisol

Cortisol is a glucocorticoid produced primarily by the zona fasciculata. It serves to maintain normal metabolism and to enhance resistance to stress.

Figure 37.7 The adrenal gland. (a) Anatomy; (b) histology.

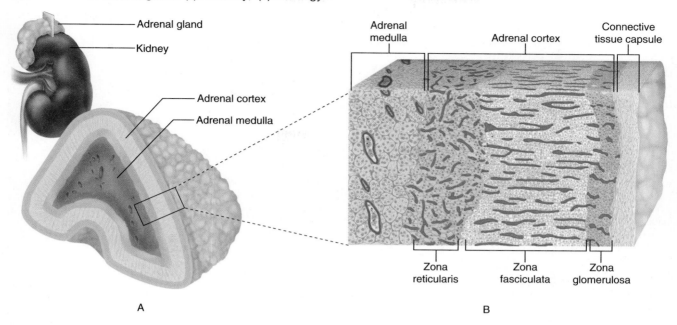

A

B

Cortisol increases the quantity of amino acids, fatty acids, and glucose available to body cells by causing (1) a decrease in protein synthesis, (2) the release of fatty acids from adipose tissue, and (3) the formation of glucose from noncarbohydrates. At therapeutic levels, it also tends to block processes that produce inflammation.

A chronic excess of glucocorticoids may result in **Cushing's syndrome,** a condition characterized by protein depletion, muscle and bone weakness, slow healing, hyperglycemia, hydration of tissues, scraggly hair, and possibly masculinization in females.

A chronic deficiency of glucocorticoids may result in **Addison's disease,** a condition characterized by hypoglycemia, dehydration, and electrolyte imbalance, which may be fatal if untreated.

The production of cortisol and other glucocorticoids is stimulated by the secretion of ACTH from the pituitary gland, which is increased in times of stress.

Sex Steroids

Small amounts of **androgens** and **estrogens** are produced by the zona reticularis. The androgens are relatively unimportant in adult males because the testes produce so much more testosterone, but in females they are responsible for the development of pubic and axillary hair at puberty and maintain the libido (sex drive) in adult life. Adrenal estrogens are relatively unimportant in women of reproductive age, but after menopause they are the only source of estrogens.

Hypersecretion of androgens often accompanies Cushing's syndrome producing **androgenital syndrome (AGS).** In children, it causes enlargement of the penis or clitoris and premature onset of puberty. In adult females, it produces masculinizing effects, such as increased body hair and deepening of the voice.

The Adrenal Medulla

Two related hormones are produced by the adrenal medulla: **epinephrine** and **norepinephrine.** The effects of these hormones resemble those of the sympathetic nervous system: increased heart and respiration rates, elevated blood pressure and blood glucose concentration, and decreased digestive action. The action of these hormones and the sympathetic nervous system prepare the body to meet energy-expending emergencies. The production of both hormones is stimulated by sympathetic nerve fibers.

The Thymus Gland

The thymus gland lies in the mediastinum above the heart and consists of two long lobes. It is relatively large in young children but after puberty diminishes in size throughout life. Histologically, the gland consists of an outer **cortex** and an inner **medulla** as shown in Figure 37.8. The cortex is composed of lymphoid-like tissue that contains massive numbers of lymphocytes. The medulla contains fewer lymphocytes and *thymic corpuscles* whose function is unknown. The thymus secretes a hormone,

Figure 37.8 The thymus and pineal glands.

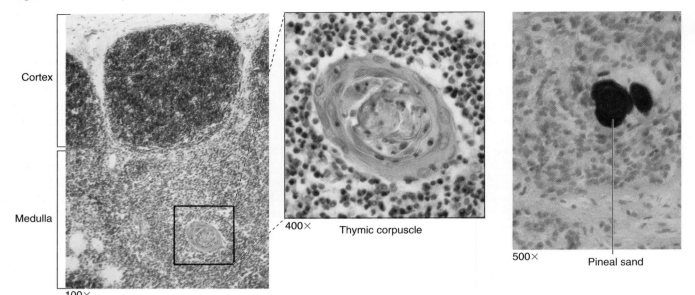

Cortex

Medulla

100×

400×   Thymic corpuscle

500×   Pineal sand

Thymus Gland

Pineal Gland

thymosin, which promotes the development and maturation of T-lymphocytes (*T* for thymus) within the thymus gland. T-lymphocytes develop within the cortex and migrate into the medulla where maturation is completed. Then, mature T-cells enter the blood or lymph and exit the thymus.

The Pineal Gland

The pineal gland is located in the brain under the posterior end of the corpus callosum. At puberty, the gland begins to form small calcium deposits called **pineal sand** whose significance is not known. See Figure 37.8.

The pineal gland's major secretion is the hormone **melatonin,** whose production varies with the day-night diurnal cycle and is regulated by light-generated impulses reaching the pineal via a retina-hypothalmic pathway. Peak levels occur at night and lowest levels occur during midday. Further, seasonal variation in day length alters melatonin secretion.

In animals that reproduce seasonally, reproduction is associated with relative changes in day length, and changes in melatonin production regulate these events. The reproductive function of melatonin in humans is unclear, but it may inhibit precocious sexual development. Variations in melatonin secretion may be involved with normal circadian (roughly 24-hour) rhythms, such as sleep and body temperature. Variations also may play a role in seasonal mood changes experienced by some people.

The Testes

The male gonads, the **testes,** contain numerous coiled seminiferous tubules that produce spermatozoa. Cells located in spaces between these tubules, the **interstitial cells,** secrete the male hormone, **testosterone.** Interstitial cells are shown between sectioned seminiferous tubules in Figure 37.9.

Figure 37.9 Testicular tissue.

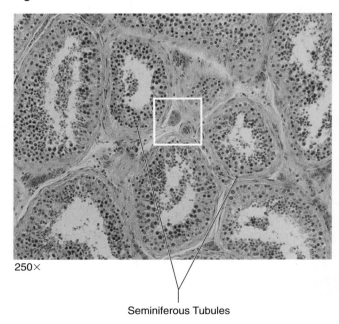

250×

Seminiferous Tubules

Chorionic gonadotropin, secreted by the placenta, stimulates testosterone production during embryonic and fetal development. Testosterone secretion before birth is responsible for the development of the male sex organs and descent of the testes into the scrotum. From birth to puberty, little testosterone is produced.

At puberty, the secretion of interstitial cell stimulating hormone (ICSH) by the anterior pituitary triggers increased testosterone secretion which, in turn, brings about sexual maturation. Testosterone exerts a negative feedback control on ICSH production.

In addition to stimulating the maturation of the male sex organs, increased testosterone levels promote a spurt in overall body growth, especially in height, and the development of secondary sexual characteristics. Male secondary sexual characteristics include (1) male distribution of body hair, (2) larynx enlargement that deepens the voice, (3) an increase in bone thickness and muscle development, and (4) development of broad shoulders.

Testosterone maintains the male sex organs, sex drive, and secondary sexual characteristics in adult life.

The Ovaries

The **germinal epithelium** of the ovary is formed early in embryological development. It produces numerous **primordial follicles** that reside in the ovarian cortex. Some of these follicles migrate inward to become **primary follicles,** which consist of a primary oocyte enclosed in a sphere of follicular cells. Starting at puberty, follicle stimulating hormone (FSH) from the anterior lobe of the pituitary stimulates some of the primary follicles to develop further. Usually, only one will become a **mature,** or **graafian, follicle** containing a secondary oocyte that is released at ovulation in each ovarian cycle. See Figure 37.10. The empty follicle then becomes a **corpus luteum** under the stimulation of luteinizing hormone (LH) from the anterior pituitary.

The ovaries produce the female sex hormones, estrogens and progesterone. FSH stimulates the production of **estrogens** by developing follicles. After ovulation, LH promotes the secretion of **progesterone** by the corpus luteum.

Estrogens

Estrogens consist of several related compounds and are produced primarily by the developing ovarian follicles. But estrogens are also formed by the corpus luteum, by the placenta during pregnancy, and, in minute quantities, by the adrenal cortex.

Estrogens promote the lengthening of long bones, the maturation of female sex organs, and the development of secondary sexual characteristics. Female secondary sexual characteristics include (1) feminization of the skeleton, such as broadening of the pelvic girdle and enlargement of the pelvic opening, (2) development of breasts, (3) increased fat

Figure 37.10 Ovarian tissue.

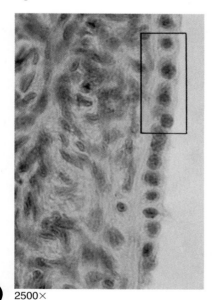

2500×

Germinal epithelium
(boxed cells)

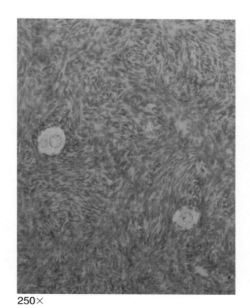

250×

Primordial follicles

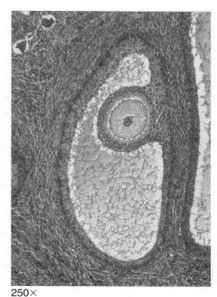

250×

Mature graafian follicle

deposition under the skin, especially in the hips, buttocks, and thighs, and (4) female distribution of body hair. Estrogens also work with progesterone to promote the development of the uterine lining, and they help to regulate the ovarian and uterine cycles.

Progesterone

Progesterone is primarily secreted by a corpus luteum that is formed from an empty follicle after ovulation. Progesterone continues the development of the uterine lining that was begun under the stimulation of estrogens, preparing the uterus to receive an early embryo.

Without pregnancy, degeneration of the corpus luteum causes a decrease in progesterone production, which results in a breakdown of the endometrium, which results in menstruation. If pregnancy occurs, chorionic gonadotropin stimulates the corpus luteum to form increased amounts of progesterone and some estrogen to maintain the uterine lining

until this role is taken over by placental estrogens and progesterone.

Progesterone works with estrogen in promoting the development of secretory cells in the mammary glands of the breasts in preparation for milk production after childbirth.

ASSIGNMENT

1. Label Figures 37.1, 37.4, and 37.11.
2. Complete Sections A through E of the laboratory report.
3. Examine prepared slides of endocrine glands. Compare your observations with the photomicrographs. Make labeled drawings in Section F of the laboratory report that will help you identify the glands from microscope slides.
4. Complete the laboratory report.

Figure 37.11 The pituitary regulatory mechanism.

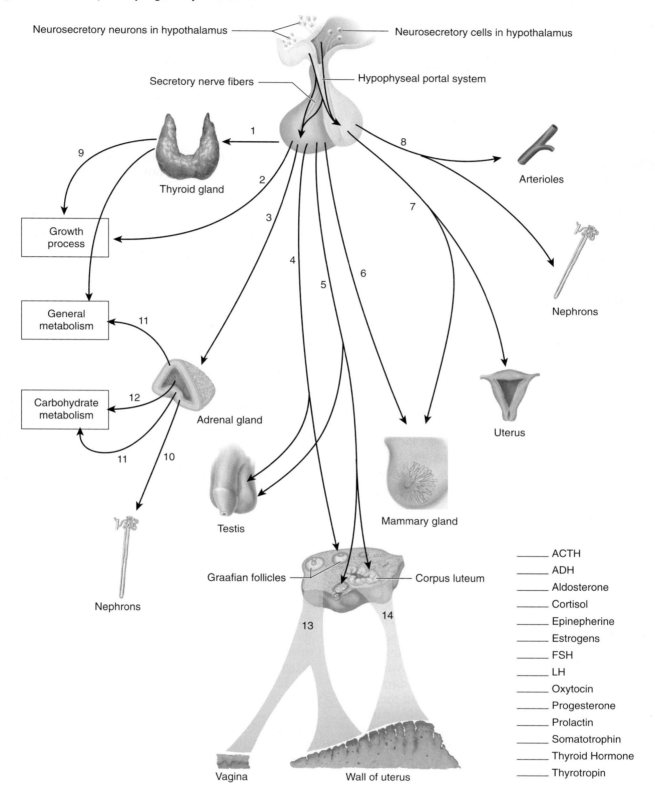

Neurosecretory neurons in hypothalamus

Neurosecretory cells in hypothalamus

Secretory nerve fibers

Hypophyseal portal system

1

9

Thyroid gland

Growth process

2

3

8

Arterioles

7

Nephrons

General metabolism

11

4

5

6

Carbohydrate metabolism

12

Adrenal gland

11

10

Uterus

Nephrons

Testis

Mammary gland

Graafian follicles

Corpus luteum

13

14

Vagina

Wall of uterus

_____ ACTH
_____ ADH
_____ Aldosterone
_____ Cortisol
_____ Epinepherine
_____ Estrogens
_____ FSH
_____ LH
_____ Oxytocin
_____ Progesterone
_____ Prolactin
_____ Somatotrophin
_____ Thyroid Hormone
_____ Thyrotropin

Exercise 38

THE REPRODUCTIVE ORGANS

OBJECTIVES

After completing this exercise, you should be able to:

1. Identify the parts of the male and female reproductive systems on charts or models and describe their functions.
2. Describe spermatogenesis and oogenesis.
3. Identify histological structures in prepared slides of ovary and testis.
4. Define all terms in bold print.

Materials

    Models of human reproductive organs
    Prepared slides of human testis, ovary, uterine wall, and uterine tube

This study of the human reproductive systems includes (1) labeling anatomical illustrations, (2) locating the reproductive organs on classroom models, and (3) microscopic study of spermatogenesis and the tissues of selected organs.

The Male Organs

Figure 38.1 illustrates a sagittal section of the male reproductive system. The primary sex organs (gonads) are the **testes** (testicles), which are located in an exterior pouch, the **scrotum.** They produce the spermatozoa and the male sex hormones. In a male fetus, the testes develop within the abdominal cavity and descend into the scrotum 1 to 2 months before birth.

Partially surrounding each testis is a downward extension of the peritoneum, the **tunica vaginalis** (label 27), that is carried into the scrotum during the descent of the testes. Between a testis and the scrotum is connective tissue and the **cremaster muscle** (label 21). Only a small portion of this muscle is shown in the illustration of testicular detail. The cremaster muscle contracts and relaxes to raise or lower the testis to maintain a testicular temperature of 94° to 95° F. Higher temperatures prevent the production of viable sperm.

A fibrous capsule forms the outer wall of each testis. **Testicular septa,** formed of connective tissue, divide the testis into several lobules. Each lobule contains **seminiferous tubules** that produce **spermatozoa.** All seminiferous tubules merge to form the **rete testis,** a complex network of tubules. Cilia within the tubules of the rete testis move the spermatozoa through several **vasa efferentia** (label 23) and on into the highly coiled **epididymis** (label 6) where the spermatozoa mature. Note that the epididymis lies over the superior and posterior portion of the testis. From the epididymis, spermatozoa pass into the **vas deferens,** the central canal within the **spermatic cord** (label 22). Note the histology of the vas deferens, x.s., especially the smooth muscle layers, in Figure HA-21.

Trace a vas deferens from the epididymis and note that it passes over the pubic bone and urinary bladder into the pelvic cavity. The distal end of each vas deferens is enlarged to form an **ampulla.** The ampulla merges with the duct from a **seminal vesicle** (label 2) to form the **ejaculatory duct.** The ejaculatory duct on each side empties into the **prostatic urethra** (label 11), the portion of the urethra that is within the **prostate gland.** The small **bulbourethral glands** (label 4) empty their secretions into the urethra at the base of the penis.

Semen is formed of spermatozoa and the alkaline secretions of the accessory glands. Two-thirds of the semen is derived from secretions of the seminal vesicles, and about one-third comes from prostatic fluid. Secretions of the bulbourethral glands and spermatozoa account for very little of the semen volume. The alkaline secretions protect the spermatozoa from acid environments, provide nutrients for the sperm, and activate their swimming movements.

Erection of the penis occurs when its three cylinders of erectile tissue fill with blood in response to sexual stimulation. The **corpus spongiosum** is the cylinder of tissue that surrounds the **penile urethra.** Its distal end is enlarged to form the **glans penis,** and its proximal end is enlarged forming the **bulb of the penis.** The two **corpora cavernosa** are located in the dorsal part of the penis and are separated by a medial septum of connective tissue called the **septum penis** (label 15). Note in the cross section of the penis how the cylinders of erectile tissue are separately enveloped by connective

Figure 38.1 Male reproductive organs.

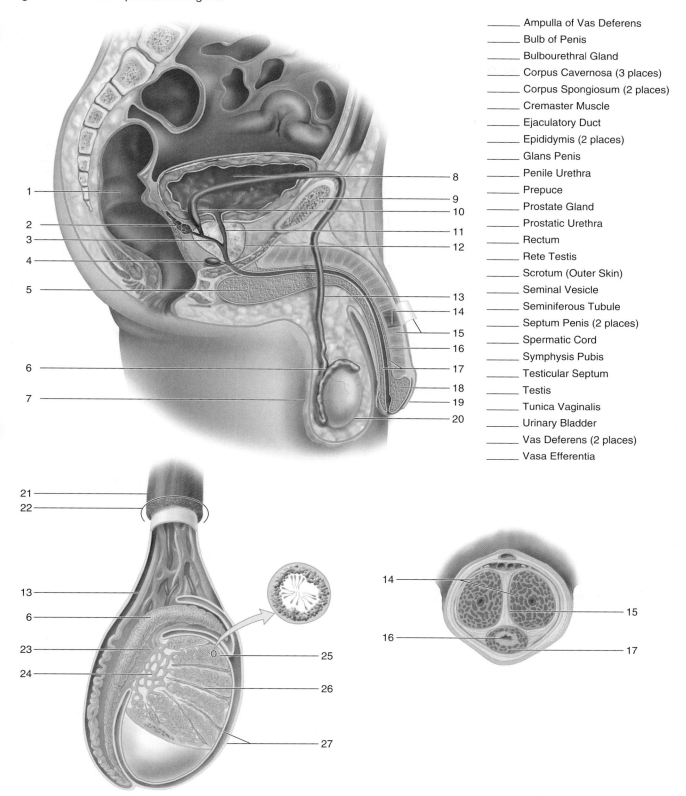

_____ Ampulla of Vas Deferens
_____ Bulb of Penis
_____ Bulbourethral Gland
_____ Corpus Cavernosa (3 places)
_____ Corpus Spongiosum (2 places)
_____ Cremaster Muscle
_____ Ejaculatory Duct
_____ Epididymis (2 places)
_____ Glans Penis
_____ Penile Urethra
_____ Prepuce
_____ Prostate Gland
_____ Prostatic Urethra
_____ Rectum
_____ Rete Testis
_____ Scrotum (Outer Skin)
_____ Seminal Vesicle
_____ Seminiferous Tubule
_____ Septum Penis (2 places)
_____ Spermatic Cord
_____ Symphysis Pubis
_____ Testicular Septum
_____ Testis
_____ Tunica Vaginalis
_____ Urinary Bladder
_____ Vas Deferens (2 places)
_____ Vasa Efferentia

tissue and bound together by additional connective tissue. Observe the histology of the penile urethra and erectile tissue in Figure HA-21.

A sheath of skin, the **prepuce,** begins just behind the glans and extends to cover it. Circumcision is a surgical procedure that removes the prepuce to facilitate sanitation.

During sexual stimulation, the secretion from the bulbourethral glands provides an alkaline environment within the urethra prior to the passage of the spermatozoa. Peristaltic contractions move the spermatozoa from the epididymis into the urethra. The secretions of the seminal vesicles are mixed with sperm in the ejaculatory ducts, and prostatic fluids are added in the proximal part of the urethra. During ejaculation, rhythmic contractions of the bulbospongiosus muscle at the base of the penis propel the semen through the urethra and out of the body.

Sperm motility requires an alkaline environment and nutrients as an energy source. The alkaline prostatic fluid increases the pH of the female vagina from about 4.0 to around 7.5 and activates the sperm. Fructose and other carbohydrates in seminal vesicle fluid serve as an energy source for sperm. Prostaglandins in seminal vesicle fluid stimulate wavelike contractions of the uterus and uterine tubes, helping to move sperm throughout the female reproductive tract.

Spermatogenesis

The process by which spermatozoa are produced in the seminiferous tubules is called **spermatogenesis.** It begins at puberty and continues throughout life. Figure 38.2 illustrates a section of a seminiferous tubule and a diagram of the developmental stages of spermatozoa formation. Both mitosis and meiosis are involved.

Figure 38.2 Spermatogenesis.

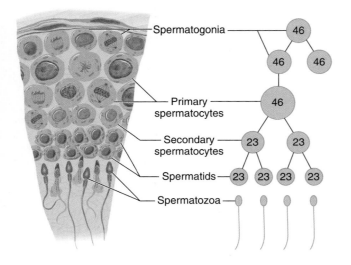

All spermatozoa originate from **spermatogonia** that are located at the periphery of a seminiferous tubule. These cells contain 23 pairs of chromosomes, a total of 46 chromosomes, the same as all body cells. Since they contain both members of each chromosome pair, they are said to be diploid. The mitotic division of a spermatogonium forms one replacement spermatogonium and one **primary spermatocyte.**

Each diploid primary spermatocyte divides by **meiotic cell division,** a type of cell division that consists of one chromosome replication and two successive cell divisions. Meiotic division produces four haploid spermatids from the single diploid primary spermatocyte.

The first meiotic division forms two haploid **secondary spermatocytes.** Each secondary spermatocyte contains only 23 chromosomes, one member of each chromosome pair. These chromosomes are already replicated, so each chromosome consists of two chromatids joined at a centromere.

The second meiotic division occurs as each secondary spermatocyte divides to yield two haploid **spermatids,** each with 23 chromosomes. In this division, the chromatids separate to provide a haploid set of chromosomes for each spermatid. The spermatids subsequently mature to become spermatozoa.

ASSIGNMENT

1. Label Figure 38.1.
2. Locate the male reproductive organs on the models available.

The Female Organs

Refer to Figures 38.3, 38.4, and 38.5 as you study this section.

External Genitalia

Two folds of skin lie on each side of the **vaginal orifice,** the labia majora and labia minora as shown in Figure 38.3. The larger exterior folds are the **labia majora,** which consist of rounded folds of adipose tissue covered by skin. They merge anteriorly with the **mons pubis,** an elevation of fatty tissue over the symphysis pubis. The outer surfaces of the labia majora possess hair; their inner surfaces are smooth and moist. The smaller interior folds, the **labia minora,** are devoid of hair and merge posteriorly with the labia majora. Anteriorly, they join to form a hoodlike covering, the **prepuce** of the clitoris. The **clitoris** is a small protuberance of erectile tissue located at the anterior junction of the labia minora. It is homologous to

Figure 38.3 Female genitalia.

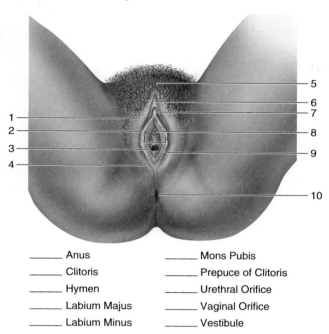

_____ Anus	_____ Mons Pubis
_____ Clitoris	_____ Prepuce of Clitoris
_____ Hymen	_____ Urethral Orifice
_____ Labium Majus	_____ Vaginal Orifice
_____ Labium Minus	_____ Vestibule

the penis in the male and is highly sensitive to sexual stimulation. Collectively, the external female reproductive organs are called the **vulva.**

The area between the labia minora is the **vestibule.** The vagina opens into the posterior portion of the vestibule. The **urethral orifice** is anterior to the vaginal opening and posterior to the clitoris. On either side of the vaginal orifice are the openings of the vestibular glands, which provide a mucous secretion for vaginal lubrication. The **hymen** (label 4) is a thin mucous membrane that partially covers the vaginal opening. Its condition or absence is not a determiner of virginity.

Internal Organs

The internal female reproductive organs are the ovaries, uterine tubes, uterus, and vagina.

The Ovaries

The primary sex organs (gonads) of the female reproductive system are the **ovaries** (label 5 in Figure 38.4). They produce the **ova,** or egg cells, and the female sex

Figure 38.4 Posterior view of female internal reproductive organs.

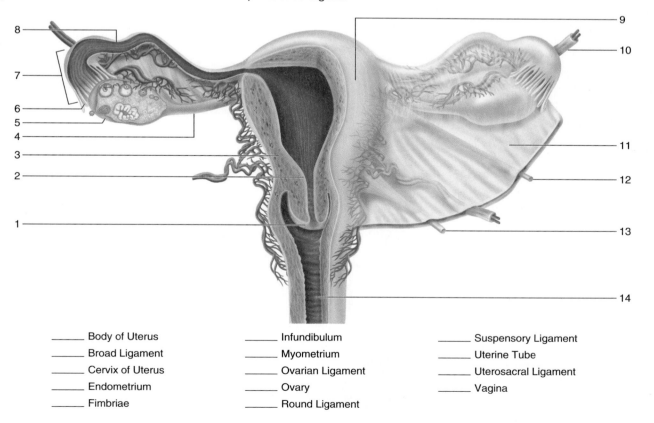

_____ Body of Uterus	_____ Infundibulum	_____ Suspensory Ligament
_____ Broad Ligament	_____ Myometrium	_____ Uterine Tube
_____ Cervix of Uterus	_____ Ovarian Ligament	_____ Uterosacral Ligament
_____ Endometrium	_____ Ovary	_____ Vagina
_____ Fimbriae	_____ Round Ligament	

hormones. The ovaries are ovoid organs located against either side of the lateral walls of the pelvic cavity.

The Uterine Tubes

The **uterine tubes** (oviducts or fallopian tubes) extend from the ovaries to the uterus. Near the ovary, each tube is expanded to form a funnel-shaped **infundibulum** (label 7 in Figure 38.4) that bears a number of fingerlike extensions, the **fimbriae.** The fimbriae and infundibula receive the oocytes released from the ovaries. The oocytes are carried toward the uterus by peristaltic contractions of the uterine tubes and the beating cilia of the ciliated columnar cells that line the tubes. Fertilization usually occurs within the upper third of a uterine tube. Observe the histology of a uterine tube in Figure HA-22.

The Uterus

The **uterus** is a hollow, pear-shaped, thick-walled organ within which fetal development takes place. It is located medially over the vagina and is bent anteriorly over the urinary bladder. The **fundus** of the uterus is a broad curvature that is superior/anterior to the junction with the uterine tubes. The **body** is the middle portion that forms most of the uterus, and the

cervix is the lower third that extends into the upper portion of the vagina. The **cervical orifice** is the uterine opening into the vagina.

The uterine wall is composed of three layers. The outer **perimetrium** is a layer of the peritoneum. The thick middle layer, the **myometrium** (label 3 in Figure 38.4), is composed of smooth muscle. The inner **endometrium** forms the mucosal lining and consists of two parts. The basal layer is attached to the myometrium. The functional layer is closest to the uterine cavity, is built up and shed during each menstrual cycle, and is covered with columnar epithelium on its free surface. Note the histology of the myometrium and endometrium in Figure HA-22.

Ligaments

The ovaries, uterine tubes, and uterus are held in place and supported by several ligaments. The **broad ligament** is a fold of the peritoneum that supports the uterus, uterine tubes, and ovaries. It extends from the lateral surfaces of the uterus to the lateral pelvic walls. Two **uterosacral ligaments** (label 13, Figure 38.4; label 13, Figure 38.5) extend from the

---

Figure 38.5 Midsagittal section of female reproductive organs.

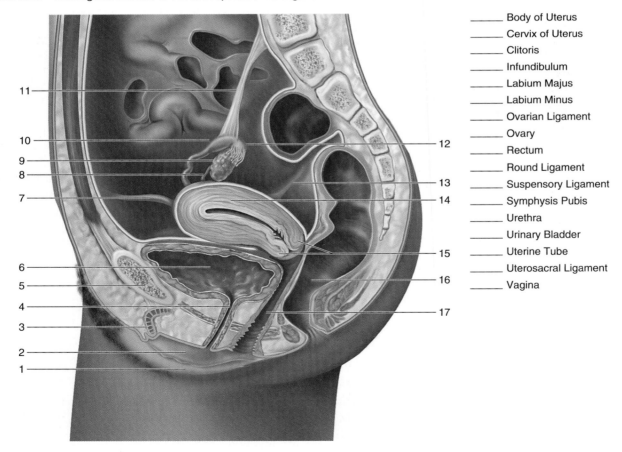

_____ Body of Uterus
_____ Cervix of Uterus
_____ Clitoris
_____ Infundibulum
_____ Labium Majus
_____ Labium Minus
_____ Ovarian Ligament
_____ Ovary
_____ Rectum
_____ Round Ligament
_____ Suspensory Ligament
_____ Symphysis Pubis
_____ Urethra
_____ Urinary Bladder
_____ Uterine Tube
_____ Uterosacral Ligament
_____ Vagina

Figure 38.6 Oogenesis.

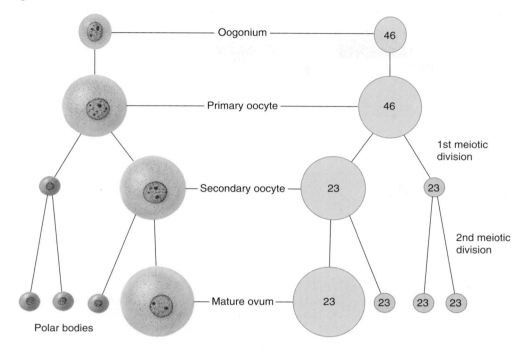

Oogenesis

cervix to the sacral wall of the pelvic cavity. A **round ligament** (label 12, Figure 38.4; label 7, Figure 38.5) extends from each side of the uterus to the anterior body wall.

In addition to the broad ligament, each ovary is held in place by two ligaments. An **ovarian ligament** extends from the uterus to the medial surface of the ovary. The **suspensory ligament** (label 10, Figure 38.4 label 11, Figure 38.5) extends from the lateral surface of the ovary to the pelvic wall.

Vagina

The **vagina** is a fibromuscular canal extending from the uterus to the vestibule. It receives the penis during sexual intercourse and serves as the birth canal.

Oogenesis

The development of the egg cells is called **oogenesis,** and it involves both mitotic and meiotic divisions. See Figure 38.6. The **germinal epithelium** is formed early in the prenatal development of the ovaries. It occurs on the outer surface of the ovaries and consists of as many as 400,000 **oogonia** that form by mitotic division. By the end of the third month of development, mitotic division has ceased and some oogonia have migrated inward to become **primary oocytes.** Each is surrounded by a sphere of cells forming a **primary follicle.** Each primary oocyte is diploid since it contains 23 pairs of chromosomes, or a total of 46 chromosomes.

Starting at puberty, several primary oocytes are stimulated to develop further by FSH. Usually, only one of them will undergo meiotic division in each ovarian cycle. The first meiotic division forms one large **secondary oocyte** and one small nonfunctional **polar body.** Both of these cells are haploid since they contain only 23 chromosomes, one member of each chromosome pair. The chromosomes are already replicated and consist of two chromatids joined at the centromere. Each secondary oocyte is located within a **secondary,** or **developing, follicle** that enlarges and fills with fluid to become a **mature,** or **graafian, follicle.** Ovulation occurs with the rupture of the mature follicle. The extruded secondary oocyte and first polar body enter the infundibulum and are carried toward the uterus by the uterine tube.

If a secondary oocyte is penetrated by a spermatozoan (activation), it undergoes the second meiotic division, which forms the ovum and a polar body. The first polar body may also divide to form two polar bodies. Thus, the meiotic division of a diploid primary oocyte forms four haploid cells: one ovum and three polar bodies. Note that the bulk of the cytoplasm passes first to the secondary oocyte and then to the ovum. After the formation of the ovum, the egg nucleus and the sperm nucleus unite (fertilization) to form a diploid zygote. The polar bodies disintegrate.

Note that the orderly process of gametogenesis (spermatogenesis and oogenesis) results in the zygote receiving one member of each chromosome pair from each parent.

ASSIGNMENT

1. Label Figures 38.3, 38.4, and 38.5.
2. Complete Sections A through E of the laboratory report.
3. Locate the female reproductive organs on the models available.

Microscopic Study

Examine the prepared slides noted below and compare your observations with Figures HA-21 and HA-22 of the Histology Atlas. Make labeled drawings of all your observations in Section F of the laboratory report.

ASSIGNMENT

1. Examine a prepared slide of human testis with the low power and high-dry objectives. Compare your observations with Figures 37.9, 38.2, and HA-21.

   a. Note the connective tissue forming the walls of the seminiferous tubules. Locate a few interstitial cells. What is their function?
   b. Locate maturing spermatozoa and the stages of spermatogenesis.

2. Examine a prepared slide of human ovary with the low power and high-dry objectives. Compare your observations with Figure 37.10.

   a. Locate the cuboidal cells of the germinal epithelium at the outer edge of the ovary. Are these cells haploid or diploid?

   b. Locate the different developmental stages of ovarian follicles and find a secondary oocyte within a mature follicle.

3. Examine prepared slides of the uterine wall and compare your observations with Figure HA-22A and B. In what tissue layer are uterine glands most abundant? What type of epithelium forms the surface lining of the endometrium?

4. Examine a prepared slide of uterine tube, x.s., and compare your observations with Figure HA-22C and D. Note the irregularly shaped lumen. What type of epithelium lines the lumen?

5. Complete the laboratory report.

HISTOLOGY ATLAS

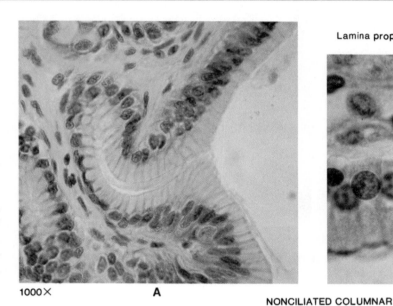

1000✕   **A**

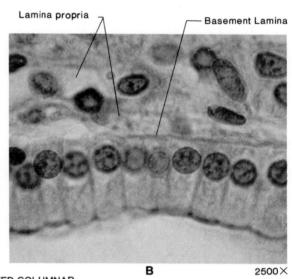

Lamina propria   Basement Lamina

B   2500✕

NONCILIATED COLUMNAR

Brush Border

Goblet Cell

2500✕   **C**

PLAIN COLUMNAR WITH BRUSH BORDER

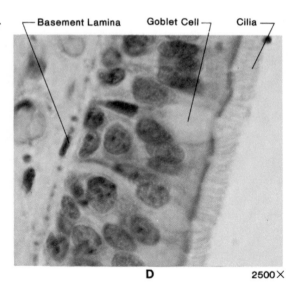

Basement Lamina   Goblet Cell   Cilia

D   2500✕

CILIATED PSEUDOSTRATIFIED COLUMNAR

Columnar epithelium, forming the internal lining of the digestive and respiratory systems, contains numerous goblet cells that secrete mucus. They are present but unlabeled in illustration A, as well as in illustrations C and D.

Note the difference in the orientation of the nuclei in simple columnar (A, B, and C) and pseudostratified columnar (D). A brush border, formed of microvilli on columnar cells lining the small intestine (C), appears distinctively different than cilia on pseudostratified columnar cells lining the trachea (D).

Note the distinctive basement lamina, consisting largely of proteins and reticular fibers, that attaches all epithelium to the underlying lamina propria. The lamina propria underlies all epithelial tissues, and it consists of areolar connective tissue containing nerves, blood vessels, and lymphatic vessels. Lymphocytes, mast cells, plasma cells, and eosinophils also occur in the lamina propria, a site of immune reactions.

Figure HA-1   Columnar epithelium.

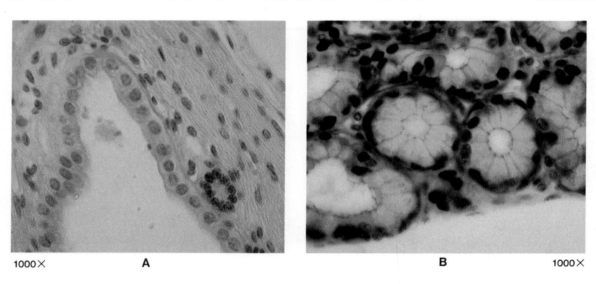

1000× **A**

B 1000×

CUBOIDAL EPITHELIUM

Lamina propria ⏐ ⏐ Basement Lamina ⏐ Binucleate Cells ⏐

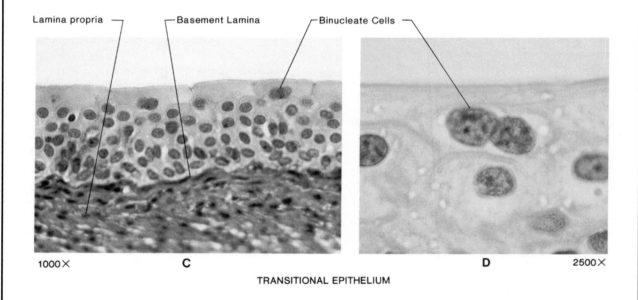

1000× **C**

D 2500×

TRANSITIONAL EPITHELIUM

Cells forming a sheet of simple cuboidal epithelium typically have a boxlike shape as shown in illustration A. However, when they form a tubule or a duct of a gland they often take on a pyramidal shape as shown in illustration B.

Transitional epithelium, illustrations C and D, is a stratified epithelium whose surface cells are large and dome-shaped rather than squamous, cuboidal, or columnar when the tissue is not stretched. Some surface cells are binucleate, and the basal cells are somewhat columnar in shape. The loose arrangement of cells allows stretching of the tissue, and only then are surface cells cuboidal or squamous in shape, depending upon the degree of stretching.

Figure HA-2 Cuboidal and transitional epithelium.

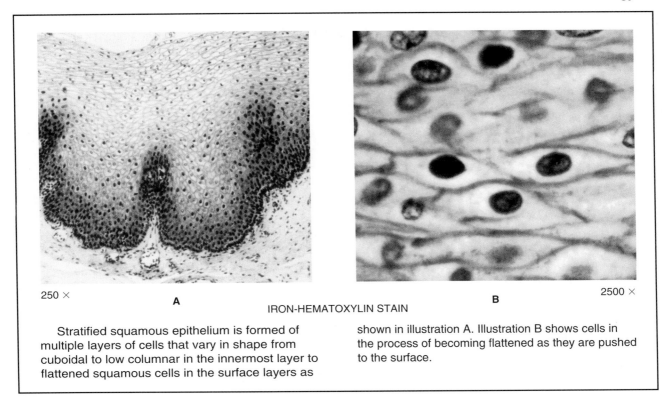

250 ×

A

B

2500 ×

IRON-HEMATOXYLIN STAIN

Stratified squamous epithelium is formed of multiple layers of cells that vary in shape from cuboidal to low columnar in the innermost layer to flattened squamous cells in the surface layers as shown in illustration A. Illustration B shows cells in the process of becoming flattened as they are pushed to the surface.

Figure HA-3 Stratified squamous epithelium.

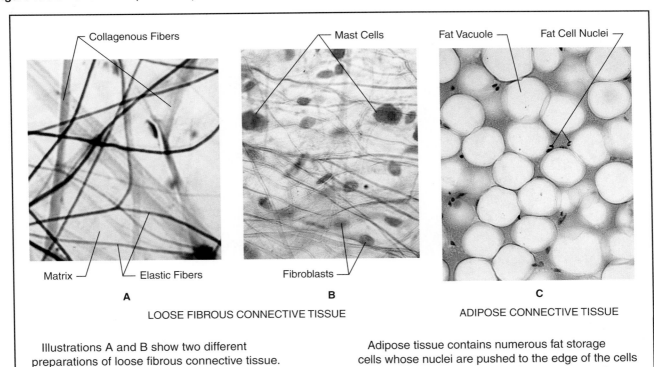

Collagenous Fibers

Mast Cells

Fat Vacuole

Fat Cell Nuclei

Matrix

Elastic Fibers

Fibroblasts

A

B

C

LOOSE FIBROUS CONNECTIVE TISSUE

ADIPOSE CONNECTIVE TISSUE

Illustrations A and B show two different preparations of loose fibrous connective tissue. Collagenous and elastic fibers are more distinct in A, but fibroblasts are more clearly shown in B. A site of immune reactions, areolar connective tissue usually contains mast cells, macrophages, lymphocytes, and plasma cells.

Adipose tissue contains numerous fat storage cells whose nuclei are pushed to the edge of the cells by fat-filled vacuoles. Adipose tissue occurs under the skin and around organs where it serves as a protective cushion and a storage site of reserve energy.

Figure HA-4 Areolar and adipose connective tissue.

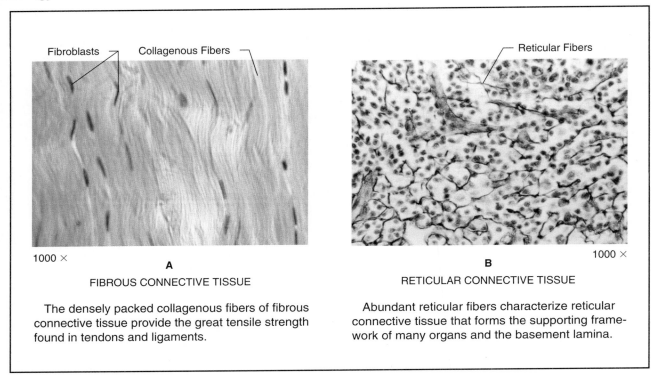

A
FIBROUS CONNECTIVE TISSUE

The densely packed collagenous fibers of fibrous connective tissue provide the great tensile strength found in tendons and ligaments.

B
RETICULAR CONNECTIVE TISSUE

Abundant reticular fibers characterize reticular connective tissue that forms the supporting framework of many organs and the basement lamina.

Figure HA-5   Fibrous and reticular connective tissue.

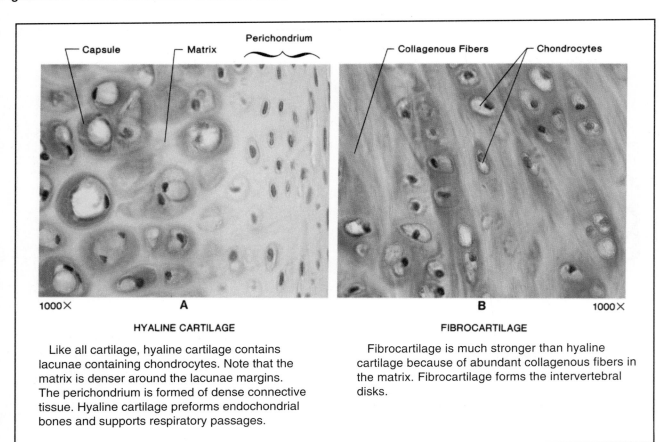

A
HYALINE CARTILAGE

Like all cartilage, hyaline cartilage contains lacunae containing chondrocytes. Note that the matrix is denser around the lacunae margins. The perichondrium is formed of dense connective tissue. Hyaline cartilage preforms endochondrial bones and supports respiratory passages.

B
FIBROCARTILAGE

Fibrocartilage is much stronger than hyaline cartilage because of abundant collagenous fibers in the matrix. Fibrocartilage forms the intervertebral disks.

Figure HA-6   Hyaline cartilage and fibrocartilage.

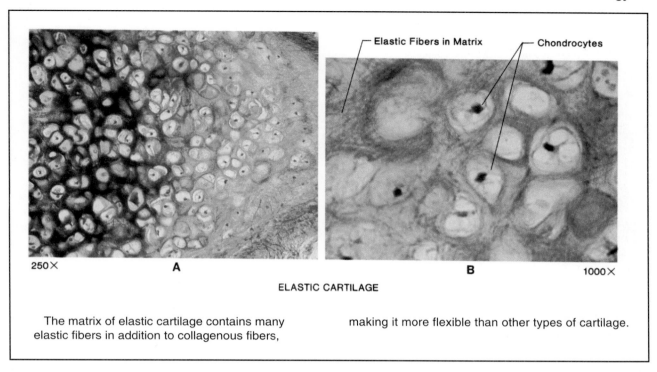

ELASTIC CARTILAGE

The matrix of elastic cartilage contains many elastic fibers in addition to collagenous fibers, making it more flexible than other types of cartilage.

Figure HA-7   Elastic cartilage.

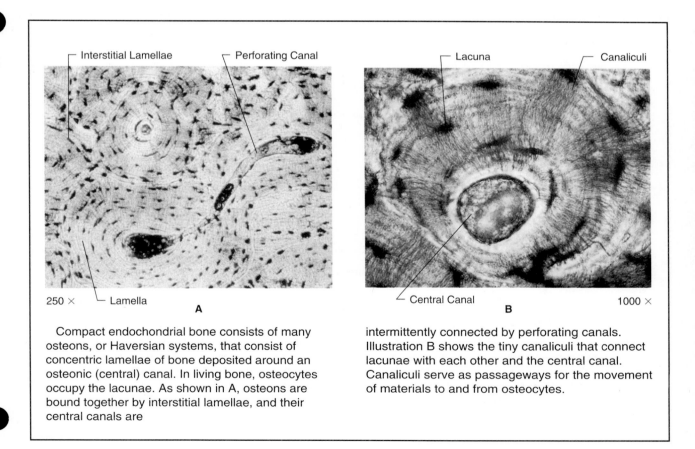

Compact endochondrial bone consists of many osteons, or Haversian systems, that consist of concentric lamellae of bone deposited around an osteonic (central) canal. In living bone, osteocytes occupy the lacunae. As shown in A, osteons are bound together by interstitial lamellae, and their central canals are intermittently connected by perforating canals. Illustration B shows the tiny canaliculi that connect lacunae with each other and the central canal. Canaliculi serve as passageways for the movement of materials to and from osteocytes.

Figure HA-8   Compact bone tissue.

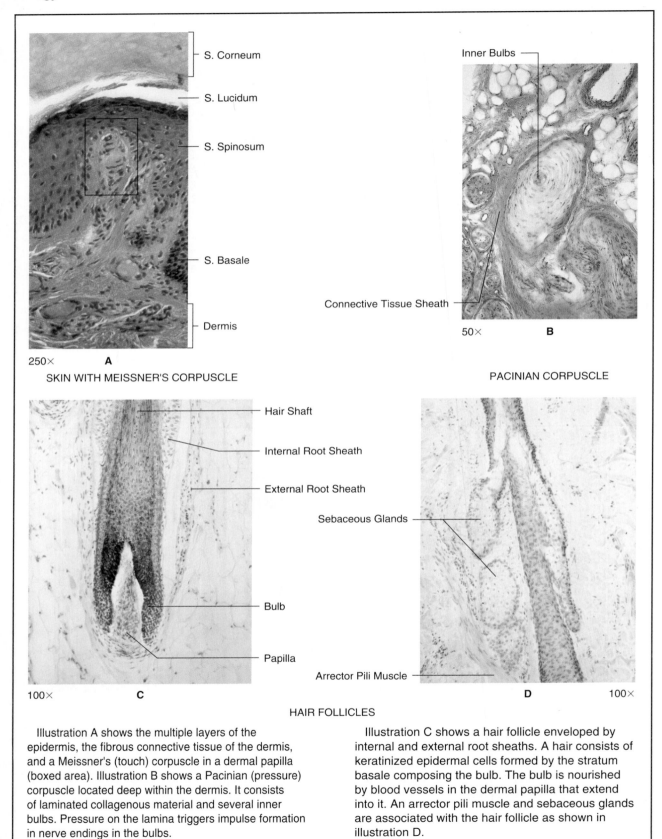

S. Corneum

S. Lucidum

S. Spinosum

S. Basale

Dermis

250× **A**

SKIN WITH MEISSNER'S CORPUSCLE

Inner Bulbs

Connective Tissue Sheath

50× **B**

PACINIAN CORPUSCLE

Hair Shaft

Internal Root Sheath

External Root Sheath

Bulb

Papilla

100× **C**

Sebaceous Glands

Arrector Pili Muscle

D 100×

HAIR FOLLICLES

Illustration A shows the multiple layers of the epidermis, the fibrous connective tissue of the dermis, and a Meissner's (touch) corpuscle in a dermal papilla (boxed area). Illustration B shows a Pacinian (pressure) corpuscle located deep within the dermis. It consists of laminated collagenous material and several inner bulbs. Pressure on the lamina triggers impulse formation in nerve endings in the bulbs.

Illustration C shows a hair follicle enveloped by internal and external root sheaths. A hair consists of keratinized epidermal cells formed by the stratum basale composing the bulb. The bulb is nourished by blood vessels in the dermal papilla that extend into it. An arrector pili muscle and sebaceous glands are associated with the hair follicle as shown in illustration D.

Figure HA-9   The integument.

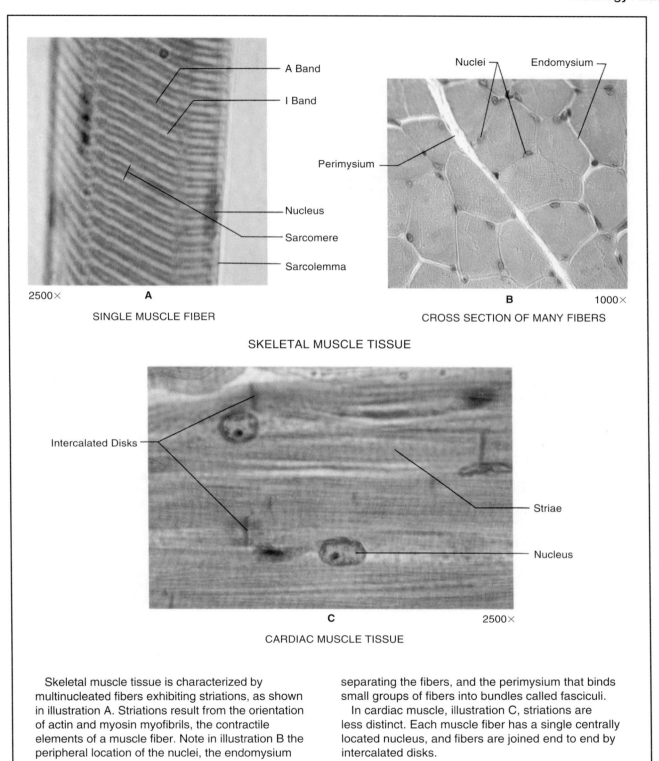

A Band

I Band

Nucleus

Sarcomere

Sarcolemma

2500× **A**

SINGLE MUSCLE FIBER

Nuclei **Endomysium**

Perimysium

B 1000×

CROSS SECTION OF MANY FIBERS

SKELETAL MUSCLE TISSUE

Intercalated Disks

Striae

Nucleus

C 2500×

CARDIAC MUSCLE TISSUE

Skeletal muscle tissue is characterized by multinucleated fibers exhibiting striations, as shown in illustration A. Striations result from the orientation of actin and myosin myofibrils, the contractile elements of a muscle fiber. Note in illustration B the peripheral location of the nuclei, the endomysium separating the fibers, and the perimysium that binds small groups of fibers into bundles called fasciculi.

In cardiac muscle, illustration C, striations are less distinct. Each muscle fiber has a single centrally located nucleus, and fibers are joined end to end by intercalated disks.

Figure HA-10   Skeletal and cardiac muscle tissues.

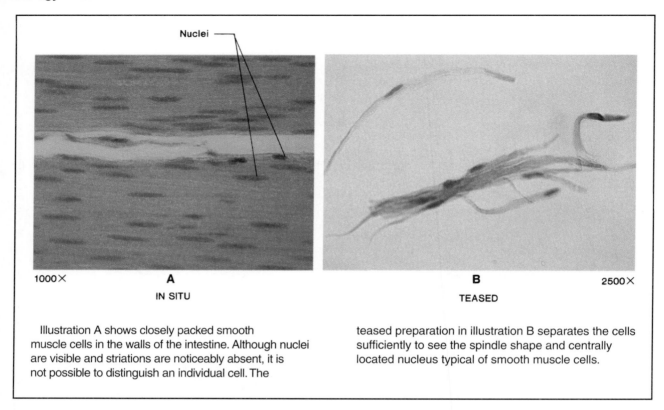

Nuclei

1000×   **A**   **B**   2500×

IN SITU   TEASED

Illustration A shows closely packed smooth muscle cells in the walls of the intestine. Although nuclei are visible and striations are noticeably absent, it is not possible to distinguish an individual cell. The

teased preparation in illustration B separates the cells sufficiently to see the spindle shape and centrally located nucleus typical of smooth muscle cells.

Figure HA-11   Smooth muscle tissue.

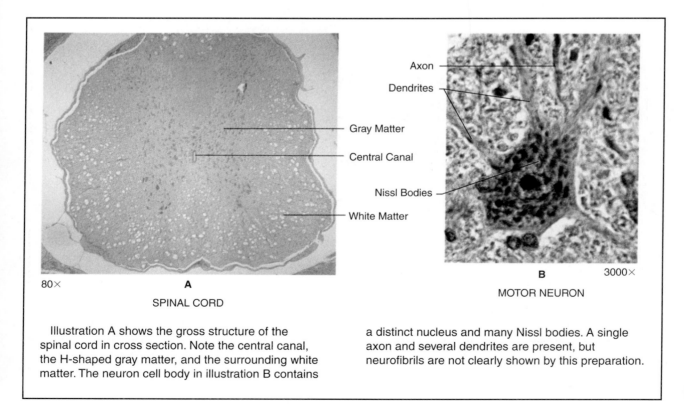

Axon

Dendrites

Gray Matter

Central Canal

Nissl Bodies

White Matter

80×   **A**   **B**   3000×

SPINAL CORD   MOTOR NEURON

Illustration A shows the gross structure of the spinal cord in cross section. Note the central canal, the H-shaped gray matter, and the surrounding white matter. The neuron cell body in illustration B contains

a distinct nucleus and many Nissl bodies. A single axon and several dendrites are present, but neurofibrils are not clearly shown by this preparation.

Figure HA-12   Nerve tissue.

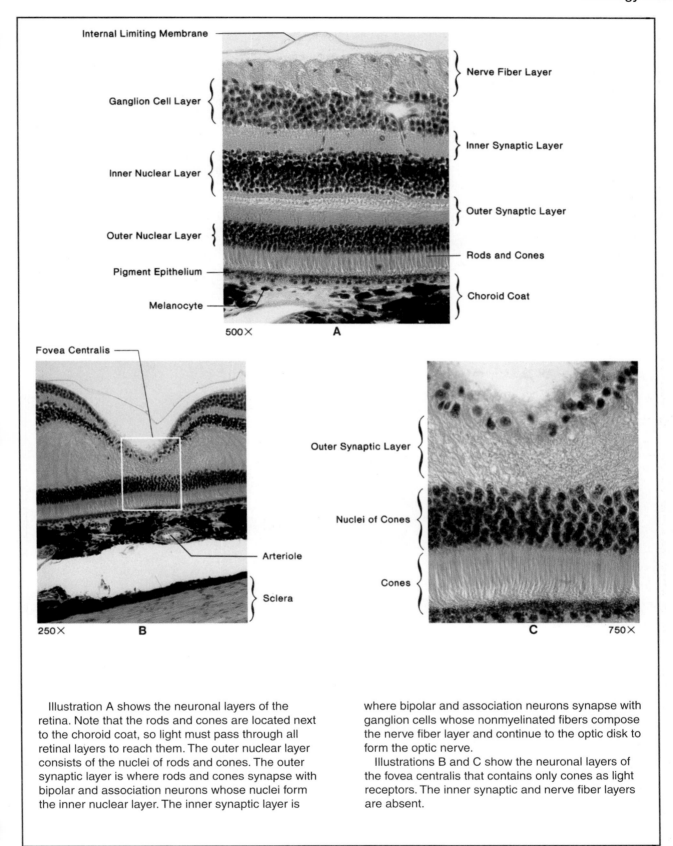

Illustration A shows the neuronal layers of the retina. Note that the rods and cones are located next to the choroid coat, so light must pass through all retinal layers to reach them. The outer nuclear layer consists of the nuclei of rods and cones. The outer synaptic layer is where rods and cones synapse with bipolar and association neurons whose nuclei form the inner nuclear layer. The inner synaptic layer is where bipolar and association neurons synapse with ganglion cells whose nonmyelinated fibers compose the nerve fiber layer and continue to the optic disk to form the optic nerve.

Illustrations B and C show the neuronal layers of the fovea centralis that contains only cones as light receptors. The inner synaptic and nerve fiber layers are absent.

Figure HA-13   The retina of the eye.

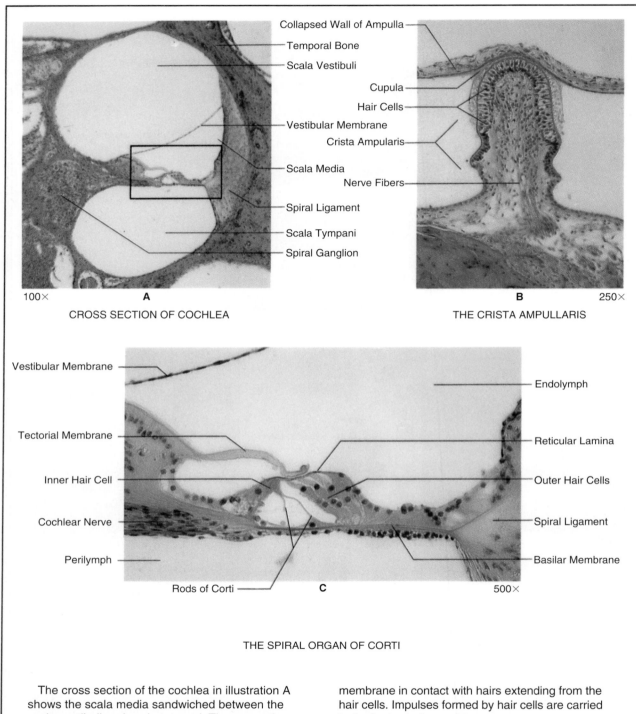

Collapsed Wall of Ampulla

Temporal Bone

Scala Vestibuli

Cupula

Hair Cells

Vestibular Membrane

Crista Ampularis

Scala Media

Nerve Fibers

Spiral Ligament

Scala Tympani

Spiral Ganglion

100× **A**

CROSS SECTION OF COCHLEA

B 250×

THE CRISTA AMPULLARIS

Vestibular Membrane

Endolymph

Tectorial Membrane

Reticular Lamina

Inner Hair Cell

Outer Hair Cells

Cochlear Nerve

Spiral Ligament

Perilymph

Basilar Membrane

Rods of Corti **C** 500×

THE SPIRAL ORGAN OF CORTI

The cross section of the cochlea in illustration A shows the scala media sandwiched between the scala vestibuli and scala tympani. The supporting spiral ligament is continuous with the basilar membrane and is derived from the periosteum of the temporal bone. The spiral organ of Corti (boxed area) is enlarged in illustration C. Note the basilar membrane, hair cells, and the flaplike tectorial

membrane in contact with hairs extending from the hair cells. Impulses formed by hair cells are carried to the brain via the cochlear nerve.

The crista ampullaris in illustration B is a receptor for dynamic equilibrium found in the ampulla of each semicircular canal. The cupula was compressed during slide preparation but the other components are normal.

Figure HA-14 The cochlea and crista ampullaris.

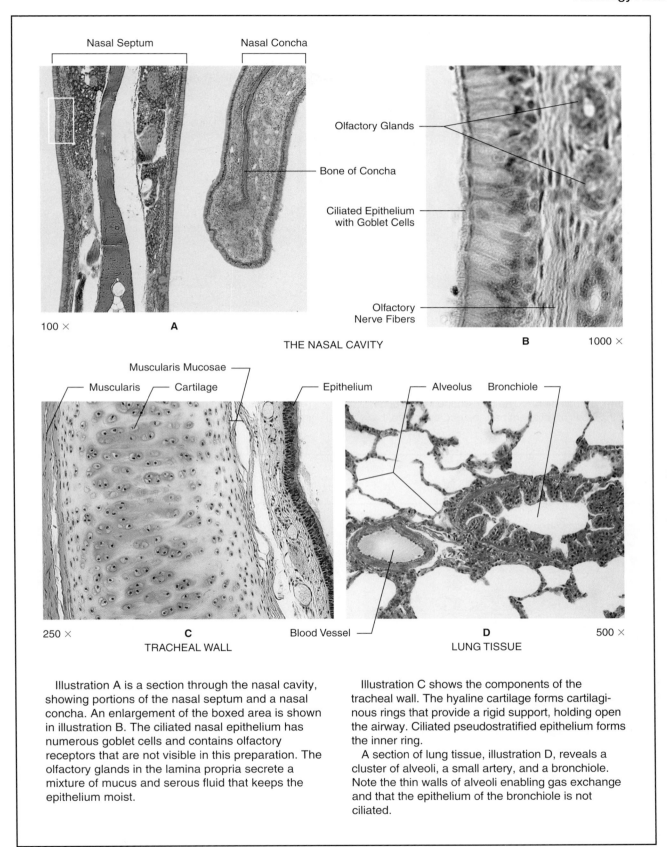

Nasal Septum

Nasal Concha

Olfactory Glands

Bone of Concha

Ciliated Epithelium
with Goblet Cells

Olfactory
Nerve Fibers

100 ×   **A**

THE NASAL CAVITY

B   1000 ×

Muscularis Mucosae

Muscularis   Cartilage

Epithelium

Alveolus   Bronchiole

250 ×   **C**

TRACHEAL WALL

Blood Vessel

D   500 ×

LUNG TISSUE

Illustration A is a section through the nasal cavity, showing portions of the nasal septum and a nasal concha. An enlargement of the boxed area is shown in illustration B. The ciliated nasal epithelium has numerous goblet cells and contains olfactory receptors that are not visible in this preparation. The olfactory glands in the lamina propria secrete a mixture of mucus and serous fluid that keeps the epithelium moist.

Illustration C shows the components of the tracheal wall. The hyaline cartilage forms cartilaginous rings that provide a rigid support, holding open the airway. Ciliated pseudostratified epithelium forms the inner ring.

A section of lung tissue, illustration D, reveals a cluster of alveoli, a small artery, and a bronchiole. Note the thin walls of alveoli enabling gas exchange and that the epithelium of the bronchiole is not ciliated.

Figure HA-15   Histology of respiratory structures.

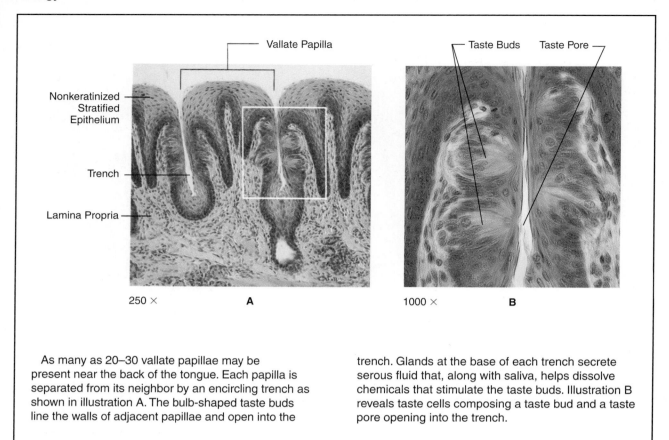

As many as 20–30 vallate papillae may be present near the back of the tongue. Each papilla is separated from its neighbor by an encircling trench as shown in illustration A. The bulb-shaped taste buds line the walls of adjacent papillae and open into the trench. Glands at the base of each trench secrete serous fluid that, along with saliva, helps dissolve chemicals that stimulate the taste buds. Illustration B reveals taste cells composing a taste bud and a taste pore opening into the trench.

Figure HA-16   Taste buds.

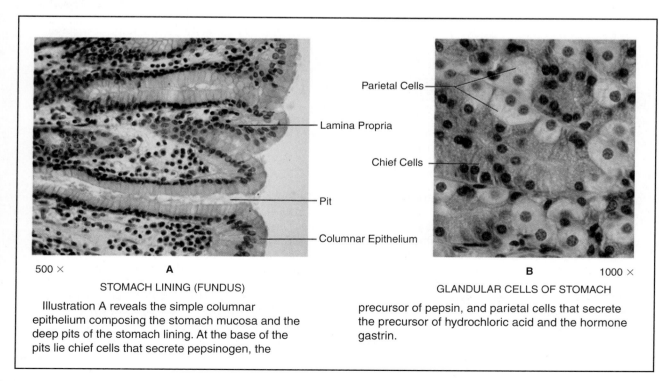

STOMACH LINING (FUNDUS)

GLANDULAR CELLS OF STOMACH

Illustration A reveals the simple columnar epithelium composing the stomach mucosa and the deep pits of the stomach lining. At the base of the pits lie chief cells that secrete pepsinogen, the precursor of pepsin, and parietal cells that secrete the precursor of hydrochloric acid and the hormone gastrin.

Figure HA-17   Stomach mucosa.

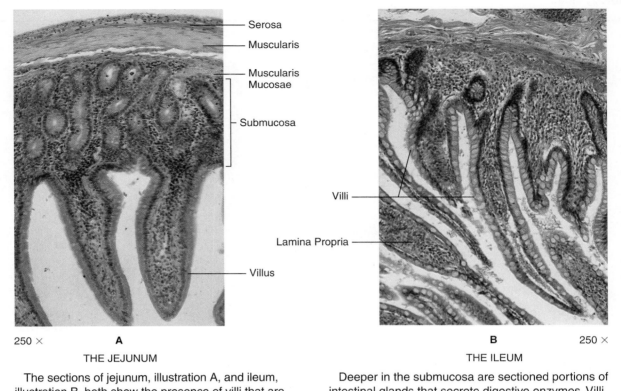

Serosa
Muscularis
Muscularis Mucosae
Submucosa
Villus

Villi
Lamina Propria

250 ×   **A**
THE JEJUNUM

B   250 ×
THE ILEUM

The sections of jejunum, illustration A, and ileum, illustration B, both show the presence of villi that are covered with simple columnar epithelium containing numerous goblet cells.

Deeper in the submucosa are sectioned portions of intestinal glands that secrete digestive enzymes. Villi greatly increase the surface area of the absorptive epithelial surface.

Figure HA-18   The jejunum and ileum.

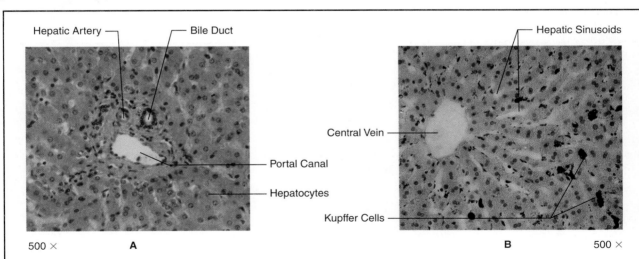

Hepatic Artery
Bile Duct
Hepatic Sinusoids

Central Vein

Portal Canal
Hepatocytes

Kupffer Cells

500 ×   **A**

B   500 ×

The liver consists of many lobules containing a central vein as shown in illustration B. A branch of the portal vein, a portal canal, and a bile duct located at a "corner" of a lobule are in illustration A. A lobule is mostly composed of hepatocytes (liver cells) that radiate from a central vein in irregular rows. Between the rows are sinusoids, which are vascular channels for blood flowing between an artery and portal vein to the central vein.

Hepatocytes synthesize bile components and certain blood proteins. Stationary phagocytic cells, Kupffer cells, remove bacteria, other foreign material, and cellular debris from blood as it flows through the sinusoids. Kupffer cells are darkly stained in illustration B because they have removed carbon particles injected into the animal prior to sacrifice and tissue preparation.

Figure HA-19   Liver tissue.

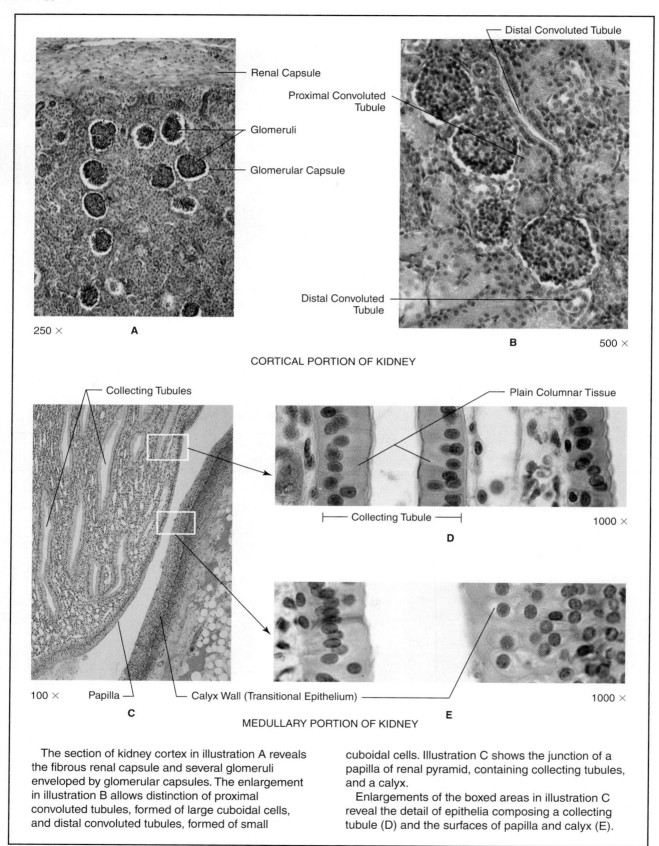

CORTICAL PORTION OF KIDNEY

MEDULLARY PORTION OF KIDNEY

The section of kidney cortex in illustration A reveals the fibrous renal capsule and several glomeruli enveloped by glomerular capsules. The enlargement in illustration B allows distinction of proximal convoluted tubules, formed of large cuboidal cells, and distal convoluted tubules, formed of small cuboidal cells. Illustration C shows the junction of a papilla of renal pyramid, containing collecting tubules, and a calyx.

Enlargements of the boxed areas in illustration C reveal the detail of epithelia composing a collecting tubule (D) and the surfaces of papilla and calyx (E).

Figure HA-20 Kidney histology.

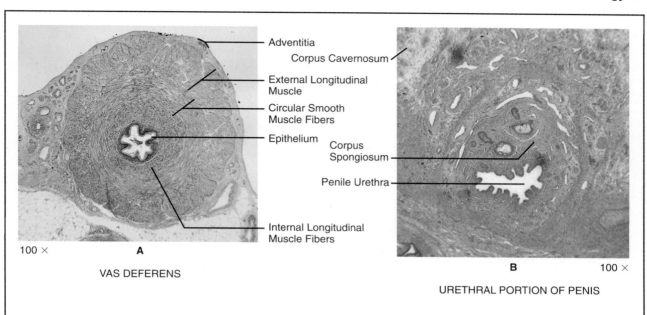

A 100 ×

VAS DEFERENS

- Adventitia
- Corpus Cavernosum
- External Longitudinal Muscle
- Circular Smooth Muscle Fibers
- Epithelium
- Corpus Spongiosum
- Penile Urethra
- Internal Longitudinal Muscle Fibers

B 100 ×

URETHRAL PORTION OF PENIS

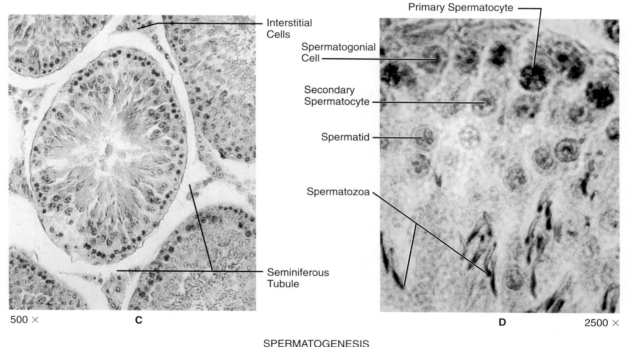

500 × **C**

- Interstitial Cells
- Primary Spermatocyte
- Spermatogonial Cell
- Secondary Spermatocyte
- Spermatid
- Spermatozoa
- Seminiferous Tubule

D 2500 ×

SPERMATOGENESIS

Illustration A reveals that the vas deferens is a thick-walled tube consisting mostly of circular and longitudinal muscle layers. Contraction of these muscles helps to propel spermatozoa during ejaculation. Illustration B shows urethra surrounded by the corpus spongiosum and a portion of one of the corpora cavernosa. Illustration C reveals a cross section of seminiferous tubules in a testis. Note the mature sperm in the lumen of the tubule and the interstitial cells between the tubules that produce testosterone. Stages of spermatogenesis are shown in illustration D.

Figure HA-21 Male reproductive tissue.

233

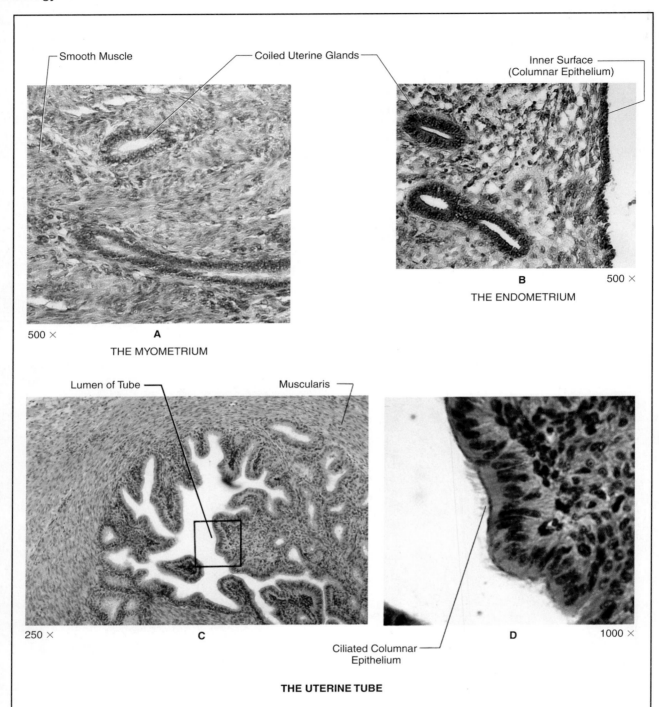

Smooth Muscle — Coiled Uterine Glands — Inner Surface — (Columnar Epithelium)

500 ×   **A**

THE MYOMETRIUM

B   500 ×

THE ENDOMETRIUM

Lumen of Tube — Muscularis —

250 ×   **C**   **D**   1000 ×

Ciliated Columnar — Epithelium

THE UTERINE TUBE

The wall of the uterus consists of three layers: endometrium, myometrium (the muscular layer forming most of the uterine wall), and perimetrium (a serous membrane covering). As shown in illustration A, some uterine glands extend into the myometrium. The simple columnar epithelial cells of the endometrium, illustration B, overlay highly vascular tissue with many uterine glands. The functional layer of the endometrium is shed at each menses and rebuilt during the proliferative stage of the uterine cycle.

A uterine tube, illustration C, consists of an inner mucosa, a middle muscle layer, and an outer serosa membrane. The epithelium, illustration D, consists of a mixture of ciliated and nonciliated columnar cells. After ovulation, cilia propel the secondary oocyte toward the uterus.

Figure HA-22   The uterus and uterine tube.

Laboratory Report 1

Student _____

Lab Section _____

INTRODUCTION TO HUMAN ANATOMY

A. Figures

Write the labels for Figures 1.1 through 1.6 in the spaces provided.

Figure 1.1

1. _____
2. _____
3. _____
4. _____
5. _____
6. _____
7. _____
8. _____
9. _____
10. _____
11. _____
12. _____
13. _____
14. _____
15. _____
16. _____
17. _____
18. _____
19. _____

Figure 1.2

Anterior View

1. _____
2. _____
3. _____
4. _____
5. _____
6. _____
7. _____
8. _____
9. _____
10. _____
11. _____
12. _____
13. _____
14. _____
15. _____
16. _____
17. _____
18. _____
19. _____
20. _____

Posterior View

1. _____
2. _____
3. _____
4. _____
5. _____
6. _____
7. _____
8. _____

9. _____

10. _____

11. _____

12. _____

Figure 1.3

Abdominal Regions

1. _____

2. _____

3. _____

4. _____

5. _____

6. _____

7. _____

8. _____

9. _____

Abdominal Quadrants

1. _____

2. _____

3. _____

4. _____

Figure 1.4

Midsagittal Section

1. _____

2. _____

3. _____

4. _____

5. _____

6. _____

7. _____

8. _____

Frontal Section

1. _____

2. _____

3. _____

4. _____

5. _____

6. _____

7. _____

Figure 1.5

1. _____

2. _____

3. _____

4. _____

5. _____

6. _____

Figure 1.6

1. _____

2. _____

3. _____

4. _____

5. _____

6. _____

B. Directional Terms, Sections, and Surfaces

Write in the answer column the term from the list below that corresponds to the statements that follow.

Directional Terms		Sections
anterior	medial	frontal
distal	posterior	midsagittal
inferior	proximal	parasagittal
lateral	superior	transverse

Directional Terms

1. The hand is _____ to the elbow.
2. The knee is _____ to the ankle.
3. The mouth is _____ to the nose.
4. The ear is _____ to the nose.
5. The teeth are _____ to the cheeks.

Sections

6. Divides the body into equal left and right halves.
7. Divides the body into superior and inferior parts.
8. Divides the body into anterior and posterior parts.
9. Divides the body into unequal left and right parts.

Surfaces

Write the surface on which the following are located.

10. Kneecap
11. Ear
12. Umbilicus (navel)
13. Buttocks
14. Palm of hand
15. Nose

1. _____
2. _____
3. _____
4. _____
5. _____
6. _____
7. _____
8. _____
9. _____
10. _____
11. _____
12. _____
13. _____
14. _____
15. _____

C. Body Regions and Surface Features

Write the term from the list below that corresponds to the following statements.

antebrachial   olecranal   hypogastric
antecubital   digital   inguinal
axillary   epigastric   lumbar
brachial   femoral   pectoral
calcaneal   gluteal   popliteal
carpal   hypochondriac   tarsal

1. The ankle.
2. The forearm.
3. Heel of the foot.
4. Lateral chest region.
5. The posterior surface of the knee.
6. Depression at junction of thigh with abdomen.
7. The armpit.
8. Abdominal region overlying the urinary bladder.
9. Abdominal region superior to the lumbar region.
10. Abdominal region directly above the umbilical region.
11. The lower back lateral to the vertebral column.
12. The fingers and toes.
13. The buttocks.
14. The proximal portion of the upper extremity.
15. The anterior surface of the elbow joint.

1. _____
2. _____
3. _____
4. _____
5. _____
6. _____
7. _____
8. _____
9. _____
10. _____
11. _____
12. _____
13. _____
14. _____
15. _____

D. Membranes

Write the names of the membranes that match the following statements.

Membranes

mesentery   peritoneum, visceral
pericardium, parietal   pleura, parietal
pericardium, visceral   pleura, visceral
peritoneum, parietal

1. Double-layered membrane around the heart.
2. Serous membrane attached to surface of lungs.
3. Serous membrane attached to surface of stomach.
4. Double-layered membrane supporting intestines.
5. Serous membrane lining wall of abdominal cavity.
6. Serous membrane tightly attached to the exterior surface of the heart.
7. Serous membrane lining the inside of the thoracic cavity.

1. _____
2. _____
3. _____
4. _____
5. _____
6. _____
7. _____

Student _____

Lab Section _____

BODY ORGANIZATION

A. Complete the Chart

Organ System	Functions	Components
Integumentary		
Skeletal		
Muscular		
Nervous		
Endocrine		

Organ System	Functions	Components
Cardiovascular		
Lymphatic		
Respiratory		
Digestive		
Urinary		
Reproductive		

B. Terms

Define these terms and give examples of each.

Organian _____

Organ system _____

C. Matching

Write the name of the organ system(s) corresponding to the words or phrases that follow.

cardiovascular   lymphatic   respiratory
digestive   muscular   skeletal
endocrine   nervous   urinary
integumentary   reproductive

Functions

1. Protective covering of the body.
2. Transports materials throughout the body.
3. Enables gas exchange between blood and atmosphere.
4. Secretes hormones.
5. Digests food to form absorbable nutrients.
6. Provides supporting framework for the body.
7. Contractions enable body movements.
8. Removes wastes from the blood to form urine.
9. Enables rapid responses to environmental changes.
10. Collects tissue fluid from intercellular spaces.
11. Provides chemical control of body functions.
12. Perpetuates the species.
13. Produces blood cells.
14. Shields body from ultraviolet radiation.
15. Cleanses lymph and returns it to the bloodstream.

1. _____
2. _____
3. _____
4. _____
5. _____
6. _____
7. _____
8. _____
9. _____
10. _____
11. _____
12. _____
13. _____
14. _____
15. _____

Components

1. Stomach and intestines.
2. Pituitary, adrenal, and thyroid glands.
3. Trachea, bronchi, and lungs.
4. Arteries, veins, and capillaries.
5. Cartilages and ligaments.
6. Kidneys, ureters, and urethra.
7. Spleen, tonsils, and adenoids.
8. Epidermis and dermis.
9. Uterus, oviducts, and vagina.
10. Liver and pancreas.
11. Nasal cavity, pharynx, and larynx.
12. Brain and spinal cord.
13. Sensory receptors and nerves.
14. Pancreas, ovaries, and testes.
15. Vasa deferentia, urethra, and penis.

1. _____
2. _____
3. _____
4. _____
5. _____
6. _____
7. _____
8. _____
9. _____
10. _____
11. _____
12. _____
13. _____
14. _____
15. _____

NOTES

Laboratory Report 3

THE MICROSCOPE

A. Parts of the Microscope

Record in the answer column the microscope part described by the statements below.

arm   mechanical stage
base   objective
condenser   ocular
diaphragm   revolving nosepiece
focusing knob, coarse   stage
focusing knob, fine   stage aperture
lamp

1. Platform on which slides are placed for viewing.
2. Lens that is closest to the microscope slide.
3. Controls amount of light entering the condenser.
4. Focusing knob used with low power objective only.
5. Concentrates light on object being observed.
6. Light source.
7. Serves as a "handle" in carrying a microscope.
8. Lens you look through to view the image.
9. Part to which objectives are attached.
10. Opening in center of the stage.
11. Enables precise movement of the slide.
12. Bottom of the microscope.
13. Focusing knob used with all objectives.
14. Focusing knob with the larger diameter.
15. May be rotated to bring different objectives into viewing position.

1. _____
2. _____
3. _____
4. _____
5. _____
6. _____
7. _____
8. _____
9. _____
10. _____
11. _____
12. _____
13. _____
14. _____
15. _____

B. Magnification

Record the magnification of the ocular(s) and objectives on your microscope and calculate the total magnification obtained when using each objective.

Ocular	×	Objective	=	Total Magnification
_____ ×		_____ ×		_____ ×
_____ ×		_____ ×		_____ ×
_____ ×		_____ ×		_____ ×

C. True–False

Record your answer to these statements as true or false.

1. A microscope should be carried with two hands.
2. You should start your observations with the low power objective.
3. Blue light gives better resolution than white light.
4. Malfunctions should be reported to your instructor.
5. If necessary, you should disassemble the ocular to clean it.

1. _____
2. _____
3. _____
4. _____
5. _____

D. Observations

1. Diagram the appearance of the letter *e* as observed with the following:

unaided eye

low-power objective

high-dry objective

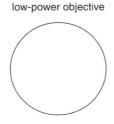

2. Draw the appearance of the crossed hairs with the following:

low-power objective

high-dry objective

E. Completion

Write your responses to the following statements in the answer column.

1. Tissue used to clean the lenses.
2. Objective with the greatest working distance.
3. Objective with the least working distance.
4. Usually, the condenser should be kept at its _____ position.
5. What effect (increase, no change, or decrease) does an increase in magnification of an objective have on the following?
    a. diameter of field
    b. working distance
    c. light intensity
6. The contrast of the image is _____ (increased, unchanged, or decreased) when light intensity is reduced.
7. Controls light intensity in microscopes with a constant light source.
8. Circle of light seen while looking into the ocular.
9. When the slide is moved to the left, the image moves to the _____.
10. Maximum resolution with a light microscope.
11. Three liquids that may be used to clean lenses.
12. When an object is in focus with one objective and is still in focus when another objective is rotated into position, the microscope is said to be _____.
13. Only focusing knob used with high-dry and oil immersion objectives.

1. _____

2. _____

3. _____

4. _____

5a. _____

5b. _____

5c. _____

6. _____

7. _____

8. _____

9. _____

10. _____

11a. _____

11b. _____

11c. _____

12. _____

13. _____

Laboratory Report 4

Student _____

Lab Section _____

CELL ANATOMY

A. Figure

List the labels for Figure 4.1.

1. _____
2. _____
3. _____
4. _____
5. _____
6. _____

7. _____
8. _____
9. _____
10. _____
11. _____

12. _____
13. _____
14. _____
15. _____
16. _____

B. Cell Structure and Organelle Function

1. Select the structure(s) from the following list that are described by the statements below.

centrioles
centrosome
chromatin granules
cilia
cytoplasm
flagellum
Golgi complex
lysosomes
microvilli

mitochondria
nuclear membrane
nucleolus
nucleus
plasma membrane
RER
ribosomes
secretory vesicles
SER

1. Short, hairlike projections on a cell surface.
2. Short bundles of microtubules oriented at right angles to each other.
3. Increase absorptive surface area of certain cells.
4. Provides motility of sperm cells.
5. A stack of membranous sacs near the ER.
6. Packages materials for extracellular transport.
7. Sites of ATP production via cellular respiration.
8. Structures in nucleus composed of DNA and protein.
9. A double membrane surrounding the nucleus.
10. A spherical structure in the nucleus composed of RNA and protein.
11. ER studded with ribosomes.
12. Organelle with an inner membrane having many folds.
13. Controls passage of materials into and out of cells.
14. ER without ribosomes.
15. Sacs of digestive enzymes.
16. Sites of protein synthesis.
17. Control center of the cell.
18. Assembles RNA and protein that will form ribosomes.
19. Sacs containing materials destined for export.
20. Contains the chromosomes of the cell.
21. Parts of chromosomes visible in nondividing cells.
22. Tiny cytoplasmic organelles composed of RNA and protein.

1. _____
2. _____
3. _____
4. _____
5. _____
6. _____
7. _____
8. _____
9. _____
10. _____
11. _____
12. _____
13. _____
14. _____
15. _____
16. _____
17. _____
18. _____
19. _____
20. _____
21. _____
22. _____

2. Write the terms that complete the sentences in the paragraphs below.

The plasma membrane consists of a bilayer of 1 molecules in which 2 molecules are embedded. Its 3 permeability regulates the passage of materials into and out of the cell. In a similar manner, the nuclear envelope regulates the passage of materials between the nucleus and the 4.

The nucleus is the control center of the cell because the genetic information encoded in the 5 molecules of the 6 controls cellular functions. The nucleolus assembles 7 and 8, which will move into the cytoplasm to become ribosomes.

The 9 provides passageways for the movement of materials within the cytoplasm. Proteins synthesized on 10 of the RER are carried to the 11, which packages them in condensing vesicles that carry the proteins toward the cell surface. Near the plasma membrane, condensing vesicles become 12 that release their contents outside the cell.

Energy for cellular work is made available by cellular respiration occurring in the 13, which contain their own RNA and 14 and are able to 15 themselves.

1. _____
2. _____
3. _____
4. _____
5. _____
6. _____
7. _____
8. _____
9. _____
10. _____
11. _____
12. _____
13. _____
14. _____
15. _____

3. Summarize your understanding of cell structure by completing the following table.

Cell Component	Structure/Composition	Function
Plasma Membrane		
Nucleus		
Nucleolus		
RER		
SER		

Cell Component	Structure/Composition	Function
Ribosome		
Mitochondrium		
Golgi Complex		
Secretory Vesicle		
Lysosome		
Microtubule		
Microfilament		
Centriole		
Cilia		

C. Microscopic Study

1. Make a series of sketches showing how an *Amoeba* moves by amoeboid movement. Use arrows to show the direction of movement and the flowing of cytoplasm. Label the nucleus, plasmasol, plasmagel, and plasma membrane.

2. Draw a few cheek epithelial cells. Label the nucleus, cytoplasm, and plasma membrane.

3. Are human sperm larger or smaller than cheek epithelial cells? _____
   Draw a few sperm and label the head and the flagellum.

4. Draw the appearance of the different types of blood cells that you observed. Label the nucleus, cytoplasm, and plasma membrane.

5. Draw a few microorganisms that you observed in pond water. Label the nucleus, cytoplasm, plasma membrane, and cilia or flagella (if present).

Laboratory Report 5

Student _____

Lab Section _____

MITOTIC CELL DIVISION

A. Matching

Write the name of the correct response in the answer column.

anaphase   interphase   prophase
cytokinesis   metaphase   telophase

1. Chromosomes line up at equator of spindle.
2. Chromosomes become visible as rodlike structures.
3. Nuclear envelope is intact.
4. Chromatin granules are present.
5. Each pair of centrioles moves to opposite ends of cell.
6. Cleavage furrow forms.
7. Centrioles replicate.
8. Nuclear envelope disappears.
9. Starts with formation of cleavage furrow.
10. Chromosomes are replicated.
11. Sister chromatids separate and move to opposite poles of the spindle.
12. New nuclear envelopes are formed.
13. Daughter chromosomes uncoil and become less distinct.
14. Spindle is formed of microtubules.
15. Forms daughter cells.

1. _____
2. _____
3. _____
4. _____
5. _____
6. _____
7. _____
8. _____
9. _____
10. _____
11. _____
12. _____
13. _____
14. _____
15. _____

B. Completion

1. Briefly describe the process of mitotic cell division. _____

_____

_____

_____

_____

_____

_____

2. There are 23 pairs of chromosomes in human body cells. How many chromosomes are in a parent cell? _____ in each daughter cell? _____

3. How many daughter chromosomes are in a human cell at anaphase? _____

4. Compare the genetic composition of parent cell and daughter cells. _____

_____

Why is this relationship important? _____

_____

C. Microscopic Study

Diagram the appearance of the whitefish cells in each phase of the cell cycle and label the aster, astral fiber, spindle fiber, chromatin granules, nuclear envelope, replicated chromosome, sister chromatids, and daughter nuclei. List the key features of each stage below your drawings.

Interphase

Prophase

Metaphase

Anaphase

Telophase

Daughter Cells

Laboratory Report 6

Student _____

Lab Section _____

DIFFUSION AND OSMOSIS

A. Concepts

1. Define diffusion. _____

_____

_____

2. Define osmosis. _____

_____

_____

3. Describe the relationship between solute concentration and the osmotic pressure of an aqueous solution. _____

_____

_____

4. Consider two sodium chloride (NaCl) solutions separated by a semipermeable membrane. Solution A is a 5% solution; solution B is a 10% solution. Both water and NaCl can readily pass through the membrane. Indicate the following:

The direction of net water movement. _____

The direction of net NaCl movement. _____

When this movement will stop. _____

The hypotonic solution. _____

5. If a 1.5% NaCl solution is separated by a semipermeable membrane from another NaCl solution that is isotonic, indicate the following:

Salt concentration of the isotonic solution. _____

Direction of net water movement. _____

B. Brownian Movement

1. Describe the movement of the ink particles observed at 1000×. _____

_____

2. Explain the cause of Brownian movement. _____

_____

C. Diffusion and Temperature

1. For each beaker, record the temperature of the water and the time required for molecules of potassium permanganate to diffuse throughout the water.

Beaker A: Temp. _____   Diffusion Time _____

Beaker B: Temp. _____   Diffusion Time _____

Beaker C: Temp. _____   Diffusion Time _____

2. Describe the relationship between the rate of diffusion and temperature. _____

_____

D. Diffusion and Molecular Weight

1. Record the diameter of the colored area around each granule.

   Potassium permanganate: _____ mm Methylene blue: _____ mm

2. Describe the relationship between the rate of diffusion and molecular weight. _____

_____

E. Osmosis

1. Record the height (in mm) of the water-sucrose columns during a 30-minute interval.

   A: At start: _____ mm   B: At start: _____ mm

       At end: _____ mm   At end: _____ mm

       Change: _____ mm   Change: _____ mm

2. Which osmometer has the higher concentration of sucrose? _____

   Explain your response. _____

_____

_____

F. Cell Membrane Integrity

1. Record the degree of transparency and the effect on the red blood cells (lysis, crenation, normal) of the solutions.

2. Indicate the number of the solution that is:

   isotonic _____   hypertonic _____   hypotonic _____

Solution	Transparency	Cell Shape
1. 2.0% glucose		
2. 5.0% glucose		
3. 0.3% NaCl		
4. 0.9% NaCl		
5. 2.0% NaCl		

Laboratory Report 7

Student _____

Lab Section _____

EPITHELIAL AND CONNECTIVE TISSUES

A. Epithelial Tissues

Select from the list of tissues those that are described by the following statements.

columnar, simple ciliated
columnar, simple nonciliated
columnar, stratified
columnar, pseudostratified ciliated
columnar, pseudostratified
  nonciliated

cuboidal, simple
cuboidal, stratified
squamous, simple
squamous, stratified
transitional

1. Epidermis of the skin.
2. Lining of the intestine.
3. Lining of the blood vessels.
4. Lining of the urinary bladder.
5. Lining of anal canal.
6. Lining of uterine tubes.
7. Secretory epithelium in the pancreas.
8. Lining of large ducts in salivary glands.
9. Lining of nasal cavity and trachea.
10. Single layer of elongate cells without cilia.
11. Single layer of cubelike cells.
12. Multilayered; outer layer of thin, flat cells.
13. Ciliated cells appear multilayered but are not.
14. Multilayered; cells flatten when stretched.
15. Multilayered; outer layer of elongate cells.

1. _____
2. _____
3. _____
4. _____
5. _____
6. _____
7. _____
8. _____
9. _____
10. _____
11. _____
12. _____
13. _____
14. _____
15. _____

B. Connective Tissues

Select from the list of tissues those that are described by the statements that follow.

adipose tissue
bone
dense fibrous tissue
elastic cartilage

fibrocartilage
hyaline cartilage
loose fibrous tissue
reticular tissue

1. Superficial fascia.
2. Framework of lymph nodes.
3. Tendons and ligaments.
4. Supporting rings in trachea.
5. Supports pinna of external ear.
6. Occurs on ends of long bones.
7. Intervertebral disks.
8. Fat storage.
9. Most flexible supporting tissue.
10. Solid matrix of calcium salts.
11. Most rigid supporting tissues.
12. Cartilage with many collagenous fibers.
13. White, glassy appearance.
14. Serves as insulation material.
15. Preforms most bones in fetal skeleton.

1. _____
2. _____
3. _____
4. _____
5. _____
6. _____
7. _____
8. _____
9. _____
10. _____
11. _____
12. _____
13. _____
14. _____
15. _____

C. Terminology

Write the term defined below in the answer column.

1. Protein in white fibers.
2. Fiber-producing cell.
3. Mucus-secreting cell.
4. Lines medullary cavity.
5. Intercellular substance.
6. Produces bone matrix.
7. Cartilage cell.
8. Plates in spongy bone.
9. Rings of compact bone.
10. Space around bone cell.

1. _____
2. _____
3. _____
4. _____
5. _____
6. _____
7. _____
8. _____
9. _____
10. _____

D. Figure

Record the labels for Figure 7.9 in the answer column.

E. Microscopic Study

Use the space below and additional sheets of paper as necessary to draw the appearance of each tissue observed. Add labels and make notations that will help you recognize these tissues in a lab practicum.

Figure 7.9

1. _____
2. _____
3. _____
4. _____
5. _____
6. _____
7. _____
8. _____
9. _____
10. _____
11. _____
12. _____
13. _____

Laboratory Report 8

Student _____

Lab Section _____

THE INTEGUMENT

A. Matching

Select the structure from the list below that is described by the following statements and record it in the answer column.

apocrine sweat gland
arrector pili muscle
dermal papilla
dermis
eccrine sweat gland
epidermis
hypodermis

Meissner's corpuscle
Pacinian corpuscle
sebaceous gland
stratum basale
stratum corneum
stratum granulosum
stratum spinosum

1. Layer of epidermis containing melanocytes.
2. Contains blood vessels that nourish hair follicle.
3. Touch receptor in dermal papillae.
4. Sweat gland emptying secretion into hair follicle.
5. Inner layer of skin formed of connective tissue.
6. Outermost layer of epidermis.
7. Epidermal layer just interior to stratum granulosum.
8. Most abundant type of sweat gland.
9. Pressure receptor located deep in dermis.
10. Secretes oily lubricant into hair follicle.
11. Superficial fascia.
12. Activation produces goose bumps.
13. Formed of loose connective and adipose tissues.
14. Five layers of stratified squamous epithelium.
15. Has numerous collagen fibers that give strength to skin.
16. Forms watery secretion containing waste materials.
17. Epidermal layer in which cell division occurs.
18. Secretion is responsible for body odor.
19. Epidermal layer with granule precursors of keratin.
20. Responsible for the pattern of fingerprints.
21. Sweat gland activated primarily by thermal stimuli.
22. Epidermal layer producing cells that form a hair.
23. Contains insulating fatty tissue.
24. Epidermal layer preventing evaporative water loss.
25. Skin layer containing nerves and blood vessels.
26. Epidermal layer of flat, dead, keratinized cells.
27. Contains numerous sensory receptors.
28. Larger, but less numerous, sweat glands.
29. Secretes sebum.
30. Contains coiled portion of eccrine sweat glands.

1. _____
2. _____
3. _____
4. _____
5. _____
6. _____
7. _____
8. _____
9. _____
10. _____
11. _____
12. _____
13. _____
14. _____
15. _____
16. _____
17. _____
18. _____
19. _____
20. _____
21. _____
22. _____
23. _____
24. _____
25. _____
26. _____
27. _____
28. _____
29. _____
30. _____

B. Figures

Record the labels for Figures 8.1 and 8.2 in the answer column.

C. Completion

1. Describe the function of melanin. _____

   _____

   _____

   _____

2. Explain why a "tan" is temporary. _____

   _____

   _____

   _____

   _____

   _____

   _____

D. Microscopic Study

Use the space below and additional sheets of paper to make drawings and notations that will help you understand skin structure and recognize skin and its components on a lab practicum.

Figure 8.1

1. _____
2. _____
3. _____
4. _____
5. _____
6. _____
7. _____
8. _____
9. _____
10. _____
11. _____
12. _____
13. _____
14. _____
15. _____
16. _____
17. _____

Figure 8.2

1. _____
2. _____
3. _____
4. _____
5. _____
6. _____
7. _____
8. _____
9. _____
10. _____
11. _____
12. _____

E. Density of Touch Receptors

Indicate the minimum distance between the tips of the dividers that produced a two-point sensation.

Back of neck _____ mm   Inner surface of forearm _____ mm

Palm of hand _____ mm   Tip of index finger _____ mm

Is there a correlation between the density of touch receptors and the importance of touch sensations for the body parts tested?
_____

Laboratory Report 9

Student ____________________

Lab Section ____________________

THE SKELETAL PLAN

A. Figures

List the labels for Figures 9.1, 9.2, and 9.3.

Figure 9.1

1. ____________________
2. ____________________
3. ____________________
4. ____________________
5. ____________________
6. ____________________
7. ____________________
8. ____________________
9. ____________________
10. ____________________
11. ____________________
12. ____________________

Figure 9.2

1. ____________________
2. ____________________
3. ____________________
4. ____________________
5. ____________________
6. ____________________
7. ____________________
8. ____________________
9. ____________________
10. ____________________
11. ____________________
12. ____________________
13. ____________________
14. ____________________
15. ____________________
16. ____________________
17. ____________________
18. ____________________
19. ____________________
20. ____________________
21. ____________________
22. ____________________
23. ____________________
24. ____________________

Figure 9.3

A. ____________________

B. ____________________

C. ____________________

D. ____________________

E. ____________________

F. ____________________

G. ____________________

H. ____________________

I. ____________________

J. ____________________

B. Long Bone Structure

Select the response that matches the statements that follow and write it in the answer column.

articular cartilage	epiphyseal line
cancellous bone	epiphysis
compact bone	medullary cavity
diaphysis	periosteum
endosteum	red marrow
epiphyseal disk	yellow marrow

1. Shaft portion of a long bone.
2. Hollow chamber in the bone shaft.
3. Type of marrow in the medullary cavity.
4. Enlarged end of a long bone.
5. Lining of the medullary cavity.
6. Type of bone tissue forming most of the diaphysis.
7. Fibrous membrane covering the bone.
8. Site of linear growth.
9. Protective surface on ends of a long bone.
10. Type of bone tissue forming interior of epiphyses.
11. Reduces friction in joints formed by long bones.
12. Type of marrow in cancellous bone of the femur.
13. Formed by fusion of epiphysis and diaphysis.

1. _____
2. _____
3. _____
4. _____
5. _____
6. _____
7. _____
8. _____
9. _____
10. _____
11. _____
12. _____
13. _____

Provide these answers on the basis of your study of the split long bones.

1. Describe the distribution of compact and spongy bone. _____

   Compact bone _____

   _____

   Spongy bone _____

   _____

2. Is the periosteum elastic or nonelastic? _____ weak or strong? _____

   Is the periosteum firmly or loosely attached to the bone? _____

3. How is a tendon or ligament attached to the bone? _____

   _____

   _____

4. Feel the surface of an articular cartilage. Describe how it feels. _____

   _____

5. How do articular cartilages aid movement at a joint formed by two long bones? _____

   _____

   _____

6. Had the beef bone attained full growth? _____ Explain. _____

   _____

C. Parts of the Skeleton

Select the response that matches the statements that follow and write it in the answer column.

clavicle	patella	sternum
coxal bone	radius	symphysis pubis
femur	rib	thoracic cage
fibula	sacrum	tibia
humerus	scapula	ulna
hyoid	skull	vertebrae

1. Shoulder blade.
2. Collarbone.
3. Breastbone.
4. Shinbone.
5. Kneecap.
6. Upper arm bone.
7. Bones of spinal column.
8. Thighbone.
9. Lateral bone of forearm.
10. Joint between coxal bones.
11. Horseshoe-shaped bone under lower jaw.
12. Hipbone.
13. Medial bone of forearm.
14. Thin bone lateral to tibia.
15. Bony framework of the head.
16. Two bones forming the shoulder girdle.
17. Three bones forming the elbow joint.
18. Two bones forming the knee joint.
19. Bones with which ribs articulate posteriorly.
20. Limb bone articulating with a scapula.
21. Limb bone articulating with a coxal bone.
22. Forms protective framework for thorax.
23. Part of vertebral column attached to coxal bones.
24. Articulates with humerus and clavicle.

1. _____
2. _____
3. _____
4. _____
5. _____
6. _____
7. _____
8. _____
9. _____
10. _____
11. _____
12. _____
13. _____
14. _____
15. _____
16a. _____
16b. _____
17a. _____
17b. _____
17c. _____
18a. _____
18b. _____
19. _____
20. _____
21. _____
22. _____
23. _____
24. _____

D. Surface Features

Select the term that matches the descriptions that follow.

condyle	sinus
crest	spine
fissure	sulcus
foramen	trochanter
fossa	tubercle
head	tuberosity
meatus	

1. An enlarged end of a bone that is supported by a narrow neck.
2. A passageway for nerves or blood vessels.
3. An air-filled cavity.
4. A shallow depression.
5. A sharp, slender process.
6. A very large process.
7. A groove or furrow.
8. A knucklelike process that articulates with another bone.
9. A narrow slit.
10. A large canal or passageway.
11. A narrow ridge.
12. A low, roughened process to which a muscle attaches.
13. A small, rounded process.

1. _____
2. _____
3. _____
4. _____
5. _____
6. _____
7. _____
8. _____
9. _____
10. _____
11. _____
12. _____
13. _____

E. Bone Fractures

Select the type of fracture that matches the statements that follow and write it in the answer column.

comminuted	oblique
compacted	segmental
fissured	spiral
greenstick	transverse

1. Results from twisting of the bone.
2. Complete fracture at 90° to bone axis.
3. Break on one side only due to bending of the bone.
4. One part of a bone is forced into another part of the same bone.
5. A linear, incomplete splitting of the bone.
6. Complete fracture *not* at 90° to bone axis.
7. Bone broken into several pieces.
8. A single piece is broken out of the bone.

Write in the answer column the type of fracture that is described by the statements below.

9. Fractured bone breaks through the skin.
10. Fractured bone is not broken clear through.
11. Fracture caused by excessive vertical force.
12. Fractured bone does not break through the skin.

1. _____
2. _____
3. _____
4. _____
5. _____
6. _____
7. _____
8. _____
9. _____
10. _____
11. _____
12. _____

Laboratory Report 10

Student ______________________________

Lab Section ______________________________

THE SKULL

A. Figures

List the labels for Figures 10.1 through 10.7 and Figure 10.9 in the answer columns.

Figure 10.1

1. ______________________________
2. ______________________________
3. ______________________________
4. ______________________________
5. ______________________________
6. ______________________________
7. ______________________________
8. ______________________________
9. ______________________________
10. ______________________________
11. ______________________________
12. ______________________________
13. ______________________________
14. ______________________________
15. ______________________________
16. ______________________________
17. ______________________________
18. ______________________________
19. ______________________________
20. ______________________________
21. ______________________________
22. ______________________________

Figure 10.2

1. ______________________________
2. ______________________________
3. ______________________________
4. ______________________________
5. ______________________________
6. ______________________________
7. ______________________________
8. ______________________________
9. ______________________________
10. ______________________________
11. ______________________________
12. ______________________________
13. ______________________________
14. ______________________________
15. ______________________________
16. ______________________________
17. ______________________________
18. ______________________________
19. ______________________________

Figure 10.3

1. ______________________________
2. ______________________________
3. ______________________________
4. ______________________________
5. ______________________________
6. ______________________________
7. ______________________________
8. ______________________________
9. ______________________________
10. ______________________________
11. ______________________________
12. ______________________________
13. ______________________________
14. ______________________________

Figure 10.4

1. _____
2. _____
3. _____
4. _____
5. _____
6. _____
7. _____
8. _____
9. _____
10. _____
11. _____
12. _____
13. _____
14. _____
15. _____
16. _____
17. _____

Figure 10.5

1. _____
2. _____
3. _____
4. _____
5. _____
6. _____
7. _____
8. _____
9. _____
10. _____

Figure 10.6

1. _____
2. _____
3. _____
4. _____

Figure 10.7

1. _____
2. _____
3. _____
4. _____
5. _____
6. _____
7. _____
8. _____
9. _____
10. _____
11. _____
12. _____
13. _____
14. _____
15. _____

Figure 10.9

1. _____
2. _____
3. _____
4. _____
5. _____
6. _____
7. _____
8. _____
9. _____
10. _____
11. _____
12. _____

B. Bones

Select the bones that are described or that have the structures noted in the statements below.

ethmoid   nasal   sphenoid
frontal   nasal conchae   temporal
lacrimal   occipital   vomer
mandible   palatine   zygomatic
maxilla   parietal

1. Upper bone of nasal septum.
2. Foramen magnum.
3. Lower jaw.
4. Cheekbone.
5. Mental foramen.
6. Scroll-like bones in nasal cavity.
7. Coronoid process.
8. Zygomatic process.
9. Internal auditory meatus.
10. Carotid canal.
11. Mandibular fossa.
12. Mastoid process.
13. Ramus.
14. Sella turcica.
15. Styloid process.
16. Supraorbital foramen.
17. Upper jaw.
18. Cribriform plates.

1. _____
2. _____
3. _____
4. _____
5. _____
6. _____
7. _____
8. _____
9. _____
10. _____
11. _____
12. _____
13. _____
14. _____
15. _____
16. _____
17. _____
18. _____

C. Sutures

Select the sutures described below.

coronal   median palatine   squamosal
lambdoidal   sagittal

1. Joins parietal bones.
2. Joins frontal and parietal bones.
3. Joins palatine processes of maxillae.
4. Joins parietal and occipital bones.
5. Joins parietal and temporal bones.

1. _____
2. _____
3. _____
4. _____
5. _____

D. Completion

Write your response in the answer column.

1. Two bones forming the hard palate.
2. Two bones forming the zygomatic arch.
3. Allow compression of skull during birth.
4. Four bones that contain sinuses.
5. Point of attachment of skull with first vertebra.
6. Allow sound waves to enter skull.
7. Form external bridge of nose.
8. Contain groove for tear duct.
9. Forms lateral margin of eye orbit.
10. Forms posterior wall of eye orbit.
11. Forms superior margin of eye orbit.
12. Depression into which the pituitary gland fits.
13. Two bones forming cheekbones.
14. Two bones forming nasal septum.
15. Membranous areas at junctions of cranial bones in a fetal skull.
16. Any sharp or rounded projection of a bone.
17. Any shallow depression of a bone.
18. An air-filled cavity in a bone.
19. A fused articulation between bones.
20. A knucklelike projection of a bone.
21. A passageway for nerves or blood vessels into a bone.

1. _____
2. _____
3. _____
4. _____
5. _____
6. _____
7. _____
8. _____
9. _____
10. _____
11. _____
12. _____
13. _____
14. _____
15. _____
16. _____
17. _____
18. _____
19. _____
20. _____
21. _____

Laboratory Report 11

Student ______________________

Lab Section ______________________

THE VERTEBRAL COLUMN AND THORAX

A. Figures

List the labels for Figures 11.1, 11.2, and 11.3.

Figure 11.1

1. ______
2. ______
3. ______
4. ______
5. ______
6. ______
7. ______
8. ______
9. ______
10. ______
11. ______
12. ______
13. ______
14. ______
15. ______
16. ______
17. ______
18. ______
19. ______
20. ______
21. ______
22. ______
23. ______
24. ______
25. ______
26. ______
27. ______
28. ______
29. ______

Figure 11.2

1. ______
2. ______
3. ______
4. ______
5. ______
6. ______
7. ______
8. ______

Figure 11.3

1. ______
2. ______
3. ______
4. ______
5. ______
6. ______
7. ______
8. ______
9. ______
10. ______
11. ______
12. ______
13. ______
14. ______
15. ______
16. ______

B. Parts of the Vertebral Column

Select the response from the list at the right that matches the statement. Answers may be used more than once.

1. Facet on transverse processes.
2. Transverse foramina.
3. First cervical vertebra.
4. Floating rib.
5. Inferior tip of breastbone.
6. Midpart of breastbone.
7. Second cervical vertebra.
8. Superior part of breastbone.
9. Vertebrae of the neck.
10. True rib.
11. False rib with costal cartilage.
12. Breastbone.
13. Tailbone.
14. Possesses odontoid process.
15. Has largest vertebral foramen.

atlas
axis
cervical vertebrae
coccyx
gladiolus
lumbar vertebrae
manubrium
sacrum
sternum
thoracic vertebrae
vertebral rib
vertebrochondral rib
vertebrosternal rib
xiphoid

1. _____
2. _____
3. _____
4. _____
5. _____
6. _____
7. _____
8. _____
9. _____
10. _____
11. _____
12. _____
13. _____
14. _____
15. _____

C. Numbers

Indicate the number of components of the following.

1. Cervical vertebrae.
2. Fused vertebrae composing coccyx.
3. Fused vertebrae composing sacrum.
4. Lumbar vertebrae.
5. Pairs of false ribs.
6. Pairs of floating ribs.
7. Pairs of true ribs.
8. Pairs of vertebrochondral ribs.
9. Spinal curvatures.
10. Thoracic vertebrae.

1. _____
2. _____
3. _____
4. _____
5. _____
6. _____
7. _____
8. _____
9. _____
10. _____

D. Completion

Provide the terms or phrases that match the statements.

1. Cushion between vertebral bodies.
2. Articulates with inferior articulating surfaces.
3. Bony framework around vertebral foramen.
4. Part of vertebra bearing most load.
5. Vertebral type with largest bodies.
6. Forms posterior wall of pelvis.
7. Vertebrae with ribs.
8. Attach ribs to sternum.
9. Articulate with facets on transverse processes.
10. Openings through which spinal nerves exit.

1. _____
2. _____
3. _____
4. _____
5. _____
6. _____
7. _____
8. _____
9. _____
10. _____

Laboratory Report 12

Student _____

Lab Section _____

THE APPENDICULAR SKELETON

A. Figures

List the labels for Figures 12.1 through 12.4.

Figure 12.1

1. _____
2. _____
3. _____
4. _____
5. _____
6. _____
7. _____
8. _____

Figure 12.2

1. _____
2. _____
3. _____
4. _____
5. _____
6. _____
7. _____
8. _____
9. _____
10. _____
11. _____
12. _____
13. _____
14. _____
15. _____
16. _____
17. _____

18. _____
19. _____
20. _____
21. _____
22. _____
23. _____
24. _____
25. _____
26. _____
27. _____
28. _____

Figure 12.3

1. _____
2. _____
3. _____
4. _____
5. _____
6. _____
7. _____
8. _____
9. _____
10. _____
11. _____
12. _____
13. _____
14. _____
15. _____

Figure 12.4

1. _____
2. _____
3. _____
4. _____
5. _____
6. _____
7. _____
8. _____
9. _____
10. _____
11. _____
12. _____
13. _____

14. _____
15. _____
16. _____
17. _____
18. _____
19. _____
20. _____
21. _____
22. _____
23. _____
24. _____
25. _____

B. True–False

Compare the male and female pelvic girdles, side by side, to determine if the following statements are true or false.

1. The opening of the female pelvis is larger and more oval than that of the male pelvis.
2. The acetabula of the female pelvis face more anteriorly than those of the male pelvis.
3. The iliac crests of the female pelvis protrude laterally more than those of the male pelvis.
4. The distance between the ischial spines is greater in the female pelvis.
5. The angle of the pubic arch below the symphysis pubis is wider in the female pelvis.
6. The sacrum curves more posteriorly in the female pelvis than in the male pelvis.

1. _____
2. _____
3. _____
4. _____
5. _____
6. _____

C. Completion

Provide the bones or structures described by the statements.

1. Fossa on scapula that articulates with humerus.
2. Condyle of humerus that articulates with the radius.
3. Scapular process that articulates with clavicle.
4. Large process lateral to head of humerus.
5. Process forming point of elbow.
6. Fossa of ulna that articulates with the humerus.
7. Wrist bones.
8. Bones of palm of the hand.
9. Scapular process projecting forward under clavicle.
10. Fossa on distal anterior surface of humerus.
11. Process on distal lateral margin of radius.
12. Bones of the fingers and toes.
13. Tuberosity just below neck of radius.
14. Fossa on distal posterior surface of humerus.
15. Part of radius that articulates with the humerus.
16. Upper bone of a coxal bone.
17. Process at angle of ischium.
18. Heelbone.
19. Large foramen in a coxal bone formed by pubis and ischium.
20. Joint between the pubic bones.
21. Large process lateral to neck of femur.
22. Fossa of a coxal bone that articulates with head of femur.
23. Joint between a coxal bone and sacrum.
24. Uppermost process on posterior edge of ischium.
25. Superior edge of ilium.
26. Carpal bone that articulates with tibia and fibula.
27. Process below neck on medial surface of femur.
28. Proximal anterior process on tibia below kneecap.
29. Process on medial surface at distal end of tibia.
30. Forms the instep of the foot.
31. A sharp ridge on the surface of a bone.
32. A process just above a condyle.
33. A slender process on a bone.
34. A roughened elevation of a bone to which a muscle attaches.

1. _____
2. _____
3. _____
4. _____
5. _____
6. _____
7. _____
8. _____
9. _____
10. _____
11. _____
12. _____
13. _____
14. _____
15. _____
16. _____
17. _____
18. _____
19. _____
20. _____
21. _____
22. _____
23. _____
24. _____
25. _____
26. _____
27. _____
28. _____
29. _____
30. _____
31. _____
32. _____
33. _____
34. _____

NOTES

Laboratory Report 13

Student _____

Lab Section _____

ARTICULATIONS

A. Figures

List the labels for Figures 13.1 through 13.4.

Figure 13.1

1. _____
2. _____
3. _____
4. _____
5. _____
6. _____
7. _____
8. _____
9. _____
10. _____

Figure 13.2

1. _____
2. _____
3. _____
4. _____
5. _____
6. _____
7. _____
8. _____
9. _____
10. _____
11. _____

Figure 13.3

1. _____
2. _____
3. _____
4. _____
5. _____
6. _____
7. _____
8. _____
9. _____
10. _____
11. _____
12. _____
13. _____
14. _____

Figure 13.4

1. _____
2. _____
3. _____
4. _____
5. _____
6. _____
7. _____
8. _____
9. _____

B. Types and Subtypes

Identify the basic articulation type to which the subtypes listed belong.

amphiarthrosis   diarthrosis   synarthrosis

1. Ball-and-socket	6. Saddle
2. Condyloid	7. Suture
3. Gliding	8. Symphysis
4. Hinge	9. Synchondrosis
5. Pivot	10. Syndesmosis

1. _____
2. _____
3. _____
4. _____
5. _____
6. _____
7. _____
8. _____
9. _____
10. _____

C. Joint Characteristics

Select the joints that have the characteristics described by the statements.

ball-and-socket	saddle
condyloid	suture
gliding	symphysis
hinge	synchondrosis
pivot	syndesmosis

1. Bone ends are flat or slightly convex.
2. Slightly movable; fibrocartilaginous pad.
3. Slightly movable; bones joined by interosseous ligament.
4. Immovable; bones bonded by cartilaginous disk.
5. Rotational movement only.
6. Freely movable in one direction only.
7. Freely movable in two planes.
8. Angular movement in all directions.
9. Immovable; bonded by fibrous connective tissue.
10. Bone ends are convex in one direction and concave in the other.

1. _____
2. _____
3. _____
4. _____
5. _____
6. _____
7. _____
8. _____
9. _____
10. _____

D. Joint Classification

By examining and manipulating the bones of an articulated skeleton, classify the following joints according to the list in Section C.

1. Femur–tibia.
2. Humerus–ulna.
3. Vertebra–vertebra.
4. Phalanges–phalanges.
5. Phalanges–metacarpals.
6. Humerus–scapula.
7. Atlas–axis.
8. Carpals–carpals.
9. Tibia–fibula, distal ends.
10. Left pubis–right pubis.
11. Parietal–frontal.
12. Trapezium–metacarpal of thumb.
13. Diaphysis–epiphysis of femur in child.
14. Radius–ulna, distal ends.
15. Femur–coxal bone.

1. _____
2. _____
3. _____
4. _____
5. _____
6. _____
7. _____
8. _____
9. _____
10. _____
11. _____
12. _____
13. _____
14. _____
15. _____

Laboratory Report 14

Student _____

Lab Section _____

MUSCLE ORGANIZATION AND BODY MOVEMENTS

A. Figures

List the labels for Figure 14.1 and the types of body movements in Figure 14.2.

Figure 14.1

1. _____
2. _____
3. _____
4. _____
5. _____
6. _____
7. _____
8. _____
9. _____
10. _____
11. _____
12. _____
13. _____

Figure 14.2

A. _____
B. _____
C. _____
D. _____
E. _____
F. _____
G. _____
H. _____
I. _____
J. _____
K. _____
L. _____
M. _____

B. Muscle Structure

Match the terms below with the statements that follow.

aponeurosis   insertion
deep fascia   ligament
endomysium   origin
epimysium   perimysium
fasciculus   tendon

1. Bundle of muscle fibers.
2. Connective tissue around each fasciculus.
3. Movable end of a muscle.
4. Thin layer of connective tissue around each muscle fiber.
5. Broad sheet of connective tissue attaching a muscle to another muscle or bone.
6. Connective tissue that surrounds a group of fasciculi.
7. Immovable end of a muscle.
8. Outer layer of connective tissue that envelops entire muscle.
9. Narrow band of connective tissue attaching muscle to bone or other muscle.
10. Not part of a muscle.

1. _____
2. _____
3. _____
4. _____
5. _____
6. _____
7. _____
8. _____
9. _____
10. _____

C. Muscle Types

Match the terms in the list below with the statements that follow.

agonists   fixators

antagonists   synergists

1. Muscles that assist prime movers.
2. Prime movers.
3. Muscles opposing prime movers.
4. Muscles holding structures steady to allow prime movers to act smoothly.

1. _____

2. _____

3. _____

4. _____

D. Movements

Match the terms in the list below with the statements that follow.

abduction   hyperextension

adduction   inversion

circumduction   plantar flexion

dorsiflexion   pronation

eversion   rotation

extension   supination

flexion

1. Movement of hand from palm down to palm up.
2. Movement of leg away from midline of the body.
3. Movement of 200° around a central point.
4. Flexion of foot upward (specific term).
5. Decrease in the angle of bones at a joint.
6. Movement of hand from palm up to palm down.
7. Movement of limb subscribing a circle.
8. Turning the sole of the foot inward.
9. Increase in the angle of bones at a joint.
10. Backward movement of head.
11. Movement of arm toward midline of the body.
12. Moving sole of foot downward (specific term).
13. Turning the sole of the foot outward.

1. _____

2. _____

3. _____

4. _____

5. _____

6. _____

7. _____

8. _____

9. _____

10. _____

11. _____

12. _____

13. _____

Laboratory Report 15

Student _____

Lab Section _____

HEAD AND TRUNK MUSCLES

A. Figures

List the labels for Figures 15.1 through 15.3.

Figure 15.1

1. _____
2. _____
3. _____
4. _____
5. _____
6. _____
7. _____
8. _____
9. _____
10. _____
11. _____

Figure 15.2

1. _____
2. _____
3. _____
4. _____
5. _____
6. _____
7. _____
8. _____
9. _____
10. _____
11. _____
12. _____

Figure 15.3

1. _____
2. _____
3. _____
4. _____
5. _____
6. _____
7. _____

B. Head Muscles

Select the muscle that matches the following statements.

buccinator	platysma
frontalis	sternocleidomastoid
masseter	temporalis
orbicularis oculi	triangularis
orbicularis oris	zygomaticus

1. Primary muscle of mastication.
2. Draws corner of mouth downward only.
3. Wrinkles forehead and raises eyebrows.
4. Draws corner of mouth upward and backward.
5. Compresses cheeks.
6. Draws corner of mouth downward and backward.
7. Draws head downward toward chest.
8. Acts synergistically with the masseter.
9. Closes eyelids.
10. Closes and puckers lips.
11. Pulls scalp backward.
12. Assists in opening the mouth.
13. Holds food between teeth during chewing.
14. Contraction of muscle on right side turns face to the left.

1. _____
2. _____
3. _____
4. _____
5. _____
6. _____
7. _____
8. _____
9. _____
10. _____
11. _____
12. _____
13. _____
14. _____

C. Trunk Muscles

Select the muscle described in the following statements.

deltoid	pectoralis major
external intercostals	rectus abdominis
external oblique	serratus anterior
infraspinatus	supraspinatus
internal intercostals	teres major
internal oblique	transversus abdominis
latissimus dorsi	trapezius

1. Innermost abdominal muscle.
2. Rotates upper arm medially only.
3. Adducts and elevates scapula.
4. Rotates upper arm laterally only.
5. Extends, adducts, and rotates upper arm medially.
6. Elevates rib cage in breathing.
7. Draws scapula downward and forward.
8. Adducts and flexes upper arm medially.
9. Abducts and extends upper arm.
10. Draws head backward.
11. Primary muscle of the shoulder.
12. Large surface muscle of the upper chest.
13. Lowers rib cage during breathing.
14. Triangular surface muscle of upper back.
15. Assists deltoid in abducting arm.
16. Segmented muscle extending from ribs to pubis.
17. Hyperextends head.
18. Outermost abdominal muscle.
19. Superficial muscle covering lower back.
20. Lies just under the external oblique.

1. _____
2. _____
3. _____
4. _____
5. _____
6. _____
7. _____
8. _____
9. _____
10. _____
11. _____
12. _____
13. _____
14. _____
15. _____
16. _____
17. _____
18. _____
19. _____
20. _____

Laboratory Report 16

Student _____

Lab Section _____

MUSCLES OF THE UPPER LIMB

A. Figures

List the labels for Figures 16.1 and 16.2.

Figure 16.1

A. _____

B. _____

C. _____

D. _____

E. _____

Figure 16.2

1. _____

2. _____

3. _____

4. _____

5. _____

6. _____

7. _____

8. _____

9. _____

10. _____

11. _____

12. _____

13. _____

14. _____

B. Arm Movements

Select the muscle described in the statements below.

biceps brachii
brachialis
brachioradialis
coracobrachialis

pronator quadratus
pronator teres
supinator
triceps brachii

1. Extends the forearm.
2. Flexes and adducts the upper arm.
3. Rotates forearm laterally.
4. Muscle at distal end of forearm that is a synergist of the pronator teres.
5. Three muscles that flex the forearm.
6. Antagonist of the brachialis.
7. Flexes and rotates the forearm laterally.
8. Superficial muscle on anterior surface of upper arm.
9. Covers posterior surface of humerus.
10. Superficial muscle on lateral surface of forearm.
11. Located just under biceps brachii.
12. Primary muscle rotating forearm medially.

1. _____
2. _____
3. _____
4. _____
5a. _____
5b. _____
5c. _____
6. _____
7. _____
8. _____
9. _____
10. _____
11. _____
12. _____

C. Hand Movements

Select the muscle described in the statements below.

abductor pollicis
extensor carpi radialis longus
extensor carpi ulnaris
extensor digitorum communis
extensor pollicis longus

flexor carpi radialis
flexor carpi ulnaris

flexor digitorum profundus
flexor digitorum superficialis
flexor pollicis longus

1. Flexes and adducts the hand.
2. Flexes and abducts the hand.
3. Abducts the thumb.
4. Extends distal phalanges of fingers 2–5.
5. Flexes distal phalanges of fingers 2–5.
6. Extends and abducts the hand.
7. Extends and adducts the hand.
8. Flexes middle phalanges of fingers 2–5.
9. Extends the thumb.
10. Flexes the thumb.

1. _____
2. _____
3. _____
4. _____
5. _____
6. _____
7. _____
8. _____
9. _____
10. _____

Laboratory Report 17

Student ______________________________

Lab Section ______________________________

Muscles of the Lower Limb

A. Figures

List the labels for Figures 17.1 through 17.4.

Figure 17.1

1. ______________________________
2. ______________________________
3. ______________________________
4. ______________________________
5. ______________________________
6. ______________________________
7. ______________________________
8. ______________________________

Figure 17.2

1. ______________________________
2. ______________________________
3. ______________________________
4. ______________________________
5. ______________________________
6. ______________________________
7. ______________________________
8. ______________________________
9. ______________________________
10. ______________________________
11. ______________________________

Figure 17.3

1. ______________________________
2. ______________________________
3. ______________________________
4. ______________________________
5. ______________________________

Figure 17.4

1. ______________________________
2. ______________________________
3. ______________________________
4. ______________________________
5. ______________________________
6. ______________________________
7. ______________________________

B. Thigh Movements

Select the muscle(s) from the list that is (are) described by the statements below.

adductor brevis   gluteus maximus
adductor longus   gluteus medius
adductor magnus   gluteus minimus
                       piriformis

1. Extends and rotates thigh laterally.
2. Abducts and rotates thigh laterally.
3. Three muscles that adduct, flex, and rotate thigh laterally.
4. Two muscles that abduct and rotate thigh medially.
5. Strongest adductor muscle.
6. Superficial buttocks muscle.
7. Innermost gluteus muscle.
8. Longest adductor muscle.

1. _____

2. _____

3a. _____

3b. _____

3c. _____

4a. _____

4b. _____

5. _____

6. _____

7. _____

8. _____

C. Thigh and Leg Movements

Select the muscle(s) that apply to the following statements.

biceps femoris   semitendinosus
gracilis   tensor fasciae latae
rectus femoris   vastus intermedius
sartorius   vastus lateralis
semimembranosus   vastus medialus

1. Four muscles extending the leg.
2. Adducts thigh and flexes leg.
3. Flexes the leg and thigh.
4. Flexes and abducts thigh.
5. Three hamstrings flexing the leg.
6. Deepest quadriceps muscle.
7. Anterior central quadriceps muscle.
8. Lateral hamstring muscle.
9. Medial hamstring muscle.

1a. _____

1b. _____

1c. _____

1d. _____

2. _____

3. _____

4. _____

5a. _____

5b. _____

5c. _____

6. _____

7. _____

8. _____

9. _____

D. Leg and Foot Movements

Select the muscle(s) that apply to the following statements.

extensor digitorum longus
extensor hallucis longus
flexor digitorum longus
flexor hallucis longus
gastrocnemius

peroneus brevis
peroneus longus
peroneus tertius
soleus
tibialis anterior
tibialis posterior

1. Two muscles with a common Achilles tendon.
2. Plantar flexes and inverts the foot.
3. Flexes the leg and plantar flexes the foot.
4. Two muscles that extend and evert the foot.
5. Flexes the great toe.
6. Extends the great toe.
7. Flexes toes 2–5.
8. Extends toes 2–5.
9. Dorsiflexes and everts the foot.
10. Dorsiflexes and inverts the foot.
11. Two calf muscles that plantar flex the foot.
12. Supports the transverse arch of the foot.

1a. _____
1b. _____
2. _____
3. _____
4a. _____
4b. _____
5. _____
6. _____
7. _____
8. _____
9. _____
10. _____
11a. _____
11b. _____
12. _____

E. Figure

List the labels for Figure 17.5.

Anterior

1. _____
2. _____
3. _____
4. _____
5. _____
6. _____
7. _____
8. _____
9. _____
10. _____
11. _____
12. _____
13. _____
14. _____
15. _____
16. _____
17. _____
18. _____
19. _____
20. _____

Posterior

1. _____
2. _____
3. _____
4. _____
5. _____
6. _____
7. _____
8. _____
9. _____
10. _____
11. _____
12. _____
13. _____
14. _____
15. _____

NOTES

Student _____

Lab Section _____

MUSCLE AND NERVE TISSUES

A. Muscle Tissue Characteristics
Select the type of muscle tissue described by the statements below.

cardiac   skeletal   smooth

1. Attached to bones.
2. Rhythmic contractions.
3. Voluntary in function.
4. Cells tapered to a point at each end.
5. Cells without striations.
6. Cells with striations and many nuclei.
7. Occurs in walls of blood vessels.
8. Cells separated by intercalated disks.
9. Striated cells with single nucleus.
10. Spontaneous contractions without neural activation.
11. Fibers are grouped in fasciculi.
12. Cells often branched to form network.
13. Forms muscle of heart.

B. Drawings
Make drawings of the three types of muscle tissue from your microscopic studies. Label pertinent parts.

1. _____
2. _____
3. _____
4. _____
5. _____
6. _____
7. _____
8. _____
9. _____
10. _____
11. _____
12. _____
13. _____

C. Figure
List the labels for Figure 18.4.

1. _____
2. _____
3. _____
4. _____
5. _____
6. _____
7. _____
8. _____
9. _____

10. _____
11. _____
12. _____
13. _____
14. _____
15. _____
16. _____
17. _____

D. Neuron Structure

Select the structure from the list that is described in the statements that follow.

axon   neurilemma
cell body   neurofibrils
collateral   Nissl bodies
dendrite   nodes of Ranvier
myelin sheath   Schwann cells

1. Contains neuron nucleus.
2. Side branch of an axon.
3. Myelinated process of a motor neuron.
4. Myelinated process of a sensory neuron.
5. Process receiving impulses.
6. Process transmitting impulses to an effector.
7. Thin filaments in the cell body and processes.
8. Clumps of RER in the cell body.
9. Gaps between Schwann cells.
10. Formed by inner wrappings of Schwann cells.
11. Outer layer of a Schwann cell.
12. Cells covering all peripheral neuron fibers.

1. _____
2. _____
3. _____
4. _____
5. _____
6. _____
7. _____
8. _____
9. _____
10. _____
11. _____
12. _____

E. Neuron Types

Select the structural type of neuron described in the statements that follow.

bipolar   multipolar   unipolar

1. Possesses a single process that forms a T-shaped branch.
2. Possesses a single dendrite and an axon.
3. Most common structural type.
4. Possesses multiple dendrites and an axon.
5. Spinal sensory neurons are examples.

1. _____
2. _____
3. _____
4. _____
5. _____

Select the functional type of neuron described in the statements that follow.

interneuron   motor   sensory

1. An afferent neuron.
2. An efferent neuron.
3. Totally within central nervous system.
4. Carries impulses to a muscle or gland.
5. Carries impulses to central nervous system.

1. _____
2. _____
3. _____
4. _____
5. _____

F. Drawings

Make drawings of neuron cell bodies from your microscopic studies. Label pertinent parts.

Laboratory Report 19

Student _____

Lab Section _____

THE NATURE OF MUSCLE CONTRACTION

A. Figure
List the labels for Figure 19.1.

1. _____   6. _____

2. _____   7. _____

3. _____   8. _____

4. _____   9. _____

5. _____   10. _____

B. Neuromuscular Junction
Select the terms that match the statements below.

acetylcholine   synaptic cleft
axon synaptic knobs   synaptic vesicle
sarcolemma

1. Space between axon tip and sarcolemma.
2. Embedded in invaginations of the sarcolemma.
3. Secreted into synaptic cleft when action potential reaches the axon tip.
4. Membrane covering muscle fiber.
5. Membranous container of acetylcholine.
6. Reacts with receptors on sarcolemma to initiate an action potential in the muscle fiber.

1. _____

2. _____

3. _____

4. _____

5. _____

6. _____

List the sequence of events in the transmission of an action potential from an axon terminal branch to a muscle fiber.

_____

_____

_____

_____

C. Myofibril Ultrastructure
Select the terms that match the statements below.

A band   myosin
actin   sarcomere
I band   Z line

1. Part of myofibril between two Z lines.
2. Thick myofilament with cross-bridges.
3. Thin myofilament attached to Z lines.
4. Main myofilament within the A band.
5. Lies between two I bands.
6. Myofilament within I bands.

1. _____

2. _____

3. _____

4. _____

5. _____

6. _____

D. Mechanics of Contraction

Select the terms that match the statements below.

actin   cross-bridges on myosin
ATP   sarcoplasmic reticulum
calcium ions

1. Releases calcium ions when depolarized by action potential.
2. Expose the active sites on the actin myofilament.
3. Attach to active sites on actin molecules.
4. Provides energy for contraction.
5. Pull actin myofilaments toward the center of the A band.

1. _____

2. _____

3. _____

4. _____

5. _____

List the sequence of events in the contraction of a muscle fiber after it has been activated by a neural impulse. _____

_____

_____

_____

_____

_____

_____

E. Experimental Muscle Contraction

1. Indicate the length of the muscle strands before and after contraction and calculate the percentage of contraction.

   Before contraction: _____ mm

   After contraction: _____ mm

   Percentage of contraction: _____

2. Draw the pattern of the striations as they appear in relaxed and contracted muscle fibers.

Relaxed

Contracted

Student _____

Lab Section _____

THE PHYSIOLOGY OF MUSCLE CONTRACTION:
Using the BIOPAC© Student Lab System

Subject Profile

Name _____   Height _____

Age _____   Weight _____

Gender: Male/Female   Dominant arm: right/left

A. Standard and Integrated EMG

Part 1: Data and Calculations

EMG Measurements

Cluster#	Forearm 1 (Dominant)				Forearm 2			
	Min	Max	P-P	Mean	Min	Max	P-P	Mean
1								
2								
3								
4								

Use the mean measurement from the table to compute the percentage increase in EMG activity recorded between the weakest clench and the strongest clench of forearm 1.

Calculation:

Answer: _____%

Part 2: Questions

1. If you did both forearms on the same subject, are the mean measurements for the right and left maximum grip EMG cluster

the same?_____ Which one suggests the greater grip strength?_____ Explain. _____

_____

_____

_____

2. What factors in addition to gender contribute to observed differences in clench strength?_____

_____

_____

_____

3. Explain the source of signals detected by the EMG electrodes.

_____

_____

_____

4. What is meant by the term "motor unit recruitment"?

_____

_____

_____

5. Define electromyography.

_____

_____

_____

B. Motor Unit Recruitment and Fatigue

Part 1: Data and Calculations

1. Motor unit recruitment data:

Peak #	Force (kg) Increments Assigned	Forearm 1 (Dominant)			Forearm 2		
		Force at Peak [CH 1] Mean (kg)	Raw EMG [CH 3] p-p (mV)	Int. EMG [CH40] Mean (mV)	Force at Peak [CH 1] Mean (kg)	Raw EMG [CH 3] p-p (mV)	Int. EMG [CH40] Mean (mV)
1							
2							
3							
4							
5							
6							
7							
8							
9							

2. Fatigue data:

Forearm 1 (Dominant)			Forearm 2		
Maximum Clench Force	50% of Max Clench Force	Time to Fatigue	Maximum Clench Force	50% of Max Clench Force	Time to Fatigue
CH 1 Value	Calculate	CH 40 Delta T	CH 1 Value	Calculate	CH 40 Delta T

Part 2: Questions

1. Is the strength of the dominant arm different from that of the nondominant arm?_____
   How do you explain this?_____

   _____

   _____

2. Is there a difference in the absolute values of force generated by males and females in your class? _____ If so, what
   might explain the difference?_____

   _____

   _____

   _____

3. Is there a difference in fatigue times between the two arms? _____ Why might this happen?_____

   _____

   _____

   _____

4. Define fatigue.

   _____

   _____

   _____

5. Define dynamometry.

   _____

   _____

   _____

NOTES

THE PHYSIOLOGY OF MUSCLE CONTRACTION:
Using the Intelitool Physiogrip™

A. Threshold Stimulus, Spatial Summation, and Maximal Stimulus
1. Record the threshold and maximal stimuli.
    Threshold stimulus: _____ volts   Maximal stimulus: _____ volts
2. Attach a copy of your myogram and label the threshold contraction, spatial summation, and the maximal contraction.

B. Temporal Summation and Tetanizing Contractions
1. Record the frequency of stimulation that produced the maximum tetanizing contractions: _____
2. Attach a copy of your myogram and label temporal summation and tetanizing contractions.

C. Single Muscle Twitch

1. Record the duration of these phases.

   Latent period: _____ msec   Contraction phase: _____ msec   Relaxation phase: _____ msec

2. Attach a copy of your myogram and label the latent period, contraction phase, and relaxation phase.

D. Fatigue

1. Record the time required for fatigue to set in.

   Without the cuff: _____ min   With the cuff: _____ min

2. Attach a copy of your myogram and label the point where fatigue began.

Laboratory Report 22

THE SPINAL CORD AND REFLEX ARCS

A. Figures

List the labels for Figures 22.1, 22.2, and 22.4.

Figure 22.1

1. _____
2. _____
3. _____
4. _____
5. _____
6. _____
7. _____
8. _____
9. _____
10. _____
11. _____
12. _____
13. _____
14. _____
15. _____
16. _____
17. _____
18. _____
19. _____
20. _____
21. _____

Figure 22.2

1. _____
2. _____
3. _____
4. _____
5. _____
6. _____
7. _____
8. _____
9. _____
10. _____
11. _____
12. _____
13. _____
14. _____
15. _____
16. _____

Figure 22.4

1. _____
2. _____
3. _____
4. _____
5. _____
6. _____

1. _____
2. _____
3. _____
4. _____
5. _____
6. _____
7. _____

B. Spinal Nerves and Plexuses

Select the terms that match the statements below.

brachial plexus   1 pair
cervical plexus   5 pairs
lumbar plexus   8 pairs
sacral plexus   12 pairs

1. The number of pairs of cervical nerves.
2. The number of pairs of thoracic nerves.
3. The number of pairs of lumbar nerves.
4. Formed of spinal nerves C_1 through C_4.
5. Formed of spinal nerves L_4 and L_5 and S_1 through S_3.
6. Formed of spinal nerves C_5 through C_8 plus T_1.
7. Formed of spinal nerves L_1 through L_4.

C. The Spinal Cord

Select the terms that match the statements below.

arachnoid mater	pia mater
central canal	subarachnoid space
dura mater	subdural space
epidural space	vertebral canal
gray matter	white matter

1. Meninx adhered to surface of spinal cord.
2. Outermost fibrous meninx.
3. Meninx between other meninges.
4. Channel continuous with ventricles of the brain.
5. Outer portion of spinal cord.
6. Inner portion of spinal cord.
7. Space between outer meninx and vertebrae.
8. Space around spinal cord containing cerebrospinal fluid.
9. Part of spinal cord containing unmyelinated neuron processes.
10. Part of dorsal cavity containing spinal cord.

1. _____
2. _____
3. _____
4. _____
5. _____
6. _____
7. _____
8. _____
9. _____
10. _____

D. Reflex Arcs

Select the terms that match the statements below.

effector	preganglionic efferent neuron
gray matter	receptor
interneuron	sensory neuron
motor neuron	spinal ganglion
postganglionic efferent neuron	white matter

1. A muscle or gland.
2. Carries impulses from receptor to spinal cord.
3. Carries impulses from spinal cord to an effector.
4. Receives stimuli and forms impulses.
5. Part of spinal cord with many myelinated neuron processes.
6. Contains cell bodies of sensory neurons.
7. Carries impulses from spinal cord to autonomic ganglia.
8. Neuron within the gray matter.
9. Carries impulses from autonomic ganglia to viscera.
10. Where synapses occur in spinal cord.

1. _____
2. _____
3. _____
4. _____
5. _____
6. _____
7. _____
8. _____
9. _____
10. _____

E. Diagnostic Reflexes

Record the results of your responses.

Biceps reflex: left arm _____   right arm _____

Triceps reflex: left arm _____   right arm _____

Patellar reflex: left leg _____   right leg _____

Achilles reflex: left foot _____   right foot _____

Explain why reinforcement techniques improve responses. _____

_____

Laboratory Report 23

Student _____

Lab Section _____

BRAIN ANATOMY: EXTERNAL

A. Figures

List the labels for Figures 23.1, 23.2, and 23.4.

Figure 23.1

1. _____
2. _____
3. _____
4. _____
5. _____
6. _____
7. _____
8. _____
9. _____
10. _____

Figure 23.2

1. _____
2. _____
3. _____
4. _____
5. _____
6. _____
7. _____
8. _____
9. _____
10. _____
11. _____
12. _____
13. _____

Figure 23.4

1. _____
2. _____
3. _____
4. _____
5. _____
6. _____
7. _____
8. _____
9. _____
10. _____
11. _____
12. _____
13. _____
14. _____
15. _____
16. _____
17. _____
18. _____
19. _____
20. _____

B. The Meninges

Select the structures described by the statements below.

arachnoid mater   epidural space   subarachnoid space
dura mater   pia mater   subdural space

1. The fibrous meninx attached to cranial bones.
2. Meninx attached to the brain surface.
3. The middle meninx.
4. Space containing cerebrospinal fluid.
5. Meninx containing sagittal sinus.

1. _____
2. _____
3. _____
4. _____
5. _____

C. Fissures and Sulci

Select the structures described by the statements below.

central sulcus   longitudinal cerebral fissure
lateral cerebral fissure   parieto-occipital fissure

1. Between frontal and parietal lobes.
2. Separates cerebral hemispheres.
3. Between temporal and parietal lobes.
4. Between occipital and parietal lobes.

1. _____
2. _____
3. _____
4. _____

D. Brain Regions and Functions

Select the parts of the brain described by the statements below.

cerebellum   hypothalamus   midbrain
cerebrum   medulla oblongata   pons

1. Controls heart and breathing rates.
2. Functions in will, intelligence, and memory.
3. Controls body temperature and water balance.
4. Controls muscular coordination.
5. Pathway for fibers between the cerebrum and the pons, cerebellum, and medulla.

1. _____
2. _____
3. _____
4. _____
5. _____

E. Cranial Nerves

Select the cranial nerves described by the statements below.

I. Olfactory   VII. Facial
II. Optic   VIII. Vestibulocochlear
III. Oculomotor   IX. Glossopharyngeal
IV. Trochlear   X. Vagus
V. Trigeminal   XI. Accessory
VI. Abducens   XII. Hypoglossal

1. Innervates abdominal viscera.
2. Carries impulses from inner ear.
3. Carries motor impulses for swallowing reflexes.
4. Innervates teeth and gums.
5. Carries motor impulses to four muscles moving eyeball.
6. Two nerves each innervating one exterior eye muscle.
7. Innervates muscles of the tongue.
8. Carries impulses from light receptors in the eye.
9. Innervates facial muscles and salivary glands.
10. Innervates the trapezius and muscles of the pharynx and larynx.
11. Carries impulses interpreted as odors.

1. _____
2. _____
3. _____
4. _____
5. _____
6. _____
7. _____
8. _____
9. _____
10. _____
11. _____

F. Functional Areas of the Cerebrum

1. List the labels for Figure 23.5.

1. _____
2. _____
3. _____
4. _____
5. _____
6. _____
7. _____
8. _____
9. _____
10. _____
11. _____
12. _____
13. _____

2. Select the functional areas described by the statements below.

auditory association   primary somatomotor
general interpretation   primary somatosensory
motor speech   primary taste
prefrontal   primary visual
premotor   sensory speech
primary auditory   somatosensory association
primary olfactory area   visual association

1. Precentral gyrus of frontal lobe.
2. Postcentral gyrus of parietal lobe.
3. Voluntary control of body movements.
4. Occipital lobe.
5. Controls speaking.
6. Visual images.
7. Superior temporal gyrus.
8. Medial surface of temporal lobe.
9. Touch sensations of the skin.
10. Coordinates learned motor skills.
11. Odor sensations.
12. Sound sensations.
13. Provides assessment of sensory input.
14. Interprets meaning of written and spoken words.
15. Site of judgment, critical thinking, and personality.

1. _____
2. _____
3. _____
4. _____
5. _____
6. _____
7. _____
8. _____
9. _____
10. _____
11. _____
12. _____
13. _____
14. _____
15. _____

NOTES

Laboratory Report 24

Student _____

Lab Section _____

BRAIN ANATOMY: INTERNAL

A. Figures

List the labels for Figures 24.2 and 24.3.

Figure 24.2

1. _____
2. _____
3. _____
4. _____
5. _____
6. _____
7. _____
8. _____
9. _____
10. _____
11. _____
12. _____
13. _____
14. _____
15. _____
16. _____
17. _____

18. _____
19. _____
20. _____
21. _____

Figure 24.3

1. _____
2. _____
3. _____
4. _____
5. _____
6. _____
7. _____
8. _____
9. _____
10. _____
11. _____
12. _____

B. General Questions

Record the name of the brain structure described below.

1. The second largest portion of the brain.
2. Composed of the midbrain, pons, and medulla.
3. Fluid within the subarachnoid space.
4. Secretes the hormone melatonin.
5. Provides an uncritical sensory awareness.
6. Generates impulses to keep the cerebrum alert.
7. Lowest portion of the brain stem.
8. The largest part of the brain.
9. Fissure separating the cerebral hemispheres.
10. Controls heart and breathing rates.

1. _____
2. _____
3. _____
4. _____
5. _____
6. _____
7. _____
8. _____
9. _____
10. _____

C. Structures

Select the structures that are described in the statements below.

arachnoid granulations   fornix

cerebral aqueduct   hypothalamus

cerebral peduncles   infundibulum

choroid plexus   interventricular foramen

corpora quadrigemina   optic chiasma

corpus callosum   thalamus

1. Receives sensory impulses and relays them to the cerebrum.
2. Contains fiber tracts associated with the sense of smell.
3. Regulates body temperature.
4. Consists of fibers connecting cerebral hemispheres.
5. Stalk supporting the hypophysis.
6. Enables cerebrospinal fluid to diffuse into the blood.
7. Allows cerebrospinal fluid to pass from the lateral ventricles into the third ventricle.
8. Crossing of optic fibers.
9. Main pathway of motor fibers from the cerebrum.
10. Reflex center for movements in response to auditory and visual stimuli.
11. Connects third and fourth ventricles.
12. Secretes cerebrospinal fluid into the ventricles.

1. _____

2. _____

3. _____

4. _____

5. _____

6. _____

7. _____

8. _____

9. _____

10. _____

11. _____

12. _____

D. Cerebrospinal Fluid

Provide the words to correctly complete the paragraph. Cerebrospinal fluid is released into the ventricles from the ___1___ in each ventricle. It passes from the lateral ventricles into the third ventricle via an opening, the ___2___ foramen, and then into the fourth ventricle through the ___3___. From the fourth ventricle, the fluid passes via the ___4___ into the ___5___ below the cerebellum. From here, some of the fluid moves upward around the cerebellum and cerebrum and diffuses into the sagittal sinus. Most of the fluid flows downward in the subarachnoid space along the ___6___ surface of the spinal cord, upward along the ___7___ surface of the spinal cord, around the cerebrum, and finally diffuses from the ___8___ into the blood in the ___9___.

1. _____

2. _____

3. _____

4. _____

5. _____

6. _____

7. _____

8. _____

9. _____

E. Figure

List the labels for Figure 24.4B.

1. _____

2. _____

3. _____

4. _____

5. _____

6. _____

7. _____

8. _____

9. _____

10. _____

Laboratory Report 25

Student ______________________

Lab Section ______________________

The Eye

A. Figures

List the labels for Figures 25.1, 25.2, and 25.3.

Figure 25.1

1. ______________________
2. ______________________
3. ______________________
4. ______________________
5. ______________________
6. ______________________
7. ______________________
8. ______________________
9. ______________________

Figure 25.2

1. ______________________
2. ______________________
3. ______________________
4. ______________________
5. ______________________
6. ______________________
7. ______________________

Figure 25.3

1. ______________________
2. ______________________
3. ______________________
4. ______________________
5. ______________________
6. ______________________
7. ______________________
8. ______________________
9. ______________________
10. ______________________
11. ______________________
12. ______________________
13. ______________________
14. ______________________
15. ______________________

B. Lacrimal Apparatus

Provide the words that correctly complete the paragraph.

The ___1___ lines the inner surface of the eyelids and covers the anterior eye surface. A fold of the conjunctiva at the medial corner of the eye forms the ___2___, a remnant of a third eyelid. Fluid, secreted by the ___3___, keeps the eye surface and ___4___ moist. At the medial corner of the eye, the fluid flows through two tiny openings, the ___5___, into the superior and inferior ___6___. Then, the fluid flows into the ___7___ and on into the ___8___ which empties the fluid into the ___9___.

1. ______________________
2. ______________________
3. ______________________
4. ______________________
5. ______________________
6. ______________________
7. ______________________
8. ______________________
9. ______________________

C. Extrinsic Muscles

Select the muscles described by statements below.

inferior oblique   superior levator palpebrae
inferior rectus   superior oblique
lateral rectus   superior rectus
medial rectus

1. Rotates eye downward.
2. Raises the upper eyelid.
3. Inserted on the medial surface of the eye.
4. Passes through a cartilaginous loop above the eye.
5. Rotates eye outward.

1. _____
2. _____
3. _____
4. _____
5. _____

D. Structures

Select the structures described by the statements below.

aqueous humor   macula lutea
choroid coat   optic disk
ciliary body   pupil
cornea   retina
fovea centralis   scleroid coat
iris   suspensory ligaments
lens   vitreous body

1. Contains light receptors: cones and rods.
2. Clear window at front of eye that bends light rays.
3. Fills cavity anterior to the lens.
4. Outer layer of the eye.
5. Opening in center of iris.
6. Attaches lens to ciliary body.
7. Fills cavity posterior to lens.
8. Precisely focuses light on retina.
9. Area of critical vision.
10. Controls shape of the lens.
11. Controls amount of light entering the eye.
12. Yellow spot containing only cones.
13. Blackish middle coat of the eyeball.
14. Provides strength to the eyeball wall.
15. Lacks rods and cones.
16. Layer of eyeball containing blood vessels.
17. Place where nerve fibers of retina leave eyeball.
18. Place where blood vessels enter eyeball.
19. Jellylike substance within eyeball.
20. Colored ring anterior to the lens.

1. _____
2. _____
3. _____
4. _____
5. _____
6. _____
7. _____
8. _____
9. _____
10. _____
11. _____
12. _____
13. _____
14. _____
15. _____
16. _____
17. _____
18. _____
19. _____
20. _____

E. Beef Eye Dissection

1. Which layer of the eyeball is the toughest? _____

2. What is the shape of the pupil? _____

3. When you hold up the lens, look at a distant object. What is unusual about the image? _____

4. When placed on a printed page, does the lens make print appear larger or smaller? _____

5. Is the lens denser near the center or periphery? _____

6. What is the clear jellylike substance? _____

7. What holds the retina in place in an intact eyeball? _____

F. Physiology of Vision

1. Name the two types of photosensitive cells in the retina. _____

2. Which type of photosensitive cell is most abundant? _____

3. Which type of photosensitive cell provides color vision? _____

4. Which type of photosensitive cell occurs in the fovea centralis? _____

5. Name the photosensitive pigment in rods. _____

6. What are the two compounds formed by the photochemical breakdown of the photosensitive pigment in rods? _____
_____

7. What vitamin is necessary for the reformation of rhodopsin? _____

8. The three types of cones are sensitive to three different colors of light. Name the colors. _____
_____

9. Explain the cause of temporary blindness when you enter a dark room from a brightly lighted area. _____
_____

   Explain the brief time lapse before dim-light vision is restored. _____
_____

G. Visual Tests

Record the results obtained in the following tests.

Test	Right Eye	Left Eye
Blind Spot (inches)		
Near Point (inches)		
Visual Acuity (20/×)		
Astigmatism (present or absent)		

H. Light Intensity and Pupil Size

1. Did the pupil size become larger or smaller in the

   exposed left eye? _____   unexposed right eye? _____

2. Explain these responses. _____
_____
_____

3. Suggest a possible benefit resulting from this reflex pattern. _____
_____

I. Color-Blindness Test

Have your laboratory partner record your responses to each test plate. The mark *x* indicates inability to read correctly.

Plate Number	Subject's Response	Normal Response	Response If Red-Green Deficiency		Response If Totally Color-blind
1		12	12		12
2		8	3		x
3		5	2		x
4		29	70		x
5		74	21		x
6		7	x		x
7		45	x		x
8		2	x		x
9		x	2		x
10		16	x		x
11		traceable	x		x
			red cones absent	green cones absent	
12		35	5	3	x
13		96	6	9	x
14		can trace two lines	purple	red	x

Laboratory Report 26

THE EAR

A. Figure

List the labels for Figure 26.1.

1. _____   13. _____
2. _____   14. _____
3. _____   15. _____
4. _____   16. _____
5. _____   17. _____
6. _____   18. _____
7. _____   19. _____
8. _____   20. _____
9. _____   21. _____
10. _____   22. _____
11. _____   23. _____
12. _____

B. Hearing (Structures)

Select the structures described in the statements below.

basilar membrane   round window
cochlear duct   scala tympani
endolymph   scala vestibuli
hair cells   stapes
incus   tectorial membrane
malleus   tympanum
middle ear   vestibular membrane
perilymph

1. Membrane receiving sound waves via air.
2. Ossicle attached to the tympanum.
3. Ossicle inserted into oval window.
4. Cochlear chamber into which oval window opens.
5. Cochlear chamber with the round window.
6. Fluid within the cochlear duct.
7. Fluid between the bony and membranous labyrinths.
8. Receptors for sound stimuli.
9. Contains 20,000 transverse fibers.
10. Cochlear chamber containing the spiral organ.
11. Membrane in which hair tips are embedded.
12. Membrane allowing movement of perilymph.
13. Transfers vibrations from fluid to hair cells.
14. Cavity traversed by the ear ossicles.
15. Activated by the movement of the perilymph.
16. Set in motion by the movement of the stapes in the oval window.

1. _____
2. _____
3. _____
4. _____
5. _____
6. _____
7. _____
8. _____
9. _____
10. _____
11. _____
12. _____
13. _____
14. _____
15. _____
16. _____

C. Hearing (True–False)

1. Sound waves may reach the cochlea through the skull bones as well as via the tympanum.
2. The more hair cells sending impulses, the louder will be the sound sensation.
3. Pitch is determined by the part of the basilar membrane that is activated.
4. Bending of the hair cells causes the formation of impulses.
5. There is no cure for conduction deafness, but nerve deafness may often be corrected.

1. _____
2. _____
3. _____
4. _____
5. _____

D. Equilibrium

Select the structures described by the statements below.

ampulla   otoconia
crista ampullaris   saccule
macula   utricle

1. Two portions of the membranous labyrinth containing sensory receptors for static equilibrium.
2. Sensory hair cells in ampullae of semicircular canals.
3. Hair cells stimulated by the movement of the otoconia.
4. Hair cells stimulated by the force of endolymph.
5. Sensory receptors for dynamic equilibrium.
6. Enlarged base of a semicircular canal.
7. Calcium carbonate crystals.

1a. _____
1b. _____
2. _____
3. _____
4. _____
5. _____
6. _____
7. _____

E. Hearing Tests

Record your results of the hearing tests.

Rinne Test

Left ear _____

Right ear _____

State a conclusion from this test. _____

Weber Test

Left ear _____

Right ear _____

State a conclusion from this test. _____

F. Static Balance Test

Record the degree of swaying as small, moderate, or great.

Standing on both feet with eyes open _____

Standing on both feet with eyes closed _____

Standing on one foot with eyes closed _____

Rate the importance of visual input in static balance. _____

_____

Student _____

Lab Section _____

BLOOD TESTS

A. Formed Elements

Select the blood cell type described in the statements below.

basophils   lymphocytes   neutrophils
eosinophils   monocytes   thrombocytes
erythrocytes

1. Agranulocytes with spherical nucleus and little cytoplasm.
2. Biconcave, enucleate cells.
3. Cells with 2–5-lobed nucleus and tiny lavender cytoplasmic granules.
4. Smallest of the formed elements.
5. Cells with red-orange cytoplasmic granules and bilobed or U-shaped nucleus.
6. Large agranulocytes with a lobed or U-shaped nucleus.
7. Cell with bilobed or U-shaped nucleus and bluish purple cytoplasmic granules.
8. 4,000,000–6,000,000 per mm^3.
9. Initiate clotting process.
10. Produce antibodies.
11. Transport oxygen and carbon dioxide.
12. Counteract products of allergy reactions.
13. Release histamine in allergy reactions.
14. 50%–70% of the leukocytes.
15. 20%–30% of the leukocytes.
16. Two types of cells that phagocytize pathogens.
17. 250,000–400,000 per mm^3.
18. Contain hemoglobin.
19. 2%–6% of the leukocytes.

1. _____
2. _____
3. _____
4. _____
5. _____
6. _____
7. _____
8. _____
9. _____
10. _____
11. _____
12. _____
13. _____
14. _____
15. _____
16a. _____
16b. _____
17. _____
18. _____
19. _____

B. Differential White Blood Cell Count

Tabulate your identification of the various types of leukocytes in the table below. Count a total of 100 white cells. Calculate the percentage of each type of leukocyte by dividing the total count for each type by 100.

Neutrophils	Lymphocytes	Monocytes	Eosinophils	Basophils
Totals				
Percent				

C. Blood Test Results

Record your test results in the table below.

Test	Normal Values	Test Results	Evaluation (over, under, normal)
Differential WBC Count	Neutrophils: 50%–70%		
	Lymphocytes: 20%–30%		
	Monocytes: 2%–6%		
	Eosinophils: 1%–3%		
	Basophils: 0.5%–1%		
Hematocrit (VPRC)	Males: 40%–54% (Av. 47%)		
	Females: 37%–47% (Av. 42%)		
Coagulation Time	3 to 6 minutes		

D. Blood Typing

1. Indicate your blood type.   ABO group _____   Rh _____
2. Complete the table below to indicate compatibility (o) and incompatibility (+) of possible transfusions. A plus (+) indicates agglutination of the erythrocytes.

Blood Type/Antigen of Donor	Blood Type and Antibodies of Recipient			
	O a, b	A b	B a	AB none
O				
A				
B				
AB				

THE HEART

A. Figures

List the labels for Figures 28.1 and 28.2.

Figure 28.1

1. _____
2. _____
3. _____
4. _____
5. _____
6. _____
7. _____
8. _____
9. _____
10. _____
11. _____
12. _____
13. _____
14. _____
15. _____
16. _____
17. _____
18. _____
19. _____
20. _____
21. _____
22. _____
23. _____
24. _____
25. _____
26. _____
27. _____
28. _____

29. _____
30. _____
31. _____
32. _____
33. _____

Figure 28.2

1. _____
2. _____
3. _____
4. _____
5. _____
6. _____
7. _____
8. _____
9. _____
10. _____
11. _____
12. _____
13. _____
14. _____
15. _____
16. _____
17. _____
18. _____
19. _____
20. _____

B. Circulation

Provide the words to correctly complete the paragraph.
During diastole, deoxygenated blood returns to the
___1___ via the ___2___ and ___3___ and enters the right
ventricle. Simultaneously, oxygenated blood returns to the
___4___ via the ___5___ and enters the left ventricle. During
systole, the right ventricle pumps blood into the
___6___, which carries blood to the lungs. Simultaneously,
the left ventricle pumps blood into the ___7___, which
carries blood to all parts of the body except the lungs.

1. _____

2. _____

3. _____

4. _____

5. _____

6. _____

7. _____

C. Structures

Select the structures described by the statements below.

aortic semilunar valve	myocardium
bicuspid valve	papillary muscles
chordae tendineae	pulmonary semilunar valve
endocardium	right atrium
epicardium	right ventricle
left atrium	tricuspid valve
left ventricle	interventricular septum

1. Partition between right and left ventricles.
2. Muscular portion of cardiac wall.
3. Lines interior of the heart.
4. Receives deoxygenated blood from the venae cavae.
5. Receives oxygenated blood from the pulmonary veins.
6. Prevents backflow of blood into right atrium.
7. Prevents backflow of blood into left atrium.
8. Prevents backflow of blood into right ventricle.
9. Prevents backflow of blood into left ventricle.
10. Pumps blood into the pulmonary artery.
11. Pumps blood into the aorta.
12. Chamber with the thickest walls.
13. Portion of ventricular wall to which chordae
    tendineae are attached.
14. Strands of fibrous tissue attaching valve cusps to the
    inner wall of the ventricles.
15. Receives blood from the coronary sinus.

1. _____

2. _____

3. _____

4. _____

5. _____

6. _____

7. _____

8. _____

9. _____

10. _____

11. _____

12. _____

13. _____

14. _____

15. _____

D. Sheep Heart Dissection

1. Indicate the number of cusps composing the heart valves. _____

   Bicuspid _____   Tricuspid _____   Aortic _____   Pulmonary _____

2. Compare the thickness of the valve cusps. _____

   Atrioventricular valves _____

   Semilunar valves _____

3. Which ventricle has the thicker wall? _____

4. Are the chordae tendineae fragile or strong? _____

   Describe their function. _____

   _____

5. Is the ventricular septum a thin sheet of connective tissue or a thick muscular wall? _____

   _____

Student _____

Lab Section _____

BLOOD VESSELS, FETAL CIRCULATION, AND LYMPHATIC SYSTEM

A. The Circulatory Plan

Select the blood vessels described by the statements below.

aorta   pulmonary vein
hepatic artery   renal artery
hepatic portal vein   renal vein
hepatic veins   vena cava, inferior
mesenteric artery vena cava, superior
pulmonary artery

1. Carries oxygenated blood to the body, except the lungs.
2. Carries blood from a kidney to the inferior vena cava.
3. Carries deoxygenated blood to the lungs.
4. Carries blood from the head and neck to the right atrium.
5. Carries oxygenated blood to the left atrium.
6. Carries blood from the aorta to the intestines.
7. Carries blood from regions below the heart to the right atrium.
8. Carries blood from the intestines to the liver.
9. Carries blood from the liver to the inferior vena cava.
10. Carries blood from the aorta to the liver.

1. _____
2. _____
3. _____
4. _____
5. _____
6. _____
7. _____
8. _____
9. _____
10. _____

B. Completion: Arteries and Veins

Provide the name of the artery or vein described by the statements.

1. Vein commonly used for venipuncture.
2. Branches to form right subclavian and right common carotid.
3. Supplies most of small intestine and part of large intestine.
4. Arterial segment between the subclavian and the brachial arteries.
5. Veins of the neck emptying into the subclavians.
6. Large superficial vein draining the leg.
7. Supplies blood to the stomach.
8. Merge to form the inferior vena cava.
9. Branches to form the anterior and posterior tibial arteries.
10. Merge to form the superior vena cava.
11. Carries blood to the kidney.
12. Carries blood to the large intestine and rectum.
13. Branch from the inferior end of the aorta.
14. Carries blood from the aortic arch to the left side of the head.
15. Return blood from the brain to brachiocephalic veins.

1. _____
2. _____
3. _____
4. _____
5. _____
6. _____
7. _____
8. _____
9. _____
10. _____
11. _____
12. _____
13. _____
14. _____
15. _____

C. Figures

List the labels for Figures 29.2, 29.3, 29.5, and 29.6.

Figure 29.2

1. _____
2. _____
3. _____
4. _____
5. _____
6. _____
7. _____
8. _____
9. _____
10. _____
11. _____
12. _____
13. _____
14. _____
15. _____
16. _____
17. _____
18. _____
19. _____
20. _____
21. _____
22. _____
23. _____
24. _____
25. _____
26. _____

Figure 29.3

1. _____
2. _____
3. _____
4. _____
5. _____
6. _____
7. _____
8. _____
9. _____
10. _____
11. _____
12. _____
13. _____
14. _____
15. _____
16. _____
17. _____
18. _____
19. _____
20. _____
21. _____
22. _____
23. _____
24. _____

Figure 29.5

1. _____
2. _____
3. _____
4. _____
5. _____
6. _____
7. _____
8. _____
9. _____
10. _____
11. _____
12. _____
13. _____
14. _____
15. _____
16. _____
17. _____
18. _____

Figure 29.6

1. _____
2. _____
3. _____
4. _____
5. _____
6. _____
7. _____
8. _____
9. _____
10. _____
11. _____
12. _____
13. _____
14. _____
15. _____
16. _____
17. _____
18. _____

D. Microscopic Study
Make a drawing of the slides.

Normal Artery   Athersclerotic Artery   Normal Vein

E. Fetal Circulation
Select the structures described by the statements below.

ductus arteriosus   umbilical arteries
ductus venosus   umbilical vein
foramen ovale

1. Opening between the atria.
2. Extensions of the hypogastric arteries.
3. Brings oxygenated blood to the fetus.
4. Connects the pulmonary artery with the aorta.
5. Carry fetal blood to the placenta.
6. Becomes the ligamentum venosum after birth.
7. Becomes the round ligament after birth.
8. Becomes the ligamentum arteriosum after birth.

1. _____
2. _____
3. _____
4. _____
5. _____
6. _____
7. _____
8. _____

F. Lymphatic System

Select the terms described by the statements below.

cisterna chyli   lymph capillaries   lymphatic vessels
lymph   lymph nodes   right lymphatic duct
                                       thoracic duct

1. Collects extracellular fluid from tissues.
2. Fluid in lymphatic vessels.
3. Empties lymph into right subclavian vein.
4. Saclike structure at inferior end of thoracic duct.
5. Cleanses lymph as it passes through.
6. Carries lymph to left subclavian vein.

1. _____
2. _____
3. _____
4. _____
5. _____
6. _____

G. Lymphocytes and Immunity

Select the terms described by the statements below.

antibodies   helper T cells
antigen-presenting cell   lymphoid tissue
antibody-mediated immunity   memory B and T cells
B cells   phagocytosis
blood   plasma cells
cell-mediated immunity   stem cells in red bone
cytokines   marrow
cytotoxic T cells   T cells

1. Source of all lymphocytes.
2. Lymphocytes maturing in thymus gland.
3. Lymphocytes maturing in red bone marrow.
4. T cells that kill diseased cells with chemicals.
5. Lymphocytes that produce antibodies.
6. Immunity enabled by T cells.
7. Where most mature lymphocytes reside.
8. Chemicals that bind to antigens.
9. Chemicals that activate B cells.
10. Lymphocytes that start a secondary immune response.
11. Chemicals secreted by helper T cells.
12. Immunity enabled by B cells.
13. T cells that start a primary immune response.
14. Immunity that targets extracellular pathogens.
15. Destroys antigens bound to antibodies.

1. _____
2. _____
3. _____
4. _____
5. _____
6. _____
7. _____
8. _____
9. _____
10. _____
11. _____
12. _____
13. _____
14. _____
15. _____

CARDIOVASCULAR PHENOMENA

A. Heart Sounds

Select the heart sound described by the statements below.

first sound   third sound
second sound   murmur

1. Vibration of ventricular walls and valve cusps.
2. Closure of the semilunar valves.
3. Closure of the atrioventricular valves.
4. Damaged heart valves.
5. Detected best in supine subject.

1. _____
2. _____
3. _____
4. _____
5. _____

B. Auscultatory Areas

Select the auscultatory area described by the statements below.

aortic semilunar area   pulmonary semilunar area
bicuspid area   tricuspid area

1. About 2 inches left of sternum at fifth rib.
2. On left edge of sternum just below second rib.
3. On left edge of sternum at the fifth rib.
4. On right edge of sternum at the second rib.
5. Site used to detect the third heart sound.

1. _____
2. _____
3. _____
4. _____
5. _____

C. Blood Pressure

1. Complete the following statement.

    The reading on the pressure gauge at the first Korotkoff sound is the _____ pressure; the reading when the Korotkoff sounds cease is the _____ pressure.

2. Indicate your blood pressure before and after exercise.

    Systolic: before exercise _____   after exercise _____

    Diastolic: before exercise _____   after exercise _____

3. Explain the cause of a higher blood pressure after exercise.

    _____

    _____

4. Explain how chronic high blood pressure (hypertension) may cause the rupture of an artery in the brain. _____

    _____

    _____

D. The Pulse

1. Record your pulse rates (beats/min) in the chart below.

Supine	Sitting	Standing	After Exercise

2. How long did it take for your pulse rate to return to normal after exercise?

_____

3. The strength of the pulse decreases in the right radial artery when you raise your right hand above your head. Why?

_____

_____

4. What is the value of a nurse's knowing the pressure points where the pulse may be taken? _____

_____

_____

_____

5. Calculate your pulse pressure.

Systolic pressure _____ − Diastolic pressure _____ = Pulse pressure _____

E. Peripheral Blood Flow

1. What part of the brain controls peripheral circulation? _____

2. Describe the normal blood flow through a capillary. _____

_____

_____

3. Describe the observable changes in circulation in the frog's foot when histamine was applied. _____

_____

_____

4. Did histamine cause vasodilation or vasoconstriction? _____

5. Describe the observable effect of epinephrine on the circulation of the frog's foot. _____

_____

_____

6. Did epinephrine cause vasodilation or vasoconstriction? _____

7. What blood vessels did histamine and epinephrine *directly* affect? _____

_____

Laboratory Report 31

Student _____

Lab Section _____

THE RESPIRATORY ORGANS

A. Figures

List the labels for Figures 31.1 and 31.2.

Figure 31.1

1. _____
2. _____
3. _____
4. _____
5. _____
6. _____
7. _____
8. _____
9. _____
10. _____
11. _____
12. _____
13. _____
14. _____
15. _____
16. _____
17. _____
18. _____

19. _____
20. _____
21. _____
22. _____

Figure 31.2

1. _____
2. _____
3. _____
4. _____
5. _____
6. _____
7. _____
8. _____
9. _____
10. _____
11. _____
12. _____
13. _____
14. _____
15. _____

1. _____
2. _____
3. _____
4. _____
5. _____
6. _____
7. _____
8. _____
9. _____
10. _____
11. _____

B. Air Pathway

Complete the following paragraph.

During inspiration, air passes through the ___1___ into the nasal cavity, where it is warmed. The surface area of the nasal cavity is increased by the three ___2___. From the nasal cavity, air passes through the ___3___ and ___4___ and into the larynx, which contains the ___5___ folds. From the larynx, air passes down the ___6___, which divides into the primary ___7___ that enter each lung. The tubes continue to branch and ultimately form small tubules called ___8___, which lead to the alveoli. Gas exchange occurs between ___9___ in the alveoli and ___10___ in the capillaries surrounding the alveoli. During ___11___, air flow is reversed through this same pathway.

C. Respiratory Structures

Select the structures described by the statements below.

alveoli
bronchi
bronchioles
epiglottis
hard palate
larynx
lingual tonsils
nasal conchae
nasopharynx

oral cavity
oral vestibule
oropharynx
palatine tonsils
pharyngeal tonsils
pleural cavity
soft palate
trachea

1. _____
2. _____
3. _____
4. _____
5. _____
6. _____
7. _____
8. _____
9. _____
10. _____
11. _____
12. _____
13. _____
14. _____
15. _____

1. Cavity between lips and teeth.
2. Tonsils in oropharynx.
3. Increase surface area of nasal cavity.
4. Sites of gas exchange.
5. Tubes branching from inferior end of trachea.
6. Prevents food from entering larynx.
7. Potential space between visceral and parietal pleurae.
8. Small tubes leading to alveoli.
9. Tonsils in nasopharynx.
10. Separated from nasal cavity by hard palate.
11. Tiny air sacs in lung.
12. Portion of palate supported by bone.
13. Contains the vocal folds.
14. Where swallowing reflex is initiated.
15. Tonsils at base of the tongue.

D. Sheep Pluck Dissection

1. Describe the shape of the cartilaginous rings of the trachea.

_____

2. What contributes to the smooth and slimy nature of the inside of the trachea? _____

_____

3. Indicate the number of lobes in the

left lung: _____   right lung: _____

4. Lung tissue seems to be full of small spaces. What pulmonary structures contribute most to this appearance?

_____

E. Microscopic Study

1. Diagram a section of tracheal wall at 100× magnification and label the ciliated epithelial lining and hyaline cartilage.
2. Diagram the epithelial lining at 400×.
3. Diagram normal lung tissue at 100× magnification. Label the alveoli.

Tracheal Wall	Epithelial Lining	Normal Lung Tissue
100×	400×	100×

Laboratory Report 32

Student _____

Lab Section _____

RESPIRATORY PHYSIOLOGY

A. Control of Breathing

Complete the following paragraph.

Breathing is controlled by the ___1___ center located in the ___2___ of the brain. The most important stimulus for breathing is the direct stimulation of this center by an ___3___ concentration of ___4___ in the blood and cerebrospinal fluid. ___5___ also acts directly on the control center. In contrast, ___6___ acts indirectly on the control center. Its concentration is detected by ___7___ in the carotid arteries and aorta, which send impulses to the control center. Usually, ___8___ concentration is of little importance in stimulating inspiration.

1. _____

2. _____

3. _____

4. _____

5. _____

6. _____

7. _____

8. _____

B. Breathing Experiments

Hyperventilation

1. Describe how you felt after hyperventilating. _____

_____

_____

2. Explain why it was easier to breathe into the paper bag. _____

_____

_____

_____

3. Indicate how long you were able to hold your breath. _____

   After normal breathing _____ sec   After hyperventilating _____ sec

   Explain your results. _____

_____

_____

Deglutition Apnea

1. Indicate whether the urge to breathe decreased or increased when you sipped the water through the straw.

_____

2. Is this phenomenon a reflex or a voluntary response? _____

C. Spirometry and Lung Capacities

Record your lung volumes determined by spirometry in the table below. Circle those values that are 22% or more below normal.

Lung Capacities	Normal (ml)	Your Capacities
Tidal Volume (TV)	500	
Minute Respiratory Volume (MRV) **(MRV = TV × Resp. Rate)**	6,000	
Expiratory Reserve Volume (ERV)	1,100	
Vital Capacity (VC) **(VC = TV + ERV + IRV)**	See Appendix C	
Inspiratory Capacity (IC) **(IC = VC − ERV)**	3,000	
Inspiratory Reserve Volume (IRV) **(IRV = IC − TV)**	2,500	

If you did the FEV_T test, provide your results below.

FEV_1 _____ % FEV_3 _____ %

Attach your spirograms to the laboratory report.

D. Clinical Applications

1. Emphysema is characterized by the rupture of numerous alveoli and a decrease in lung elasticity.

   Explain why emphysematous patients have a below normal expiratory reserve volume. _____

   _____

   _____

   How would this affect the vital capacity? _____

   _____

2. Asthma causes inflammation of the bronchial tree and is classified as an obstructive disorder. On a FEV_1 test, would you expect an asthmatic patient to have a below normal, normal, or higher than normal result? _____

   Explain. _____

   _____

   _____

Laboratory Report 33

Student _____

Lab Section _____

THE DIGESTIVE ORGANS

A. Figures

List the labels for Figures 33.1 through 33.5.

Figure 33.1

1. _____
2. _____
3. _____
4. _____
5. _____
6. _____
7. _____
8. _____
9. _____
10. _____
11. _____

Figure 33.2

1. _____
2. _____
3. _____
4. _____
5. _____
6. _____
7. _____
8. _____
9. _____
10. _____
11. _____
12. _____

Figure 33.3

1. _____
2. _____
3. _____
4. _____
5. _____
6. _____
7. _____
8. _____
9. _____

Figure 33.4

1. _____
2. _____
3. _____
4. _____
5. _____
6. _____
7. _____
8. _____

Figure 33.5

1. _____
2. _____
3. _____
4. _____
5. _____
6. _____
7. _____
8. _____
9. _____
10. _____
11. _____
12. _____
13. _____
14. _____
15. _____

16. _____
17. _____
18. _____
19. _____
20. _____
21. _____
22. _____
23. _____
24. _____
25. _____
26. _____
27. _____
28. _____
29. _____

B. Oral Cavity

Select the structures described by the statements below.

cementum   papillae, vallate
crown of tooth   parotid gland
dentin   pulp cavity
enamel   root of tooth
gingiva   sublingual gland
oropharynx   submandibular gland
papillae, filiform tonsils, lingual
papillae, fungiform tonsils, palatine

1. Portion of a tooth covered with enamel.
2. Papillae giving rough texture to the tongue.
3. Mucous membrane covering the neck of each tooth.
4. Salivary gland located anterior to the ear.
5. Substance forming most of a tooth.
6. Bonelike substance between root and periodontal ligament.
7. 8–12 large papillae located at back of tongue.
8. Salivary gland located in anterior floor of oral cavity.
9. Contains nerves and blood vessels of a tooth.
10. Where swallowing reflex is initiated.
11. Papillae with taste buds scattered over surface of tongue.
12. Tonsils located in oropharynx.
13. Salivary gland secreting salivary amylase.
14. Salivary glands secreting mucin.

1. _____
2. _____
3. _____
4. _____
5. _____
6. _____
7. _____
8. _____
9. _____
10. _____
11. _____
12. _____
13. _____
14a. _____
14b. _____

C. Esophagus to Anus

Select the structures described by the statements below.

cardiac sphincter   ileocecal valve
cecum   ileum
colon   jejunum
common bile duct   pancreatic duct
cystic duct   pyloric sphincter
duodenum   rectum
esophagus   stomach
hepatic duct   villi

1. Valve that opens to let food into the stomach.
2. Valve that opens to let chyme out of stomach.
3. Valve between the small and large intestines.
4. Pouch from which appendix extends.
5. Tube that carries food to the stomach.
6. The first 10–12 inches of the small intestine.
7. Duct that carries bile from the liver.
8. Duct that carries bile to and from gallbladder.
9. The terminal 5–6 inches of the large intestine.
10. Duct that carries bile and pancreatic juice to the duodenum.
11. Microscopic projections through which absorption of nutrients occurs.
12. Site of bacterial decomposition of food residue.
13. Middle portion of the small intestine.
14. Where food is mixed with gastric juice.

1. _____
2. _____
3. _____
4. _____
5. _____
6. _____
7. _____
8. _____
9. _____
10. _____
11. _____
12. _____
13. _____
14. _____

D. Microscopic Study

Make drawings from these slides. Label pertinent parts.

**Vallate Papilla
with Taste Buds**

Stomach Wall

Ileum Wall   **Jejunum Wall**

Liver Lobules

DIGESTION

A. Completion

1. Explain the meaning of enzymatic hydrolysis. _____

_____

_____

_____

2. What determines the three-dimensional shape of an enzyme? _____

_____

3. Explain the importance of the shape of an enzyme. _____

_____

4. Explain how temperature and pH changes may alter the shape of an enzyme. _____

_____

_____

B. Digestive Enzymes

Select the enzyme described by the statements below.

amylase	maltase
carboxypeptidase	pepsin
chymotrypsin	peptidase
lactase	sucrase
lipase	trypsin

1. Enzyme present in saliva and pancreatic juice.
2. Converts lipids to fatty acids and glycerol.
3. Intestinal enzyme converting peptides to amino acids.
4. Converts starch to maltose.
5. Converts maltose to glucose.
6. Converts sucrose to glucose and fructose.
7. Converts lactose to glucose and galactose.
8. Gastric enzyme converting proteins to peptides.
9. Two pancreatic enzymes converting proteins to peptides only.
10. Pancreatic enzyme converting peptides to peptide residues and amino acids.
11. Enzyme found in gastric, pancreatic, and intestinal juices.

1. _____

2. _____

3. _____

4. _____

5. _____

6. _____

7. _____

8. _____

9a. _____

9b. _____

10. _____

11. _____

C. Completion

Indicate the end products of digestion for the following:

Carbohydrates _____

Lipids _____

Proteins _____

D. Starch Digestion Experiments

Record your data from the experiments in the tables below. In the starch digestion row, use a "0" to indicate no digestion, a "+" to indicate some digestion, and a "C" to indicate complete digestion.

Table I Amylase activity at 20° C

Tube No.	1	2	3	4	5	6	7	8	9	10
Color										
Starch Digestion										

Table II Amylase activity at 37° C

Tube No.	1	2	3	4	5	6	7	8	9	10
Color										
Starch Digestion										

Table III Amylase activity when boiled

Tube No.	1	2	3
Color			
Starch Digestion			

Table IV pH and amylase activity

Time	2 min			4 min			6 min		
Tube No.	1	2	3	4	5	6	7	8	9
pH	5	7	9	5	7	9	5	7	9
Color									
Starch Digestion										

E. Conclusions

1. How long did it take for starch digestion to be

    evident?   20° C _____ min   37° C _____ min

    complete?   20° C _____ min   37° C _____ min

2. Did you expect this pattern of results? _____ Why? _____

    _____

3. What effect did boiling have on amylase? _____

    _____

4. At which pH was digestion most rapid? _____ How does this compare with the normal pH of saliva? _____

    _____

5. Gastric juice has a pH of 2 or less. What will its effect be on salivary amylase? _____

    _____

    Do you expect any starch digestion in the stomach? _____

Laboratory Report **35**

Student ______________________________

Lab Section ______________________________

THE URINARY ORGANS

A. *Figures*

List the labels for Figures 35.1, 35.2, and 35.3.

Figure 35.1

1. ______________________________
2. ______________________________
3. ______________________________
4. ______________________________
5. ______________________________
6. ______________________________
7. ______________________________
8. ______________________________
9. ______________________________
10. ______________________________

Figure 35.2

1. ______________________________
2. ______________________________
3. ______________________________
4. ______________________________
5. ______________________________
6. ______________________________
7. ______________________________
8. ______________________________
9. ______________________________
10. ______________________________
11. ______________________________
12. ______________________________
13. ______________________________
14. ______________________________
15. ______________________________
16. ______________________________

Figure 35.3

1. ______________________________
2. ______________________________
3. ______________________________
4. ______________________________
5. ______________________________
6. ______________________________
7. ______________________________
8. ______________________________
9. ______________________________
10. ______________________________
11. ______________________________
12. ______________________________

B. *Microscopic Study*

Diagram a glomerulus and glomerular capsule as observed on the prepared slide of kidney tissue.

C. Anatomy

Select the structures described by the statements below.

calyces	medulla	renal pelvis
collecting duct	nephron	renal pyramids
cortex	renal capsule	ureters
glomerular capsule	renal column	urethra
glomerulus	renal papilla	

1. Tubes that drain the kidneys.
2. Tube that drains the urinary bladder.
3. Portion of kidney that contains renal corpuscles.
4. Cone-shaped portions of the medulla.
5. Part of kidney containing collecting ducts.
6. Functional unit of the kidney.
7. Capillary tuft within renal corpuscle.
8. Tip of renal pyramid.
9. Tube receiving urine from several nephrons.
10. Thin, fibrous covering of the kidney.
11. Cortical tissue between renal pyramids.
12. Short tubes receiving urine from renal pyramids.
13. Part of kidney composed of renal pyramids and renal columns.
14. Funnel-like structure receiving urine from calyces.
15. Cuplike structure enveloping a glomerulus.

1. _____
2. _____
3. _____
4. _____
5. _____
6. _____
7. _____
8. _____
9. _____
10. _____
11. _____
12. _____
13. _____
14. _____
15. _____

D. Table 35.1

Calculate the concentrations (grams/liter) of the following substances in Table 35.1 and record them below.

Substance	Plasma	Filtrate	Urine
Proteins	_____	_____	_____
Chloride ions	_____	_____	_____
Potassium ions	_____	_____	_____
Sodium ions	_____	_____	_____
Glucose	_____	_____	_____
Creatinine	_____	_____	_____
Urea	_____	_____	_____
Uric acid	_____	_____	_____

1. Why do so few proteins enter the filtrate? _____

2. What substances have different concentrations in the plasma and in the filtrate? _____

3. What substances are completely reabsorbed? _____

4. What substances are more concentrated in urine than in the filtrate? _____

5. Does urine formation remove all nitrogenous wastes or only reduce their concentrations in the plasma? _____

6. What percentage of the filtrate is reabsorbed? _____

The reabsorption of what substance accounts for most of the volume reduction and the concentration of urine solutes? _____

7. List the three processes of urine formation. _____

URINE AND URINALYSIS

A. Terminology
Write the term defined below in the answer column.

1. Inflammation of the kidney (general).
2. Albumin in the urine.
3. A measure of the concentration of solutes in urine.
4. Erythrocytes in the urine.
5. Inflammation of the urinary bladder.
6. Most abundant inorganic compound in urine.
7. Leukocytes or pus components in the urine.
8. pH range of normal urine.
9. Hemoglobin in the urine.
10. More than a trace of glucose in the urine.
11. Ketones in the urine.
12. Inflammation of the kidney involving glomeruli.
13. Accumulations of materials hardened in tubules.
14. Most abundant nitrogenous waste.
15. Excessive urine production.
16. Bile pigment in the urine.
17. Inflammation of the urethra.
18. Kidney stones.
19. Little or no urine production.
20. Most abundant inorganic solute in urine.

1. _____
2. _____
3. _____
4. _____
5. _____
6. _____
7. _____
8. _____
9. _____
10. _____
11. _____
12. _____
13. _____
14. _____
15. _____
16. _____
17. _____
18. _____
19. _____
20. _____

B. Clinical Significance
Select the name of the possible clinical condition from the list below that is indicated by the urinalysis results. Write your answer in the answer column.

Clinical Conditions

calculi

diabetes insipidus

diabetes mellitus

glomerulonephritis

hemolytic anemia

liver pathology

renal failure

urinary infection

Urinalysis Results

1. Anuria.
2. Pyuria, hematuria, kidney/ureter pain.
3. Hemoglobinuria.
4. Polyuria, very low specific gravity.
5. Urobilinogenuria and/or bilirubinuria.
6. Polyuria, glycosuria, ketonuria, high specific gravity.
7. Pyuria, bacteriuria.
8. Low pH, albuminuria, pyuria, casts.

1. _____
2. _____
3. _____
4. _____
5. _____
6. _____
7. _____
8. _____

C. Urinalysis

Record on the chart below the results of the analysis of each urine specimen. All of the following tests except color and turbidity are easily done with a 10SG-Multistix.

Test	Normal Values	Unknowns		Student's Urine
		No. _____	No. _____	
Leukocytes	None			
Nitrite	None			
Urobilinogen	<2 mg/dl			
Protein	None			
pH	4.8 to 7.5			
Blood	None			
Sp. Gravity	1.002–1.030			
Ketone	None			
Bilirubin	None			
Glucose	None or trace			
Color	Pale yellow to amber			
Turbidity	Clear			

Indicate the possible clinical conditions associated with the unknowns.

No. _____ _____

_____

No. _____ _____

_____

Laboratory Report 37

Student _____

Lab Section _____

THE ENDOCRINE GLANDS

A. Figures

List the labels for Figures 37.1 and 37.4.

1. _____
2. _____
3. _____
4. _____
5. _____
6. _____
7. _____
8. _____
9. _____

Figure 37.4

1. _____
2. _____
3. _____
4. _____
5. _____
6. _____

B. Hormone Sources

Select the glandular tissue that produces the hormones listed.

adrenal cortex
adrenal medulla
hypothalamus
ovary: corpus luteum
ovary: follicles
pancreas
parathyroid gland

pineal gland
pituitary: ant. lobe
pituitary: post. lobe
testis
thymus gland
thyroid gland

1. ACTH.
2. ADH.
3. Aldosterone.
4. Calcitonin.
5. Cortisol.
6. Epinephrine.
7. Estrogens.
8. FSH.
9. Glucagon.
10. Growth hormone.
11. Insulin.
12. LH.
13. Melatonin.
14. Oxytocin.
15. Parathyroid hormone.
16. Prolactin.
17. Progesterone.
18. Testosterone.
19. Thymosin.
20. Thyrotropin.
21. Thyroid hormone.

1. _____
2. _____
3. _____
4. _____
5. _____
6. _____
7. _____
8. _____
9. _____
10. _____
11. _____
12. _____
13. _____
14. _____
15. _____
16. _____
17. _____
18. _____
19. _____
20. _____
21. _____

C. Hormone Functions

Select the hormones that perform the functions listed below.

Group 1

aldosterone   epinephrine
cortisol   melatonin

1. Increases resistance to stress.
2. Promotes excretion of K^+.
3. Increases reabsorption of Na^+.
4. Increases blood glucose level (2).
5. Prepares body for response to emergencies.
6. Increases heart and respiration rate.

Group 2

calcitonin   parathyroid hormone
glucagon   somatostatin
insulin   thyroid hormone

1. Promotes bone deposition.
2. Promotes bone reabsorption.
3. Increases blood glucose level.
4. Decreases blood glucose level.
5. Stimulates metabolism.
6. Raises blood calcium level.
7. Lowers blood calcium level.
8. Aids entrance of glucose into cells.

Group 3

ACTH   LH
ADH   oxytocin
estrogens   progesterone
FSH   testosterone
growth hormone   thyrotropin

1. Stimulates descent of testes.
2. Promotes breast development.
3. Promotes release of milk.
4. Stimulates thyroxin production.
5. Stimulates follicle development.
6. Stimulates glucocorticoid production.
7. Stimulates growth of body cells.
8. Prepares uterus for implantation of embryo.
9. Rapid increase causes ovulation.
10. Stimulates spermatogenesis.
11. Promotes water reabsorption by kidneys.
12. Strengthens uterine contractions.

Group 1

1. _____
2. _____
3. _____
4a. _____
4b. _____
5. _____
6. _____

Group 2

1. _____
2. _____
3. _____
4. _____
5. _____
6. _____
7. _____
8. _____

Group 3

1. _____
2. _____
3. _____
4. _____
5. _____
6. _____
7. _____
8. _____
9. _____
10. _____
11. _____
12. _____

D. Hormonal Imbalance

Select the hormones involved in the maladies listed and indicate if the condition is due to an excess (+) or a deficiency (−).

glucagon

glucocorticoids

growth hormone

insulin

mineralocorticoids

parathyroid hormone

thyroid hormone

1. Acromegaly.
2. Addison's disease.
3. Cretinism.
4. Cushing's syndrome.
5. Diabetes mellitus.
6. Dwarfism.
7. Gigantism.
8. Hypoglycemia.
9. Myxedema.
10. Tetany.

1. _____
2. _____
3. _____
4. _____
5. _____
6. _____
7. _____
8. _____
9. _____
10. _____

E. Figure

List the labels for Figure 37.11.

F. Microscopic Study

Make labeled drawings of your observations of the prepared microscope slides of the endocrine glands. Use additional sheets of paper as necessary.

Figure 37.11

1. _____
2. _____
3. _____
4. _____
5. _____
6. _____
7. _____
8. _____
9. _____
10. _____
11. _____
12. _____
13. _____
14. _____

G. Endocrine Hormones and Actions

Summarize your understanding of endocrine glands by completing the chart below.

Hormone	Endocrine Gland	Action
Adrenocorticotropin (ACTH)		
Aldosterone		
Antidiuretic hormone (ADH)		
Calcitonin		
Cortisol		
Epinephrine		
Estrogen		
Follicle stimulating hormone (FSH)		
Glucagon		
Growth hormone (GH)		
Insulin		
Interstitial cell stimulating hormone (ICSH)		
Luteinizing hormone (LH)		
Melatonin		
Oxytocin		
Parathyroid hormone (PH)		
Progesterone		
Prolactin (PRL)		
Somatostatin (GH-RIH)		
Thymosin		
Thyroid hormone		
Thyrotropin (TSH)		

Laboratory Report 38

Student ______________________

Lab Section ______________________

THE REPRODUCTIVE ORGANS

A. Figures

List the labels for Figures 38.1, 38.3, 38.4, and 38.5.

Figure 38.1

1. ______
2. ______
3. ______
4. ______
5. ______
6. ______
7. ______
8. ______
9. ______
10. ______
11. ______
12. ______
13. ______
14. ______
15. ______
16. ______
17. ______
18. ______
19. ______
20. ______
21. ______
22. ______
23. ______
24. ______
25. ______
26. ______
27. ______

Figure 38.3

1. ______
2. ______
3. ______
4. ______
5. ______
6. ______
7. ______
8. ______
9. ______
10. ______

Figure 38.4

1. ______
2. ______
3. ______
4. ______
5. ______
6. ______
7. ______
8. ______
9. ______
10. ______
11. ______
12. ______
13. ______
14. ______

Figure 38.5

1. _____
2. _____
3. _____
4. _____
5. _____
6. _____
7. _____
8. _____
9. _____

10. _____
11. _____
12. _____
13. _____
14. _____
15. _____
16. _____
17. _____

B. Gametogenesis

Select the cells described by the statements below.

Spermatogenesis	*Oogenesis*
primary spermatocytes	oogonia
secondary spermatocytes	ovum
spermatids	polar bodies
spermatogonia	primary oocyte
spermatozoa	secondary oocyte

1. Haploid male gametes.
2. Haploid female gamete.
3. Divide mitotically to form primary spermatocytes.
4. Cells composing germinal epithelium in ovary.
5. Large haploid cell released in ovulation.
6. Male cells formed by first meiotic division.
7. Large female cell formed by first meiotic division.
8. Male cells in which first meiotic division occurs.
9. Female cells in which first meiotic division occurs.
10. Large female cell formed by second meiotic division.

1. _____
2. _____
3. _____
4. _____
5. _____
6. _____
7. _____
8. _____
9. _____
10. _____

C. Male Organs

Select the structures described by the statements below.

bulbourethral gland	prostate gland
corpora cavernosa	prostatic urethra
corpus spongiosum	seminal vesicle
ejaculatory duct	seminiferous tubule
epididymis	testes
penis	vas deferens
prepuce	

1. Male copulatory organ.
2. Male gonads.
3. Sheath of skin over the glans penis.
4. Carries sperm from epididymis to ejaculatory duct.
5. Tube from urinary bladder into prostate gland.
6. Empties secretions into base of penile urethra.
7. Empties secretions into prostatic urethra.
8. Empties secretions into ejaculatory duct.
9. Erectile tissue forming glans penis.
10. Two dorsally located cylinders of erectile tissue.
11. Slender tube containing sperm-forming cells.
12. Site of sperm storage and maturation.

1. _____
2. _____
3. _____
4. _____
5. _____
6. _____
7. _____
8. _____
9. _____
10. _____
11. _____
12. _____

D. Female Organs

Select the structures described by the statements below.

broad ligament   myometrium
cervix   ovarian ligament
clitoris   ovaries
fimbriae   suspensory ligament
hymen   uterine tubes
infundibulum   uterosacral ligament
labia majora   vagina
labia minora   vestibular glands
mons pubis

1. Female copulatory organ.
2. Muscular wall of the uterus.
3. Narrow neck of the uterus.
4. Female gonads.
5. Fingerlike projections around infundibulum.
6. Source of ovum.
7. Provides vaginal lubrication during intercourse.
8. Duct carrying oocyte toward uterus.
9. Inner folds of vulva surrounding vestibule.
10. Nodule of erectile tissue.
11. Outer fatty folds of vulva merging with mons pubis.
12. Funnel-like opening of uterine tube.
13. Narrow ligament attaching ovary to uterus.
14. Sheetlike ligament supporting ovary, uterine tubes, and uterus.
15. Attaches uterus to posterior pelvic wall.
16. Ligament from ovary to lateral pelvic wall.

1. _____
2. _____
3. _____
4. _____
5. _____
6. _____
7. _____
8. _____
9. _____
10. _____
11. _____
12. _____
13. _____
14. _____
15. _____
16. _____

E. Completion

1. The inner lining of the uterus is the _____.
2. The mixture of sperm and secretions of accessory glands is called _____.
3. Testes are located in a sac called the _____.
4. Fertilization usually occurs in the upper third of the _____.

1. _____
2. _____
3. _____
4. _____

F. Microscopic Study

Make labeled drawings of your observations below. Use additional sheets of paper as necessary.

NOTES

VITAL CAPACITIES

Table A-1 Predicted Vital Capacities for Females

HEIGHT IN CENTIMETERS AND INCHES

AGE	CM 152 / IN 59.8	154 / 60.6	156 / 61.4	158 / 62.2	160 / 63.0	162 / 63.7	164 / 64.6	166 / 65.4	168 / 66.1	170 / 66.9	172 / 67.7	174 / 68.5	176 / 69.3	178 / 70.1	180 / 70.9	182 / 71.7	184 / 72.4	186 / 73.2	188 / 74.0
16	3,070	3,110	3,150	3,190	3,230	3,270	3,310	3,350	3,390	3,430	3,470	3,510	3,550	3,590	3,630	3,670	3,715	3,755	3,800
17	3,055	3,095	3,135	3,175	3,215	3,255	3,295	3,335	3,375	3,415	3,455	3,495	3,535	3,575	3,615	3,655	3,695	3,740	3,780
18	3,040	3,080	3,120	3,160	3,200	3,240	3,280	3,320	3,360	3,400	3,440	3,480	3,520	3,560	3,600	3,640	3,680	3,720	3,760
20	3,010	3,050	3,090	3,130	3,170	3,210	3,250	3,290	3,330	3,370	3,410	3,450	3,490	3,525	3,565	3,605	3,645	3,695	3,720
22	2,980	3,020	3,060	3,095	3,135	3,175	3,215	3,255	3,290	3,330	3,370	3,410	3,450	3,490	3,530	3,570	3,610	3,650	3,685
24	2,950	2,985	3,025	3,065	3,100	3,140	3,180	3,220	3,260	3,300	3,335	3,375	3,415	3,455	3,490	3,530	3,570	3,610	3,650
26	2,920	2,960	3,000	3,035	3,070	3,110	3,150	3,190	3,230	3,265	3,300	3,340	3,380	3,420	3,455	3,495	3,530	3,570	3,610
28	2,890	2,930	2,965	3,000	3,040	3,070	3,115	3,155	3,190	3,230	3,270	3,305	3,345	3,380	3,420	3,460	3,495	3,535	3,570
30	2,860	2,895	2,935	2,970	3,010	3,045	3,085	3,120	3,160	3,195	3,235	3,270	3,310	3,345	3,385	3,420	3,460	3,495	3,535
32	2,825	2,865	2,900	2,940	2,975	3,015	3,050	3,090	3,125	3,160	3,200	3,235	3,275	3,310	3,350	3,385	3,425	3,460	3,495
34	2,795	2,835	2,870	2,910	2,945	2,980	3,020	3,055	3,090	3,130	3,165	3,200	3,240	3,275	3,310	3,350	3,385	3,425	3,460
36	2,765	2,805	2,840	2,875	2,910	2,950	2,985	3,020	3,060	3,095	3,130	3,165	3,205	3,240	3,275	3,310	3,350	3,385	3,420
38	2,735	2,770	2,810	2,845	2,880	2,915	2,950	2,990	3,025	3,060	3,095	3,130	3,170	3,205	3,240	3,275	3,310	3,350	3,385
40	2,705	2,740	2,775	2,810	2,850	2,885	2,920	2,955	2,990	3,025	3,060	3,095	3,135	3,170	3,205	3,240	3,275	3,310	3,345
42	2,675	2,710	2,745	2,780	2,815	2,850	2,885	2,920	2,955	2,990	3,025	3,060	3,100	3,135	3,170	3,205	3,240	3,275	3,310
44	2,645	2,680	2,715	2,750	2,785	2,820	2,855	2,890	2,925	2,960	2,995	3,030	3,060	3,095	3,130	3,165	3,200	3,235	3,270
46	2,615	2,650	2,685	2,715	2,750	2,785	2,820	2,855	2,890	2,925	2,960	2,995	3,030	3,060	3,095	3,130	3,165	3,200	3,235
48	2,585	2,620	2,650	2,685	2,715	2,750	2,785	2,820	2,855	2,890	2,925	2,960	2,995	3,030	3,060	3,095	3,130	3,160	3,195
50	2,555	2,590	2,625	2,655	2,690	2,720	2,755	2,785	2,820	2,855	2,890	2,925	2,955	2,990	3,025	3,060	3,090	3,125	3,155
52	2,525	2,555	2,590	2,625	2,655	2,690	2,720	2,755	2,790	2,820	2,855	2,890	2,925	2,955	2,990	3,020	3,055	3,090	3,125
54	2,495	2,530	2,560	2,590	2,625	2,655	2,690	2,720	2,755	2,790	2,820	2,855	2,885	2,920	2,950	2,985	3,020	3,050	3,085
56	2,460	2,495	2,525	2,560	2,590	2,625	2,655	2,690	2,720	2,755	2,790	2,820	2,855	2,885	2,920	2,950	2,980	3,015	3,045
58	2,430	2,460	2,495	2,525	2,560	2,590	2,625	2,655	2,690	2,720	2,750	2,785	2,815	2,850	2,880	2,920	2,945	2,975	3,010
60	2,400	2,430	2,460	2,495	2,525	2,560	2,590	2,625	2,655	2,685	2,720	2,750	2,780	2,810	2,845	2,875	2,915	2,940	2,970
62	2,370	2,405	2,435	2,465	2,495	2,525	2,560	2,590	2,620	2,655	2,685	2,715	2,745	2,775	2,810	2,840	2,870	2,900	2,935
64	2,340	2,370	2,400	2,430	2,465	2,495	2,525	2,555	2,585	2,620	2,650	2,680	2,710	2,740	2,770	2,805	2,835	2,865	2,895
66	2,310	2,340	2,370	2,400	2,430	2,460	2,495	2,525	2,555	2,585	2,615	2,645	2,675	2,705	2,735	2,765	2,800	2,825	2,860
68	2,280	2,310	2,340	2,370	2,400	2,430	2,460	2,490	2,520	2,550	2,580	2,610	2,640	2,670	2,700	2,730	2,760	2,795	2,820
70	2,250	2,280	2,310	2,340	2,370	2,400	2,425	2,455	2,485	2,515	2,545	2,575	2,605	2,635	2,665	2,695	2,725	2,755	2,780
72	2,220	2,250	2,280	2,310	2,335	2,365	2,395	2,425	2,455	2,480	2,510	2,540	2,570	2,600	2,630	2,660	2,685	2,715	2,745
74	2,190	2,220	2,245	2,275	2,305	2,335	2,360	2,390	2,420	2,450	2,475	2,505	2,535	2,565	2,590	2,620	2,650	2,680	2,710

From: Archives of Environmental Health, February, 1966, Vol. 12, pp. 146–89. E. A. Gaensler, MD, and G. W. Wright, MD.

Table A-2 Predicted Vital Capacities for Males

HEIGHT IN CENTIMETERS AND INCHES

AGE	CM IN	152 59.8	154 60.6	156 61.4	158 62.2	160 63.0	162 63.7	164 64.6	166 65.4	168 66.1	170 66.9	172 67.7	174 68.5	176 69.3	178 70.1	180 70.9	182 71.7	184 72.4	186 73.2	188 74.0
16		3,920	3,975	4,025	4,075	4,130	4,180	4,230	4,285	4,335	4,385	4,440	4,490	4,540	4,590	4,645	4,695	4,745	4,800	4,850
18		3,890	3,940	3,995	4,045	4,095	4,145	4,200	4,250	4,300	4,350	4,405	4,455	4,505	4,555	4,610	4,660	4,710	4,760	4,815
20		3,860	3,910	3,960	4,015	4,065	4,115	4,165	4,215	4,265	4,320	4,370	4,420	4,470	4,520	4,570	4,625	4,675	4,725	4,775
22		3,830	3,880	3,930	3,980	4,030	4,080	4,135	4,185	4,235	4,285	4,335	4,385	4,435	4,485	4,535	4,585	4,635	4,685	4,735
24		3,785	3,835	3,885	3,935	3,985	4,035	4,085	4,135	4,185	4,235	4,285	4,330	4,380	4,430	4,480	4,530	4,580	4,630	4,680
26		3,755	3,805	3,855	3,905	3,955	4,000	4,050	4,100	4,150	4,200	4,250	4,300	4,350	4,395	4,445	4,495	4,545	4,595	4,645
28		3,725	3,775	3,820	3,870	3,920	3,970	4,020	4,070	4,115	4,165	4,215	4,265	4,310	4,360	4,410	4,460	4,510	4,555	4,605
30		3,695	3,740	3,790	3,840	3,890	3,935	3,985	4,035	4,080	4,130	4,180	4,230	4,275	4,325	4,375	4,425	4,470	4,520	4,570
32		3,665	3,710	3,760	3,810	3,855	3,905	3,950	4,000	4,050	4,095	4,145	4,195	4,240	4,290	4,340	4,385	4,435	4,485	4,530
34		3,620	3,665	3,715	3,760	3,810	3,855	3,905	3,950	4,000	4,045	4,095	4,140	4,190	4,225	4,285	4,330	4,380	4,425	4,475
36		3,585	3,635	3,680	3,730	3,775	3,825	3,870	3,920	3,965	4,010	4,060	4,105	4,155	4,200	4,250	4,295	4,340	4,390	4,435
38		3,555	3,605	3,650	3,695	3,745	3,790	3,840	3,885	3,930	3,980	4,025	4,070	4,120	4,165	4,210	4,260	4,305	4,350	4,400
40		3,525	3,575	3,620	3,665	3,710	3,760	3,805	3,850	3,900	3,945	3,990	4,035	4,085	4,130	4,175	4,220	4,270	4,315	4,360
42		3,495	3,540	3,590	3,635	3,680	3,725	3,770	3,820	3,865	3,910	3,955	4,000	4,050	4,095	4,140	4,185	4,230	4,280	4,325
44		3,450	3,495	3,540	3,585	3,630	3,675	3,725	3,770	3,815	3,860	3,905	3,950	3,995	4,040	4,085	4,130	4,175	4,220	4,270
46		3,420	3,465	3,510	3,555	3,600	3,645	3,690	3,735	3,780	3,825	3,870	3,915	3,960	4,005	4,050	4,095	4,140	4,185	4,230
48		3,390	3,435	3,480	3,525	3,570	3,615	3,655	3,700	3,745	3,790	3,835	3,880	3,925	3,970	4,015	4,060	4,105	4,150	4,190
50		3,345	3,390	3,430	3,475	3,520	3,565	3,610	3,650	3,695	3,740	3,785	3,830	3,870	3,915	3,960	4,005	4,050	4,090	4,135
52		3,315	3,353	3,400	3,445	3,490	3,530	3,575	3,620	3,660	3,705	3,750	3,795	3,835	3,880	3,925	3,970	4,010	4,055	4,100
54		3,285	3,325	3,370	3,415	3,455	3,500	3,540	3,585	3,630	3,670	3,715	3,760	3,800	3,845	3,890	3,930	3,975	4,020	4,060
56		3,255	3,295	3,340	3,380	3,425	3,465	3,510	3,550	3,595	3,640	3,680	3,725	3,765	3,810	3,850	3,895	3,940	3,980	4,025
58		3,210	3,250	3,290	3,335	3,375	3,420	3,460	3,500	3,545	3,585	3,630	3,670	3,715	3,755	3,800	3,840	3,880	3,925	3,965
60		3,175	3,220	3,260	3,300	3,345	3,385	3,430	3,470	3,500	3,555	3,595	3,635	3,680	3,720	3,760	3,805	3,845	3,885	3,930
62		3,150	3,190	3,230	3,270	3,310	3,350	3,390	3,440	3,480	3,520	3,560	3,600	3,640	3,680	3,730	3,770	3,810	3,850	3,890
64		3,120	3,160	3,200	3,240	3,280	3,320	3,360	3,400	3,440	3,490	3,530	3,570	3,610	3,650	3,690	3,730	3,770	3,810	3,850
66		3,070	3,110	3,150	3,190	3,230	3,270	3,310	3,350	3,390	3,430	3,470	3,510	3,550	3,600	3,640	3,680	3,720	3,760	3,800
68		3,040	3,080	3,120	3,160	3,200	3,240	3,280	3,320	3,360	3,400	3,440	3,480	3,520	3,560	3,600	3,640	3,680	3,720	3,760
70		3,010	3,050	3,090	3,130	3,170	3,210	3,250	3,290	3,330	3,370	3,410	3,450	3,480	3,520	3,560	3,600	3,640	3,680	3,720
72		2,980	3,020	3,060	3,100	3,140	3,180	3,210	3,250	3,290	3,330	3,370	3,410	3,450	3,490	3,530	3,570	3,610	3,650	3,680
74		2,930	2,970	3,010	3,050	3,090	3,130	3,170	3,200	3,240	3,280	3,320	3,360	3,400	3,440	3,470	3,510	3,550	3,590	3,630

From: Archives of Environmental Health, February, 1966, Vol. 12, pp. 146–89. E. A. Gaensler, MD, and G. W. Wright, MD.

MICROORGANISMS

Figure B-1 Protozoans.

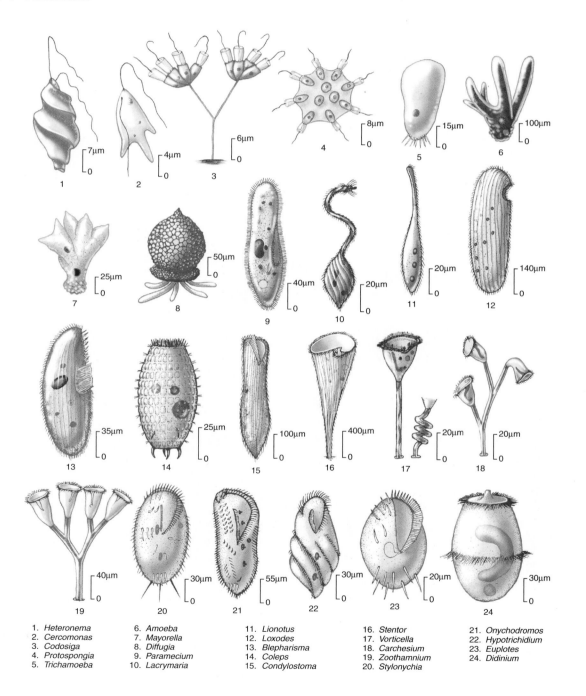

1. *Heteronema*	6. *Amoeba*	11. *Lionotus*	16. *Stentor*	21. *Onychodromos*
2. *Cercomonas*	7. *Mayorella*	12. *Loxodes*	17. *Vorticella*	22. *Hypotrichidium*
3. *Codosiga*	8. *Diffugia*	13. *Blepharisma*	18. *Carchesium*	23. *Euplotes*
4. *Protospongia*	9. *Paramecium*	14. *Coleps*	19. *Zoothamnium*	24. *Didinium*
5. *Trichamoeba*	10. *Lacrymaria*	15. *Condylostoma*	20. *Stylonychia*	

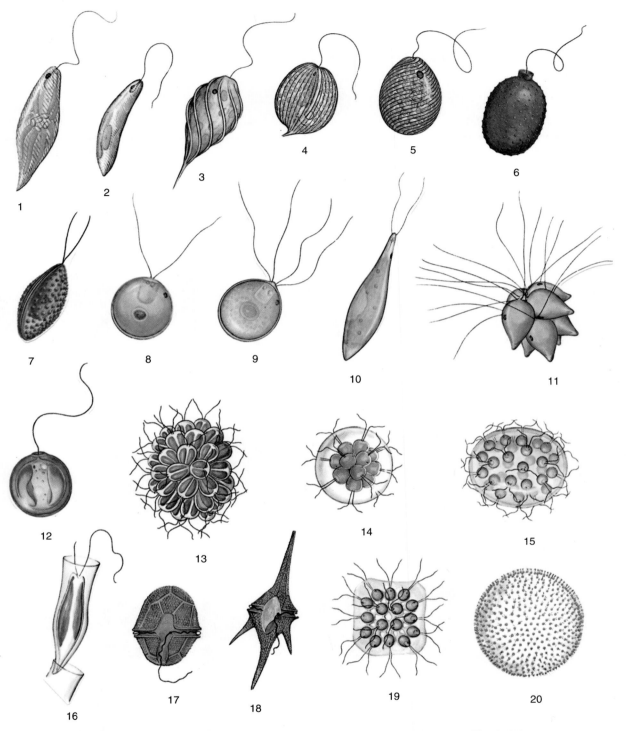

Courtesy of the U.S. Environmental Protection Agency, Office of Research & Development, Cincinnati, Ohio 45268.

1. *Euglena* (700×)
2. *Euglena* (700×)
3. *Phacus* (1000×)
4. *Phacus* (350×)
5. *Lepocinclis* (350×)

6. *Trachelomonas* (1000×)
7. *Phacotus* (1500×)
8. *Chlamydomonas* (1000×)
9. *Carteria* (1500×)
10. *Chlorogonium* (1000×)

11. *Pyrobotrys* (1000×)
12. *Chrysococcus* (3000×)
13. *Synura* (350×)
14. *Pandorina* (350×)
15. *Eudorina* (175×)

16. *Dinobyron* (1000×)
17. *Peridinium* (350×)
18. *Ceratium* (175×)
19. *Gonium* (350×)
20. *Volvox* (100×)